NANOPARTICLES IN POLYMER SYSTEMS FOR BIOMEDICAL APPLICATIONS

NANOPARTICLES IN POLYMER SYSTEMS FOR BIOMEDICAL APPLICATIONS

Edited by

Jince Thomas
Sabu Thomas
Nandakumar Kalarikkal
Jiya Jose

APPLE ACADEMIC PRESS

Apple Academic Press Inc.
3333 Mistwell Crescent
Oakville, ON L6L 0A2 Canada

Apple Academic Press Inc.
9 Spinnaker Way
Waretown, NJ 08758 USA

© 2019 by Apple Academic Press, Inc.

First issued in paperback 2021

Exclusive worldwide distribution by CRC Press, a member of Taylor & Francis Group
No claim to original U.S. Government works

ISBN 13: 978-1-77463-396-0 (pbk)
ISBN 13: 978-1-77188-703-8 (hbk)

Library and Archives Canada Cataloguing in Publication

Nanoparticles in polymer systems for biomedical applications / edited by Jince Thomas, Sabu Thomas, Nandakumar Kalarikkal, Jiya Jose.

Includes bibliographical references and index.

Issued in print and electronic formats.

ISBN 978-1-77188-703-8 (hardcover).--ISBN 978-1-351-04788-3 (PDF)

1. Nanomedicine. 2. Nanoparticles. 3. Nanotechnology. I. Thomas, Jince, editor

| R857.N34N36 2018 | 610.28'4 | C2018-904909-X | C2018-904910-3 |

Library of Congress Cataloging-in-Publication Data

Names: Thomas, Jince, editor. | Thomas, Sabu, editor. | Kalarikkal, Nandakumar, editor. | Jose, Jiya, editor.

Title: Nanoparticles in polymer systems for biomedical applications / editors, Jince Thomas, Sabu Thomas, Nandakumar Kalarikkal, Jiya Jose.

Description: Oakville, Canada ; Waretown, NJ : Apple Academic Press, [2019] | Includes bibliographical references and index.

Identifiers: LCCN 2018039873 (print) | LCCN 2018041334 (ebook) | ISBN 9781351047883 (ebook) | ISBN 9781771887038 (hardcover : alk. paper)

Subjects: | MESH: Nanoparticles | Polymers | Nanomedicine--methods

Classification: LCC R857.P6 (ebook) | LCC R857.P6 (print) | NLM QT 36.5 | DDC 610.28--dc23

LC record available at https://lccn.loc.gov/2018039873

Apple Academic Press also publishes its books in a variety of electronic formats. Some content that appears in print may not be available in electronic format. For information about Apple Academic Press products, visit our website at **www.appleacademicpress.com** and the CRC Press website at **www.crcpress.com**

CONTENTS

ABOUT THE EDITORS

Jince Thomas is currently working as a Senior Research Fellow at the International and Inter University Centre for Nanoscience and Nanotechnology, Mahatma Gandhi University, Kerala, India. His research focuses on polymer blends, polymer nanocomposites, conductive and insulating polymer composites, and biomaterials. He has two master's degrees (MTech in Polymer Science and Technology and MSc in Chemistry). He also worked at the University of Tennessee Knoxville, USA. He has authored three book chapters.

Sabu Thomas, PhD, is the Pro Vice Chancellor of Mahatma Gandhi University and Founding Director of the International and Inter University Centre for Nanoscience and Nanotechnology, Mahatma Gandhi University, Kottayam, Kerala, India. He is also a full professor of polymer science and engineering at the School of Chemical Sciences of the same university. He is a fellow of many professional bodies. Professor Thomas has authored and co-authored many papers in international peer-reviewed journals in the area of polymer science and nanotechnology. He has organized several international conferences. Professor Thomas's research group has specialized in many areas of polymers, which includes polymer blends, fiber-filled polymer composites, particulate-filled polymer composites and their morphological characterization, ageing and degradation, pervaporation phenomena, sorption and diffusion, interpenetrating polymer systems, recyclability and reuse of waste plastics and rubbers, elastomeric crosslinking, dual porous nanocomposite scaffolds for tissue engineering, etc. Professor Thomas's research group has extensive exchange programs with different industries and research and academic institutions all over the world and is performing world-class collaborative research in various fields. Professor Thomas's center is equipped with various sophisticated instruments and has established state-of-the-art experimental facilities, which cater to the needs of researchers within the country and abroad.

Dr. Thomas has more than 700 publications, with citation count of 31190. His H index is 81, and he also has six patents to his credit. Prof. Thomas has edited over 70 books. He is a reviewer for many international journals. He has received MRSI, CRSI, and nanotech medals for his outstanding work in nanotechnology. Recently Prof. Thomas has been conferred an Honoris Causa, DSc by the University of South Brittany, France, and University Lorraine, Nancy, France.

Nandakumar Kalarikkal, PhD, is an Associate Professor at the School of Pure and Applied Physics and Joint Director of the International and Inter University Centre for Nanoscience and Nanotechnology of Mahatma Gandhi University, Kottayam, Kerala, India. His research activities involve applications of nanostructured materials, laser plasma, phase transitions, etc. He is the recipient of research fellowships and associateships from prestigious government organizations such as the Department of Science and Technology and the Council of Scientific and Industrial Research of the Government of India. He has active collaboration with national and international scientific institutions in India, South Africa, Slovenia, Canada, France, Germany, Malaysia, Australia, and the United States. He has more than 130 publications in peer-reviewed journals. He has also co-edited over a dozen books of scientific interest and has co-authored many book chapters.

Jiya Jose, PhD, is a Postdoctoral Fellow of International and Inter University Centre for Nanoscience and Nanotechnology at Mahatma Gandhi University, Kottayam, Kerala, India. Her PhD is from the CSIR-National Institute of Oceanography under the guidance of Dr. Shanta Achuthankutty. Her research mainly focuses on nanomedicine, drug delivery, nanomaterials for water purification, polymer scaffolds for wound healing applications, and interaction studies on nanoparticles and microorganisms. She has written book chapters and articles in peer-reviewed journals.

CONTRIBUTORS

Ann Rose Abraham
School of Pure and Applied Physics, Centre for Nanoscience and Nanotechnology,
Mahatma Gandhi University, Kottayam 686560, Kerala, India

Bilahari Aryat
International and Inter University Centre for Nanoscience and Nanotechnology,
Mahatma Gandhi University, Kottayam 686560, Kerala, India

Arun K. Iyer
Use-inspired Biomaterials & Integrated Nano Delivery (U-BiND) Systems Laboratory,
Department of Pharmaceutical Sciences, Wayne State University, Detroit, Michigan, USA

Jiya Jose
International and Inter University Centre for Nanoscience and Nanotechnology,
Mahatma Gandhi University, Kottayam 686560, Kerala, India

Jomon Joy
School of Chemical Sciences, Mahatma Gandhi University, P D Hills, Kottayam 686560,
Kerala, India

Nandakumar Kalarikkal
School of Pure and Applied Physics, Centre for Nanoscience and Nanotechnology,
Mahatma Gandhi University, Kottayam 686560, Kerala, India
International and Inter University Centre for Nanoscience and Nanotechnology,
Mahatma Gandhi University, Kottayam 686560, Kerala, India

Veera Venkata Satyanarayana Reddy Karri
Department of Pharmaceutics, JSS College of Pharmacy, Ootacamund, Rocklands,
Udhagamandalam 643001, Tamil Nadu, India
JSS Academy of Higher Education and Research, Mysuru, Karnataka, India

Karthika M.
International and Inter University Centre for Nanoscience and Nanotechnology,
Mahatma Gandhi University, Kottayam 686560, Kerala, India

Sushil K. Kashaw
Department of Pharmaceutical Sciences, Dr. Harisingh Gour Central University, Sagar 470003,
Madhya Pradesh, India

Lizymol P. P.
Division of Dental Products, Biomedical Technology Wing, Sree Chitra Tirunal Institute for
Medical Sciences and Technology, Poojappura, Thiruvananthapuram 695012, India

Mereena Luke P.
International and Inter University Centre for Nanoscience and Nanotechnology,
Mahatma Gandhi University, Kottayam 686560, Kerala, India

Jawahar Natarajan
Department of Pharmaceutics, JSS College of Pharmacy, Ootacamund, Rocklands, Udhagamandalam 643001, Tamil Nadu, India
JSS Academy of Higher Education and Research, Mysuru, Karnataka, India

Neelakandan M. S.
International and Inter University Centre for Nanoscience and Nanotechnology, Mahatma Gandhi University, Kottayam 686560, Kerala, India

Parvathy Prasad
International and Inter University Centre for Nanoscience and Nanotechnology, Mahatma Gandhi University, Kottayam 686560, Kerala, India

Rajesh Ramasamy
Stem Cell & Immunity Research Group, Immunology Laboratory, Department of Pathology, Faculty of Medicine and Health Sciences, Universiti Putra Malaysia, 43400 Seri Kembangan, Selangor, Malaysia

Prashant Sahu
Department of Pharmaceutical Sciences, Dr. Harisingh Gour Central University, Sagar 470003, Madhya Pradesh, India

K. P. Sajesha
School of Chemical Sciences, Mahatma Gandhi University, P D Hills, Kottayam, Athirampuzha 686560, Kerala, India

Vahid Hosseinpour Sarmadi
Stem Cell & Immunity Research Group, Immunology Laboratory, Department of Pathology, Faculty of Medicine and Health Sciences, Universiti Putra Malaysia, 43400 Seri Kembangan, Selangor, Malaysia

Samaresh Sau
Use-inspired Biomaterials & Integrated Nano Delivery (U-BiND) Systems Laboratory, Department of Pharmaceutical Sciences, Wayne State University, Detroit, Michigan, USA

P. M. Sayeesh
School of Chemical Sciences, Mahatma Gandhi University, P D Hills, Kottayam, Athirampuzha 686560, Kerala, India

Nitheesha Shaji
International and Inter University Centre for Nanoscience and Nanotechnology, Mahatma Gandhi University, Kottayam 686560, Kerala, India

Sunija Sukumaran
International and Inter University Centre for Nanoscience and Nanotechnology, Mahatma Gandhi University, Kottayam 686560, Kerala, India

Jince Thomas
International and Inter University Centre for Nanoscience and Nanotechnology, Mahatma Gandhi University, Athirampuzha 686560, Kerala, India

Minu Elizabeth Thomas
School of Chemical Sciences, Mahatma Gandhi University, P D Hills, Kottayam, Athirampuzha 686560, Kerala, India

Sabu Thomas
International and Inter University Centre for Nanoscience and Nanotechnology,
Mahatma Gandhi University, Kottayam 686560, Kerala, India
School of Chemical Sciences, Mahatma Gandhi University, Kottayam 686560,
Kerala, India

Aswathy Vasudevan
International and Inter University Centre for Nanoscience and Nanotechnology,
Mahatma Gandhi University, Kottayam 686560, Kerala, India

Vibha C.
Division of Dental Products, Biomedical Technology Wing, Sree Chitra Tirunal Institute for
Medical Sciences and Technology, Poojappura, Thiruvananthapuram 695012, India

Vishnu K. A.
School of Chemical Sciences, Mahatma Gandhi University, Kottayam 686560, Kerala, India

Yadu Nath V. K.
School of Chemical Sciences, Mahatma Gandhi University, Kottayam 686560, Kerala, India
International and Inter University Centre for Nanoscience and Nanotechnology,
Mahatma Gandhi University, Kottayam 686560, Kerala, India

ABBREVIATIONS

9-NC	9-nitrocamptothecin
AFM	atomic force microscopy
AM	additive manufacturing
AMF	alternating magnetic field
API	active pharmaceutical ingredient
BBB	blood–brain barrier
BCL	B-cell lymphoma
BM	bone marrow
BMP	bone morphogenic proteins
BNNT	boron nitride nanotube
BPO	benzoyl peroxide
BSA	bovine serum albumin
BT	barium titanate
CAD	computer-aided design
CDC	Centres for Disease Control and Prevention
CHP	cholesterol bearing pullulan
CHPOA	acryl-oyl-modified cholesterol-bearing pullulan
cLABL	cyclo-(1,12)-penITDGEATDSGC
CNGs	chitin nanogels
CNS	central nervous system
CNT	carbon nanotube
CNT-NEEs	CNT nanoelectrode ensembles
CS	compressive strength
CT	computerized tomography
DDSs	drug delivery systems
DEAP	diethylaminopropyl
DF-1	dendrofullerene
DLLA	d,l-lactide
DMF	dimethylformamide
DMPT	N,N-dimethyl-p-toluidine
DMSO	dimethyl sulfoxide

DMT	N,N-dimethyl-4-toluidine
DOX	doxorubicin
DSC	differential scanning calorimetry
ECs	endothelial cells
ECM	extracellular matrix
EE	entrapment efficiency
EGDMA	ethylene glycol dimethacrylate
EGFP	enhanced green fluorescent protein
EGFR	epidermal growth factor receptor
ELPs	elastin-like polypeptides
EMT	epithelial–mesenchymal transition
EPR	enhanced permeability and retention
FAS	first apoptosis signal
FDA	Food and Drug Administration
FDM	fused deposition modeling
FGF	fibroblast growth factors
FITC	fluorescein iso-thiocyanate
FN	fibronectin
FTIR	Fourier-transform infrared spectroscopy
GALT	gut-associated lymphoid tissue
GICs	glass-ionomer cements
GMS	glycerol monostearate
GNP	gold nanoparticle
GO	graphite oxide
GPO	graphene peroxide
HA	hyaluronic acid
HAP	hydroxyapatite
HCCs	hepatocellular carcinoma cells
HEMA	hydroxyethylmethacrylate
HER	herceptin
HER-2	human epidermal receptor-2
HGF	hepatocyte growth factor
HLA-DR	human leukocyte antigen—antigen D related
hMSCs	human mesenchymal stem cells
HMWPE	high-molecular-weight polyethylene
HPE	hyperbranched poly

HuH	human hepatocyte
ICG	indo-cyanine green
IFN-γ	interferon-γ
IL-1	interleukin-1
IP	Indian Pharmacopeia
LOD	limit of detection
LOQ	limit of quantification
mAbs	monoclonal antibodies
MCL	myeloid cell leukemia
MCTSs	multicellular tumor spheroids
MIT	Massachusetts Institute of Technology
MMA	methylmethacrylate
MMPs	matrix metalloproteinases
MNPs	magnetic nanoparticles
MPA	mycophenolic acid
mPEG	methoxy(PEG)
MRI	magnetic resonance imaging
mRNA	messenger RNAs
MSCs	mesenchymal stem cells
MT1-MMP	membrane type 1 MMP
MUDFs	multiple-unit dosage forms
MWCNTs	multiwalled CNTs
NCHs	nanocomposite hydrogels
NEMS	nanoelectromechanical devices
NIPAM	N-isopropylacrylamide
NIR	near infrared
NK	natural killer
NO	nitric oxide
NPCs	nasopharyngeal carcinoma cells
NPs	nanoparticles
NTs	nanotubes
o/w	oil-in-water
P-(MAA-g-EG)	poly(methacrylic acid-g-ethylene glycol)
PAMAM	polyamidoamine
PANi	polyaniline
PARP	poly ADP ribose polymerase

PCL	poly(ε-caprolactone)
PDGF	platelet-derived growth factor
PDT	photodynamic therapy
PEAMA	poly(2-N,N-(di-ethyl-amino)ethyl methacrylate)
PEG	poly(ethylene glycol)
PEI	polyethyleneimine
PEM	piezoforce microscopy
PGA	polyglycolide
PLA	poly-lactic acid
PLGA	poly(lactic-co-glycolic acid)
PMMA	poly(methyl methacrylate)
PNIPAm	poly(N-iso-propylacrylamide)
PPF	poly(propylene fumarate)
PPy	polypyrrole
PSIs	patient-specific instruments
PSMA	prostate-specific membrane antigen
PSNPs	porous silica NPs nanoparticles nanogels
PTs	photothermal transductors
PTT	photothermal therapy
PTX	paclitaxel
PU	polyurethane
PVA	polyvinyl alcohol
PVP/PAAc	polyvinyl pyrrolidone-poly(acrylic acid)
QDs	quantum dots
RCC	radical cross-linking copolymerization
RES	reticuloendothelial system
ROS	reactive oxygen species
RP	rapid prototyping
s/o/w	solid/oil/water
SAR	specific absorption rate
SBF	simulated body fluid
SE	self-emulsifying
SEM	scanning electron microscopy
SFFs	solid free-form technologies
siRNA	small interfering RNA
SLA	stereo lithography

SLS	selective laser sintering
SMCs	smooth muscle cells
SPIONs	superparamagnetic iron oxide nanoparticles
SQUID	superconducting quantum interference device
SWCNTs	single-walled CNTs
TAFs	tumor-associated fibroblasts
TEGDMA	triethyleneglycoldimethacrylate
TEM	transmission electron microscopy
TMC	trimethylene carbonate
TNF-α	tumor necrosis factor α
TPGS	tocopheryl PEG succinate
TRAIL	TNF-related apoptosis-inducing ligand
VCAM-1	vascular cell adhesion molecule-1
VCR	vincristine sulfate
VEGF	vascular endothelial growth factor
VRP	verapamil
w/o/w	water-in-oil-in-water
XRD	X-ray diffraction
ZnO	zinc oxide

PREFACE

This book, *Nanoparticles in Polymer Systems for Biomedical Applications,* presents many of the recent technological and research accomplishments in the area of nanomedicine and nanotechnology. Within this book are presentations of the state of the art in the area, novel approaches of nanomedicine and nanotechnology, and new challenges and opportunities. Also discussed are the different types of nanomedicine drugs along with their production and commercial significance. Other topics include the use of natural and synthetic nanoparticles for the production of nanomedicine drugs, different types of nanoparticles systems, drug carriers, wound healing antimicrobial activity, the effect of natural materials in nanomedicine, and toxicity of nanoparticles.

This innovative book provides an up-to-date record on the key findings, observations, and fabrication of drugs related to nanomedicine and nanotechnology. It is intended to assist as a reference resource for medical research fields as well as in science fields. The various chapters in this book have been contributed by prominent researchers from academe, industry, and government–private research laboratories across the world. This book will be a very valuable reference source for university and college faculties, professionals, postdoctoral research fellows, senior graduate students, and researchers from R&D laboratories working in the area of health care and therapeutic fields.

Chapter 1 gives a detailed outline and an overview of the state of the art in the field and also presents new types of nanoparticles, opportunities, and the significance of nanoengineering in nanomedicine fields. Chapter 2 discusses the detailed studies of PLGA-based nanoparticles for cancer therapy. The authors explain the preparation and modification of PLGA nanoparticles and their releasing mechanism. Chapter 3 provides the benefits and application of superparamagnetic ferrite nanoparticles in nanomedicine fields.

Chapter 4 focuses on the biomedical applications of 3D printing. The mechanisms and principle of 3D printing are discussed. Chapter 5 looks at nanogels design for targeted cancer therapy. Properties, preparation, benefits, and patent status of nanogels are also correlated in this chapter. In Chapter 6, piezoelectrical materials for biomedical applications are presented. In this chapter, the authors review the piezoelectricity in biological tissues and their effects on biomedical and tissue engineering applications. Synthesis and comparison of lipophilic drugs are discussed in Chapter 7. This chapter mainly focuses on the characterization of lipophilic drugs. Chapter 8 gives the idea about the nanocomposites hydrogel for various biomedical applications. In this chapter, the authors review the benefits of carbon-based nanocomposite hydrogels, polymeric-material-based nanocomposite hydrogels, and inorganic-nanoparticle-based nanocomposite hydrogels. Chapter 9 highlights development of new bioactive multifunctional inorganic–organic hybrid resin for biomedical applications.

Mesenchymal stem cells are discussed in Chapter 10. This chapter includes interaction between MSCs and tumor cells, suppression of immune response; MSCs inhibit tumor cell proliferation and angiogenesis. Chapter 11 is about the purification of water using bionanoparticles. Nanoparticles used for transdermal delivery are discussed in Chapter 12. Chapter 13 discusses the toxicity of nanoparticles. In this chapter, the authors review the effect of toxicity of nanomaterials in drug delivery systems.

Finally, thanks to God for the successful completion of this book, and the editors would like to express their honest gratitude to all the contributors of this book who provided excellent support for the fruitful completion of this endeavor. We appreciate them for their commitment and the genuineness for their contributions to this book. We would like to acknowledge all the reviewers who have taken their valuable time to make critical comments on each chapter. We also acknowledge the support of Apple Academic Press.

—Jince Thomas
Sabu Thomas

CHAPTER 1

AN INTRODUCTION TO SPECIALLY TAILORED NANOMATERIALS FOR BIOMEDICAL APPLICATIONS

MINU ELIZABETH THOMAS[1*], K. P. SAJESHA[1], P. M. SAYEESH[1], and JINCE THOMAS[2]

[1]*School of Chemical Sciences, Mahatma Gandhi University, P D Hills, Kottayam, Athirampuzha 686560, Kerala, India*

[2]*International and Inter University Centre for Nanoscience and Nanotechnology, Mahatma Gandhi University, Athirampuzha 686560, Kerala, India*

Corresponding author. E-mail: minuputhen@gmail.com

ABSTRACT

Nanotechnology, the widely expanding branch, deals with the study of the properties, applications, and toxicity of particles of smaller dimensions particularly less than 100 nm. This chapter provides a brief introduction to the exciting world of nanoscience and its importance in study. This chapter also provides an overview to the different types of nanomaterials such as quantum dots, liposomes, dendrimers, nanoshells and nanoparticles, nanotubes, and polymeric nanoparticles that are mainly used in biomedical application. The necessity for the adaptation of green nanoparticles is also discussed in this chapter.

1.1 INTRODUCTION

The world we live in is gifted with magnificent particles that have been provoking curiosity in human minds. Since ancient times, these particles are being categorized on the basis of their dimensions. These include particles from macroscale dimension to microscale and even to lower nanoscale dimension. The particles with dimension above 10^{-3} m belong to macroscale, those between 10^{-3} and 10^{-6} m belongs to microscale and those from 10^{-7} to 10^{-9} m are classified as nanoscaled particles. Thus, according to the definition given by National Nanotechnology Initiative,[1] nanoparticles are structures having sizes ranging from 1 to 100 nm (1 nm = 10^{-9} m) in at least one dimension. However, the prefix "nano" is being commonly used for particles that extend up to several hundred nanometers in size.

The major natural sources of nanoparticles in the atmosphere includes volcanic eruptions, desert surfaces, dusts from cosmic sources located in the solar system or outside it, hurricanes, etc. A very good example of nanoparticle sources is the eruption of Krakatoa on August 27, 1883.[2] The smoke column reached 80 km above and the dust thrown into the ionosphere not only caused strange optical effects visible in North America and Europe but also acted as a solar radiation filter that decreased the global temperature with about 1.5°C in the next 2 years. Another example is the Tunguska event on June 30, 1908, whose probable cause was the collision of a gigantic meteorite or comet fragment with our Earth. In nature, the examples of nanoparticles can be seen on the wings of Morpho butterflies and on the feathers of peacocks. The brilliant coloration is due to the presence of optical nanostructures on their surfaces (Fig. 1.1).

The importance of this dimension is that when the size of the particles is below 5 nm, the properties of matter differ significantly from that at the

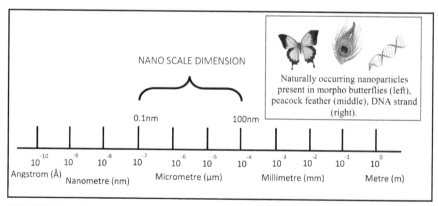

FIGURE 1.1   Dimension scale.

larger particulate scale. Macroscopic materials possess continuous physical properties. The same applies to microsized materials. But when particles are of nanoscale dimensions, the principles of classic physics become no longer capable of describing their behavior like movement, energy, etc. At these dimensions, the principles of quantum mechanics becomes applicable and can have optical, mechanical, and electrical properties which are very different from the properties of materials that are of macroscale dimension. Properties like electrical conductivity, color, strength, and weight changes when the nanolevel is reached. A metal which is a conductor at macroscale can become a semiconductor or an insulator at the nanoscale level; it helps scientists to understand the phenomena that occur naturally when matter is organized at the nanoscale. Thus, the essence is that these phenomena are based on quantum effects and other simple physical effects such as expanded surface area. In addition to this, the fact that a majority of biological processes occur at the nanolevel gives scientists new models and templates to imagine and construct new processes that can enhance their work in medicine, imaging, computing, printing, chemical catalysis, materials synthesis, and in many other fields. Working at the nanolevels enables scientists to utilize the unique physical, chemical, mechanical, and optical properties of materials that naturally occur at that scale.

The entry of nano to the field of science has led to the beginning of a new branch termed as nanosciences which in turn led to the vast field of nanotechnology. Nanoscience can thus be defined as the study of different phenomena and manipulation of materials that are of atomic, molecular, and macromolecular dimensions and whose properties differ significantly from those at larger dimensions. It is an interdisciplinary science which involves the concepts of more than one discipline, such as chemistry, physics, biology, technology, etc. (Fig. 1.2).

1.2 APPLICATIONS OF NANOSCIENCE AND NANOTECHNOLOGY

The application of nanoscience to practical devices led to the emergence of a new stream called nanotechnology. Nanotechnology is based on the manipulation, control, and integration of atoms and molecules to form materials, structures, components, devices, and systems that are of nanoscale

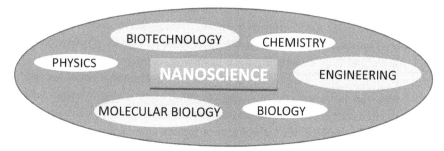

FIGURE 1.2   Nanoscience as interdisciplinary science.

dimension. Its application is highly concentrated to industrial and commercial objectives. All industrial sectors depend on materials and devices made of atoms and molecules which in turn can be improved with nanomaterials. Hence, all industries are promised to be benefited using nanotechnology. However, it was Professor Richard Feynman who gave this new perspective to nanotechnology. Nanoconcept was explained by Professor Richard Feynman with his illuminating talk "There's Plenty of Room at the Bottom"[3] on December 1959. The lecture was presented on an American Physical Society meeting held at the California Institute of Technology. He was an American theoretical physicist who received Nobel Prize in physics in 1965 along with Julian Schwinger and Sin-Itiro Tomonaga, for the contribution in developing quantum electrodynamics. Feynman proposed that it could be possible to develop a general method to manipulate matter on atomic scale with a top–bottom approach. He concluded his talk with a challenge to build a tiny motor and to write the information from a book page on a surface 1/25,000 times smaller in linear scale.

Since then nanotechnology extended its arms and at present has a wide range of applications (Fig. 1.3).

Unprecedented growth in research in the area of nanoscience has led to an increasing application of nanotechnology to medicine which brings significant advances in the diagnosis, treatment, and prevention of disease. Nanomedicine is thus a highly emerging and exploring field of research. Besides nanomedicine, nanotechnology has also led to the development of nanoelectronics which involves the combination of nanotechnology and electronic components. Its application results in increasing the capacity of electronic devices along with reduction in their weight and power con-

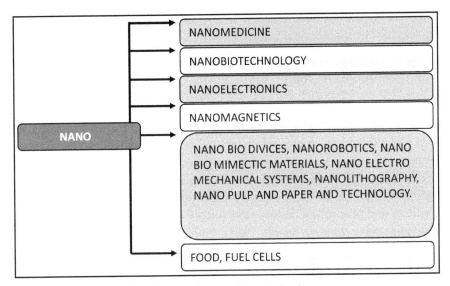

FIGURE 1.3 Applications of nanoscience and nanotechnology.

sumption, increases the density of memory chips, improves the display of certain electronic devices, reduces the size of transistors, etc. Nanobiotechnology involves the joining of molecular biology, biomedical sciences, and nanotechnology. It helps in the progress of modern medicines, regenerating tissues, etc.; with the application of nanobiotechnology, bladders have been cultured in certain American patients. Nanomagnetics[4] involve development of magnetic nanostructures for biomedical application, catalysis, biochemistry, and other industrial applications. It also involves targeted therapies and diagnostics. These materials help in understanding the physical mechanisms contributing to their function. Metal-based materials with certain nanocrystalline structures produced certain magnetic properties that were never seen before. For example, iron oxide nanoparticles known as Feridex have been used as a contrast agent in magnetic resonance imaging (MRI). It is also used in the field of food, from their production to their package. Clay nanoparticles are used in certain bottles, cartons, and thin films which provides a barrier to the gases like oxygen and carbon dioxide in the atmosphere. Silver nanoparticles are embedded in certain plastics to prevent bacterial attack. In the field of energy, research nanotechnology has provided a hand to fuel cells. The electrode and membrane electrolytes have been benefited by nanostructures and associated increase in surface area. It helps in the study of molecular electrochemical behavior.

1.3 NANOMEDICINE

Although nanotechnology has a wide range of application, its use in the field of modern medicine is the most important one. It is an amazing tool that has allowed an improved understanding of the basis of any disease, bring more refined diagnostic opportunities, and provide more effective therapeutic and preventive measures. Thus, the field of nanomedicine can be defined as the science and technology of detecting or diagnosing, treating, and preventing diseases and hurtful injury followed by preserving and improving human health, using molecular tool of nanotechnology and molecular knowledge of the human body. The availability of more durable, better prosthetics, and new systems in delivery of drugs, heat, and light to the diseased cells are of great interest and provide the hope for treatment of cancer, heart diseases, diabetes, and other deadly diseases. Particles are designed so that they are attracted only to the diseased cells and allow direct treatment of those infected cells. This technique reduces damaging of the healthy cells in the body and allows for earlier detection of diseases. For example, nanoscaled diagnostics are expected to identify the becoming of the disease prior to the appearance of symptoms on the onset of disease. Many novel nanoparticles and nanodevices are developed and reached the markets that are expected to be used with an enormous positive impact on human health. It has a vision to improve health by enhancing the efficacy and safety of nanosystems and nanodevices. It is extending its area detecting molecules associated with diseases such as cancer, diabetes mellitus, neurodegenerative diseases, detecting microorganisms and viruses associated with infections, such as pathogenic bacteria, fungi, HIV viruses, etc. The advancement in nanotechnology also helps in the treatment of neurodegenerative disorders such as Parkinson's disease and Alzheimer's disease. It is also applied in tuberculosis treatment, operative dentistry, ophthalmology, surgery, visualization, tissue engineering, antibiotic resistance, immune response, etc. For example, in cancer therapy, certain novel nanoparticles can respond to externally applied physical stimuli, making its way as suitable therapeutics or therapeutic drug delivery systems.

The nanoscaled materials that are widely used in nanomedicines to save human lives includes nanotubes, nanoshells, quantum dots, dendrimers, etc. They provide a high degree of integration between technology and biological systems. Based on the nature of application in the field of nanomedicine these nanosized materials are classified into two.

- Nanomaterials for sensory and diagnosis application
- Nanomaterials for therapeutic and drug delivery systems

Quantum dots and iron oxide nanoparticles find their application in medical diagnosis mainly for the imaging purposes. On the other hand, C_{60}, liposomes, dendrimers, and polymers find their application in drug delivery systems. However, nanomaterials like nanoshells and nanotubes have its application in both drug delivery systems and as biomedical sensors.

TABLE 1.1 Nanomaterials and Their Applications.

Diagnosis and sensory applications

Materials	Applications
Quantum dots	Imaging and tracking through fluorescence
Iron oxide particles	Imaging, MRI
Nanotubes	Biomedical sensors
Nanocore/shells	Biomedical sensors
Dendrimers	Imaging, MRI

Therapeutic and drug delivery applications

Materials	Applications
C_{60} Drug carriers	
Carbon nanotubes	Drug and gene delivery, imaging, photochemical ablation
Liposomes	Drug delivery
Dendrimers	Drug delivery
Polymers	Drug delivery
Nanocore/shells	Drug delivery

MRI, magnetic resonance imaging.

1.3.1 QUANTUM DOTS

Quantum dots are extremely small nanocrystals, usually ranging from 2 to 10 nm (10–50 atoms) in diameter and possess high semiconducting property. These small-sized materials behave differently; giving quantum dots the high degree of tunability and enabling them to have high-level applications to science and technology which had never been seen before. They have superior fluorescent properties and possess remarkable optical and electronic properties that can be precisely tuned by changing their size and composition. They are also sometimes referred to as artificial atoms, as they can be considered as a single object with bound, discrete electronic states, similar to the naturally occurring atoms or molecules. Quantum dots can be made to emit light of any wavelength with in the visible and infrared ranges and can be inserted almost everywhere, including liquid solution, dyes, etc. They can be attached to a variety of surface ligands and inserted into a variety of organisms for in vivo research.

Since the past few decades, imaging has become an important tool for disease diagnosis. The advances in magnetic resonance and computer technology have also contributed to imaging. Because of the superior photostability, narrow emission range, broad wavelength of excitation, different possibilities of modification, quantum dots have attained much attention from the science world. Due to their low expense and simple synthesis, quantum dots have already entered the market for experimental biomedical optical imaging applications. Fluorescent semiconductor quantum dots are proving to be extremely beneficial for medical applications through high-resolution cellular imaging. Their high degree of quantum yields and magnetism are the main advantages that make them an important imaging tool in in vivo imaging applications. These semiconductor quantum dots are several times brighter than normal organic dye molecules. They can produce multiple color imaging with single excitation. Fluorescent nanoparticles which include quantum dots and fluorescent dye tagged nanoparticles are being widely used in both in vitro and in vivo imaging and have become widely used for cancer cell imaging.[5–7]

Though quantum dots could revolutionize medicine, most of them are toxic. However, recent studies conducted at the University of California, Berkeley, have shown that protective coatings for quantum dots may eliminate toxicity.

1.3.2   LIPOSOMES

Liposomes are nano- or microparticles or colloidal carriers whose size ranges from 30 nm to several micrometers. They appear as spherical-shaped intracellular vesicles (Fig. 1.4) and are formed from phospholipids, steroids, etc.[8] Properties of liposomes were differing with the lipid combination, method of preparation, surface charge and mainly the size. The amphipathic nature and size make liposome as one of the good drug-delivering systems. Liposomes increase the solubility of drugs and also increase their pharmacokinetic properties like therapeutic index, fast metabolism, reduction of side effects, increases anticancer activities.[9] The drug encapsulated in the liposome is released from it depending upon the liposome composition, pH, osmotic gradient, and the surrounding medium. The benefit of using liposome as drug delivery vehicle is that it can be applied as colloidal, aerosol, or semisolid like gels or creams. Amphotericin B, doxorubicin, vasopressin, insulin, anti-inflammatory drugs, antibiotics, etc. are delivered using liposomes.

1.3.3 CARBON NANOTUBES

A long cylinder of three coordinated carbon, namely, carbon nanotube (CNT) that appeared as molecular wires were discovered in 1991 by Iijima[10] as a by-product during the synthesis of fullerene. From then, tremendous studies on CNTs are being conducted due to its unique chemical and physical properties. CNTs can be of two types: single-walled CNTs (SWCNTs) or multiwalled CNTs (MWCNTs). Pristine CNTs are insoluble in aqueous solution but surface functionalization like radicals, diazonium salts, nitrenes, and carbenes make it possible.[11–14] As a result,

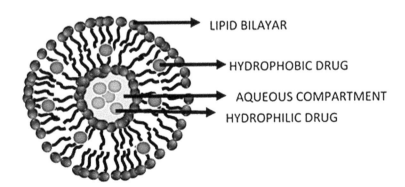

FIGURE 1.4 (See color insert.) Structure of liposome.

an extensive variety of applications mostly in biomedical field has been formulated.

Wide application of CNTs in biomedical field is due to its high surface area which is capable of adsorbing or conjugating a wide range of therapeutic agents like antibodies, drugs, genes, etc. CNTs are highly successful as an excellent vehicle for drug delivery (Fig. 1.5), which inject drugs directly to target sites without their metabolism by the body.[15] The general process of drug delivery using CNTs is that the drug is fixed on the functionalized CNTs and introduced to the animal body indirectly in a classical way such as oral or injection or directly using an external magnet to the target (like lymphatic nodes). Finally, nanotubes spill into the cells and thus the drug is delivered successfully. When the CNT drug vehicle enters the target, it delivers the drug in two possible ways: either incorporates with CNT or without CNT. However, the former is more effective as the drug molecule is released inside the cells. By using CNTs as drug vehicle, it is also possible to promote the cellular uptake of drugs, which decreases the dosage of drugs used.

Because of the unique features of CNTs, most of the anticancer drugs like epirubicin, cisplatin, methotrexate, and quercetin were successfully tested with CNTs as drug delivery vehicle.[16] Li et al. showed that SWCNTs functionalized with p-glycoprotein antibodies loaded with doxorubicin is an anticancer drug which is better than free doxorubicin. Yang et al. used MWCNTs in effective antitumor immunotherapy and applied in a mouse model bearing H22 liver tumor. SWCNTs exhibit strong absorbance of near IR region and hence is used for local antitumor hyperthermia therapy.

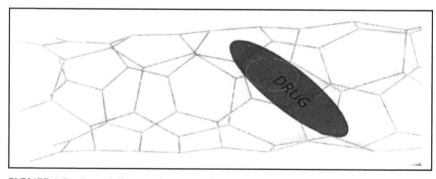

FIGURE 1.5   Drug delivery using nanotubes.

Besides anticancer drug, CNTs are used as drug vehicle for antimicrobial agents, antifungal agents, etc. They also used as a carrier for antigens.

The electrochemical property of CNT makes it highly efficient as biosensors. CNTs are used as glucose biosensors[17] for the selective detection of glucose and are known as CNT nanoelectrode ensembles (CNT-NEEs). CNT-NEEs offer membrane-free and mediator-free biosensors.

1.3.4 DENDRIMERS

Dendrimers are branched molecules. The name originates from two words "dendro" meaning tree and "meros" means part. As the name describes, dendrimers[18] are symmetric homogenous and monodispersive molecule. Dendrimer studies have started since 1978 which was first introduced by Vogtle et al. followed by Donald A. Tomalia who synthesized a family of dendrimers. Two methods, namely, convergent and divergent are most commonly used for dendrimer synthesis. In divergent method, the dendrimer grows outwards from a functionalized core molecule and the reaction proceeds to different generation. In this method, there is chance for side reaction which leads to incomplete reaction. To overcome this, more reagents are needed and face the difficulty of purification. Convergent method uses a stepwise starting from an end group and progress proceeds inward. The structure of the dendrimer makes it suitable for drug delivery. In order to use a drug, it should be water soluble. The design of dendrimer backbones makes it suitable for water solubility (Fig. 1.6).

Polyamidoamine (PAMAM) which was prepared by Tomalia et al. has a notable application in medicinal field.[19] The surface of these amine groups needs modification because of the toxic action and chances of the drug being accumulated in the liver. The drug cisplatin within the drug vehicle PAMAM exhibit a slower release of drug and is less toxic than the free cisplatin.

Dendrimers are also used in MRI contrast agents,[20] for example, diaminobutane dendrimers-Gd. The polychelated dendrimer gives a high-quality MR angiography. Apart from this, dendrimer itself acts as a medicine. Polylysine dendrimers with sulfonated naphthyl groups are used as antiviral drugs against the herpes simplex virus.[21] Vivagel is such a polylysine dendritic drug which reduces or prevents HIV virus.

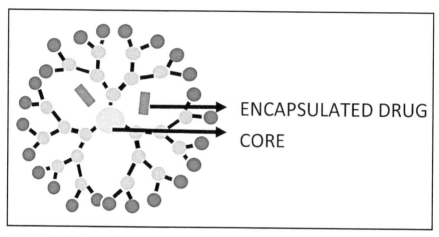

FIGURE 1.6   **(See color insert.)** Structure of dendrimers.

1.3.5 FULLERENES

Fullerenes are one of the allotropes of carbon that had put an interest to the minds of the researchers of the present time. These are molecules that are entirely made of only carbon atoms and were discovered in 1985 at Rice University by Richard Smalley, Robert Curl, and Harry Kroto.[22] They are spherical structures having pentagon and hexagonal units.

Functionalized fullerenes[23] are used for biomedical applications such as imaging and drug delivery. Fullerene's nanoscale dimensions favor passive targeting. Fullerene-based amino acids and peptides are used for peptide delivery. Paclitaxel-embedded buckysomes, composed of amphilic fullerene of range 100–200 nm in which paclitaxel is embedded in hydrophobic pocket, is used for metastatic breast cancer. Dendrofullerene (DF-1) is used for radioprotection, carboxyfullerene is used as antioxidant and neuroprotecting agent, and bisphonatefullerenes are used in bone therapy.

The endohedral metallofullerenes are used as contract agents in MRI which stimulus based on the pH. $Gd-C_{60}[C(COOH)_2]_{10}$ is a good candidate for tracking any mammalian cells. Functionalized fullerene with pyrolidinium is used for photodynamic killing of cancer cells.

1.3.6 NANOSHELL AND NANOPARTICLES

Nanoshells are a novel class of nanoparticles which are optically tunable. It consists of a core which is dielectric and a thin coating of metal shell around it. Nanoshell studies were started by Halas et al. Experimentally, first metal nanoshell was prepared by Zhou et al. using the present available techniques; there are many methods to synthesize the nanoshell. Most of them are single- and double-step synthesis. The uniform coating is the challenge that faces in the synthesis. Nanoshells can be synthesized using metal, semiconductors, and insulators. Dielectric materials can be used as core like silica and polystyrene; they are stable (Fig. 1.7). Depending on the core radius and thickness of shell, nanoshell can be designed to absorb or scatter light to a large range of spectra that can allow the maximum light penetration through the tissues. This property of nanoshells along with its chemical inertness and water solubility allows it to be used for diagnostic and therapeutic purposes. Oxide and metal nanoshells are the two different types of nanoshells. Metal nanoshells have notable application in the cancer treatment and bioimaging.

Gold nanoshell has highly tunable optical properties which are possible to probe for several antigens simultaneously by varying optical resonances.[24] Nanoshells are used as contrast agent in optical coherence tomography of breast cancer.

Nanoparticles for example gold and iron oxide used in diagnosis and drug delivery are also gaining attraction from the scientific world. Magnetic nanoparticles (MNPs) have a dimension less than 20 nm, are superparamagnetic at room temperature in presence of an external magnetic

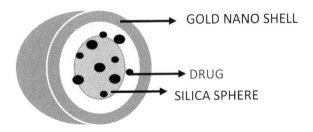

FIGURE 1.7   Structure of nanoshell.

field, but their net magnetic moment is regularly rationalized to zero by thermal agitation. This unique property makes MNPs an attractive candidate for biomedical application especially in drug delivery.[25] For example, $Au–Fe_3O_4$ dumbbell-shaped MNP targeted to breast cancer cells by conjugating cisplatin on Au and Herceptin, antibodies on with some cisplatin Fe_3O_4, etc. This targets to cancer cells more effectively than other drug vehicles. Nanoparticles incorporated with polymer shells are also introduced to the biomedical field.

1.3.7 POLYMERIC NANOPARTICLES

Polymeric nanoparticles form another set of drug delivery system with diameter ranging from 10 to 100 nm. They are obtained from synthetic polymers[1] like polycaprolactone (PCL), polyacrylamide, polyacrylate, or from natural polymers like albumin, DNA, chitosan, gelatin, etc. These polymeric nanoparticles are usually coated with nonionic surfactants to decrease the immunological interactions and intermolecular interactions among its surface chemical groups.[26] Here, drugs are immobilized on their surface through polymerization reaction or encapsulated during any polymerization step (Fig. 1.8). The drugs may be released at the target tissue by desorption, diffusion, or through nanoparticle erosion.

Some of the drugs for which polymeric nanoparticles acts as carriers include[27] Carboplatin—an antineoplastic drug, 5-flurouracil—an anticancer drug, doxorubicin—an antineoplastic agent, mitomycin C—a drug against bladder cancer, rifampicin—an antitubercolosis drug, lamivudine—an anti-HIV drug, tacrine—an anti-Alzheimer's drug, retinyl acetate—drug against skin inflammation, clotrimazole—an antifungal drug, etc. The drug, retinyl acetate is usually delivered using ethyl cellulose, which increases its aqueous stability and photostability. Similarly, chitosan-g-poly(N-vinylcaprolactam) biopolymer is generally used to deliver 5-fluorouracil that is used in cancer treatment. However, depending on the in vivo nature polymeric nanoparticles are classified as degradable polymeric nanoparticles like poly-L-lactide (PLA) and polyglycolide and non-biodegradable polymeric nanoparticles like polyurethane.

Biodegradable polymeric nanoparticles are gaining much attention in recent years as potential drug delivery systems because of their applications in the controlled release of drugs, due to their ability to target particular organs or tissues, as DNA carriers in gene therapy, in its ability to deliver proteins, peptides, etc. Perioral route of administration of the drug is used to dissolve or disperse it into the genes. The PLA, polylactide-co-glycolide, and polylactic-co-glycolic acid polymers have been used in controlled-release formulations in parenteral and in drug delivery applications in implantation due to their tissue compatible nature. Pitt et al. first reported poly($\in$-caprolactone), PCL, for the controlled release of steroids and narcotic antagonists as well as to the delivery of opthalmic drugs.

The drug delivery mechanism of polymeric nanoparticles is such that the drug is dissolved in the polymerization medium either before the monomer addition or at the end of the polymerization reaction. Then, by using ultracentrifugation or by resuspension of the particle in a surfactant-free isotonic medium, the nanoparticle suspension is purified. Various stabilizers such as dextran-70, dextran-40, dextran-10, poloxamer-188, -184, and -237, are used during polymerization. Along with this some surfactants like polysorbate-20, -40, and -80 are also added. The size of particles and its molecular mass depend upon the nature and concentration of the stabilizer and surfactants that are added.

Studies have also reported the preparation of ethyl-2-(ethoxycarbonyl) ethyl methylene malonate-co-ethylene oxide. These polymers have both hydrophilic and hydrophobic functional groups, and they can act as better polymers in preparing the long circulating nanoparticles.

The preparation of the DNA chitosan nanoparticles were reported by Mao et al. and are used as oral gene delivery systems. The chitosan nanoparticles are better carriers for loading than gelatin-based nanoparticles for loading the antineoplastic proteins. Polyesters that are novel and biodegradable and consisting of short polylactone chains attached onto PVA (polyvinyl alcohol) or charge-modified sulfobutyl-PVA were also prepared by polymerization of lactide and glycolide in the presence of core polyols of different kinds.

1.4 NANOROBOTICS

The origin of nanotechnology is most often associated with the talk given by the Nobel Prize winner Richard Feynman in 1959 which is entitled "There's Plenty of Room at the Bottom." Here Feynman discussed the possibilities of nanotechnology and how can its advancement potentially generate an enormous number of technical applications. It is the most

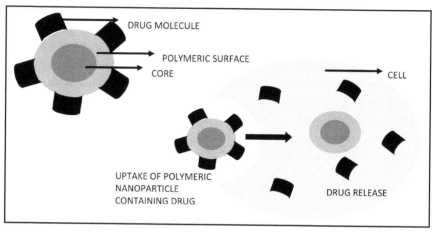

FIGURE 1.8   Polymeric nanoparticle-containing drug.

pertinent interest of the Intelligent Systems and Robotics Center to determine its role in the field of nanotechnology. Nanorobotics is an emerging branch of nanotechnology.

The first scientist to describe the medical applications of nanotechnology and nanorobots was Robert Freitas Jr. In the *Journal of American Dental Association*, he defined nanomedicine as the science and technology of diagnosing, treating, and preventing disease and traumatic injury, of relieving pain; and of preserving and improving human health, through the use of nanostructured materials, biotechnology, genetic engineering, and also complex molecular machine systems and nanorobots.[28]

Nanorobotics also referred as molecular robotics is a highly emerging and advancing multidisciplinary field that connects medical, pharmaceutical, biomedical, engineering, as well as other applied and basic sciences. Nanorobots differ from macroworld robots, due to their nanosized construction. Nanorobots are nanoelectromechanical systems and are comparable to the biological cells and organelles in their size. The technology of designing, fabricating, and programming of these nanorobots is known as nanorobotics.[29,30] Nanorobots are controllable machines having nanometer- or molecular-scale dimensions and are composed of nanoscale components. With the modern scientific technology, it has become possible to create nanorobotic devices and make them interface with the macroworld. There are many such machines existing in nature, and there is also an opportunity to build many more such devices by mimicking nature. Even though nanorobotics is fundamentally different from macrorobots, there are many similarities in its design and control techniques that could be projected and applied. Nanorobots are invisible to naked eye, and hence, modern techniques like scanning electron microscopy and atomic force microscopy are being employed to visualize the molecular structure of these devices.[31] The development of nanorobots and nanomachine components offers difficult fabrication and control challenges.

Nanorobots are of two types, bionanorobots and artificial nanorobots. Nanorobots that are designed by harnessing the properties of biological materials like peptides and DNA their designs and functionalities are bionanorobots. These are inspired not only by nature but also through machines. Artificial nanorobots are scientifically and synthetically created. However, no nanorobots have been created that is completely artificial.

Nanorobots require energy to carry out various operations like propulsion, force, actuation, communication, or any other activity in the biological system at nanoscale dimensions.[32] This energy can be generated by natural or artificial entities known as molecular motors. This motors when performs their duty at nanoscale are known as nanomachines.[33] The natural molecular motors are present in the biological system and carry out functions in the body at molecular or nanolevel. Most of these motors are made of proteins or DNA. Studies are being done on these natural motors in order to use them as efficient motors for artificial nanorobots like kinesin molecular motors,[34] flagella motors,[35] DNA scissors,[36] and DNA tweezers.[37,31] In the similar way, chemical molecular motors are used as nanomachines in artificial nanorobotics.[38] These machines are highly difficult to create but are stronger than natural machines. These machines are mostly constructed from organic compounds such as carbon, nitrogen, and hydrogen. They can be chemically, electrochemically, or photochemically controlled. Many chemical molecular machines are also constructed using interlocked organic compounds like rotaxanes and catananes.

Nanorobotics finds their applications in medicine which includes the early diagnosis and targeted drug delivery, treatment and medicine for cancer, biomedical instrumentation, surgery, pharmacokinetics, monitoring of diabetes, health care, etc. Nanorobots help in monitoring, diagnosing diseases, and help to restore lost tissue at cellular levels.[39]

In the health care field, the nanorobotics helps to perform good treatment through biomedical applications by improved techniques in treatment.

This can help in decreasing the deadly diseases in future. Nanorobotics is used in medical application, treatment of cancer, in gene therapy, for brain aneurysm, in dentistry, etc.[40] Cancer can be successfully treated with the current medical technologies and therapies using nanorobotics. Nanorobots containing embedded chemical biosensors can be used in detection of tumor cells at early stages of development inside the patient's body. Integrated nanosensors can be used for such a task in order to find its intensity. Scientists have genetically modified salmonella bacteria which are drawn to the tumors by chemicals that are secreted by cancerous cells. These bacteria carry microscopic robots which are about 3 μm in size and can automatically release capsules filled with drugs when the bacteria reach the tumor cells. By delivering drugs directly to the tumor, the nanorobot, which they named as bacteriobot, attacks the tumor leaving healthy cells apart and protects the patient from the side effects of chemotherapy. Bactiriobot can not only help in the treatment of breast cancers and colorectal but also helps in the treatment of other cancers as well. The high potential of nanorobots in the biomedical applications and the fewer side effects on its applications make nanorobots very desirable.

Nanorobots through precise chemical dosage and administration and in a similar approach could be used to deliver anti-HIV drugs.[41]

Glucose which is carried through the blood stream has the most important role in maintaining the human metabolism, and the maintenance of its correct level is a key issue in the diagnosis and treatment of diabetes. Nanorobots can thus be used to maintain the normal glucose level in the body. However, the higher the initial design cost, the more complications will be involved in designing of nanorobots, therefore making it less available. However, scientists have been succeeded to develop only biological nanorobotic systems. Artificial nanorobots are still a concept that has to be highly explored. Nanorobotics is an upcoming field that interconnects various fields of science and technology. The advantage of its application in various fields can overweight the challenges and hurdles it faces during the development process.

1.5   FUTURE PERSPECTIVES AND CONCLUSION

The widespread application of nanoparticles is due to their unique physical and chemical properties. There are a large number of processes that are widely used to synthesize these nanoparticles with desired characteristics. However, these syntheses are highly expensive, require much human labor, and are less environment friendly. Thus, the production of nanoparticles using environment-friendly processes is quite a promising area of research to the development of products that are free from unwanted toxicity. This has been attained when plants have been used as bioreactor for the synthesis of nanoparticles.[42] Example of such green nanoparticle[43] include the plants *Brassica juncea* and *Medicago sativa* which accumulates about 50 nm silver nanoparticles to a high level of 13.6% of their own weight when grown on silver nitrate substrate. Gold nanoparticles of 4 nm size in *M. sativa*, and copper nanoparticles with size 2 nm in *Iris pseudacorus* when grown on substrates containing the respective metal salts were also detected.

Although many goals have been attained with the application of nanoparticles in the field of science and medicine, more investigations and developments are required to meet the needs of our fast-running world.

KEYWORDS

- **nanoscience**
- **nanomedicine**
- **nanoparticles**
- **biosensors**
- **cancer therapy**
- **nanorobotics**
- **green nanoparticles**

REFERENCES

1. Wilczewska, A. Z., et al. Nanoparticles as Drug Delivery Systems. *Pharmacol. Rep.* **2012**, *64*(5), 1020–1037.
2. Strambeanu, N.; Demetrovici, L.; Dragos, D. Natural Sources of Nanoparticles. In *Nanoparticles' Promises and Risks*; Springer: Cham, 2015; pp 9–19.
3. Feynman, R. P. There's Plenty of Room at the Bottom. *Eng. Sci.* **1960**, *23*(5), 22–36.
4. Ralph, S. Nanomagnetics. *J. Phys.: Condens. Matter* **2003**, *15*(20), R841.
5. Sage, L. Biosphere: Finding Cancer Cells with Quantum Dots. *Anal. Chem.* **2004**, *76*(23), 453-A.
6. Yun, X., et al. Molecular Profiling of Single Cancer Cells and Clinical Tissue Specimens with Semiconductor Quantum Dots. *Int. J. Nanomed.* **2006**, *1*(4), 473.

7. Li, Z., et al. Rapid and Sensitive Detection of Protein Biomarker Using a Portable Fluorescence Biosensor Based on Quantum Dots and a Lateral Flow Test Strip. *Anal. Chem.* **2010**, *82*(16), 7008–7014.

8. Abolfazl, A., et al. Liposome: Classification, Preparation, and Applications. *Nanoscale Res. Lett.* **2013**, *8*(1), 102.

9. de Araújo Lopes, S. C., et al. Liposomes as Carriers of Anticancer Drugs. In *Cancer Treatment—Conventional and Innovative Approaches*; Intech Open: London, 2013.

10. Iijima, S. Carbon Nanotubes: Past, Present, and Future. *Physica B: Condens. Matter* **2002**, *323*(1), 1–5.

11. Holzinger, M., et al. Sidewall Functionalization of Carbon Nanotubes. *Angew. Chem. Int. Ed.* **2001**, *40*(21), 4002–4005.

12. Holzinger, M., et al. Functionalization of Single-walled Carbon Nanotubes with (R-) Oxycarbonyl Nitrenes. *J. Am. Chem. Soc.* **2003**, *125*(28), 8566–8580.

13. Holzinger, M., et al. [2+ 1] Cycloaddition for Cross-linking SWCNTs. *Carbon* **2004**, *42*(5), 941–947.

14. Bahr, J. L.; Tour, J. M. Highly Functionalized Carbon Nanotubes Using In Situ Generated Diazonium Compounds. *Chemi. Mater.* **2001**, *13*(11), 3823–3824.

15. Singh, B. G. P., et al. Carbon Nanotubes. A Novel Drug Delivery System. *Int. J. Res. Pharm. Chem.* **2012**, *2*(2), 523–532.

16. He, H., et al. Carbon Nanotubes: Applications in Pharmacy and Medicine. *BioMed Res. Int.* **2013**, *2013*, 12.

17. Lin, Y., et al. Glucose Biosensors Based on Carbon Nanotube Nanoelectrode Ensembles. *Nano Lett.* **2004**, *4*(2), 191–195.

18. Tomalia, D. A. Dendrimer Molecules. *Sci. Am.* **1995**, *272*, 62–66.

19. Svenson, S.; Tomalia, D. A. Dendrimers in Biomedical Applications—Reflections on the Field. *Adv. Drug Deliv. Rev.* **2012**, *64*, 102–115.

20. Langereis, S., et al. Multivalent Contrast Agents Based on Gadolinium−Diethylene-triaminepentaacetic Acid-terminated Poly(Propylene Imine) Dendrimers for Magnetic Resonance Imaging. *Macromolecules* **2004**, *37*(9), 3084–3091.

21. Bourne, N., et al. Dendrimers, A New Class of Candidate Topical Microbicides with Activity Against Herpes Simplex Virus Infection. *Antimicrob. Agents Chemother.* **2000**, *44*(9), 2471–2474.

22. Kroto, H. W., et al. C 60: Buckminsterfullerene. *Nature* **1985**, *318*(6042), 162–163.

23. Partha, R.; Conyers, J. L. Biomedical Applications of Functionalized Fullerene-Based Nanomaterials. *Int. J. Nanomed.* **2009**, *4*, 261

24. Bardhan, R., et al. Theranostic Nanoshells: From Probe Design to Imaging and Treatment of Cancer. *Acc. Chem. Res.* **2011**, *44*(10), 936–946.

25. Hao, R., et al. Synthesis, Functionalization, and Biomedical Applications of Multifunctional Magnetic Nanoparticles. *Adv. Mater.* **2010**, *22*(25), 2729–2742.

26. Torchilin, V., Ed. *Multifunctional Pharmaceutical Nanocarriers*; Springer Science & Business Media: New York, 2008; Vol. 4.

27. Soppimath, K. S., et al. Biodegradable Polymeric Nanoparticles as Drug Delivery Devices. *J. Control. Release* **2001**, *70*(1), 1–20.

28. Biji, B.; Narayanan, S. Nano Robotics—Its Time for Change. *IJOCR*, **2014,** *2*(5).

29. Sierra, D. P.; Weir, N. A.; Jones, J. F. *A Review of Research in the Field of Nanorobotics*. No. SAND2005-6808; Sandia National Laboratories, 2005.

30. Yoseph, B.-C. *Biomimetics: Biologically Inspired Technologies*; CRC Press: New York, 2005.
31. Ummat, A., et al. Bio-nano-robotics: State of the Art and Future Challenges. In *Invited Chapter in the Biomedical Engineering Handbook*, 3rd ed.; Yarmush, M. L., Ed.; CRC Press: New York, 2004.
32. Eric, D. K. Molecular Engineering: An Approach to the Development of General Capabilities for Molecular Manipulation. *Proc. Natl. Acad. Sci. U.S.A.* **1981**, *78*(9), 5275–5278.
33. Eric, D. K. *Nanosystems: Molecular Machinery, Manufacturing, and Computation*; John Wiley & Sons, Inc.: New York, 1992.
34. Block, S. M. Kinesin: What Gives? *Cell* **1998**, *93*(1), 5–8.
35. Berry, R. M.; Judith P. Armitage. The Bacterial Flagella Motor. *Adv. Microb. Physiol.* **1999**, *41*, 291–337.
36. Mitchell, J. C.; Yurke, B. In *DNA Scissors*, International Workshop on DNA-Based Computers; Springer: Berlin, Heidelberg, 2001.
37. Yurke, B., et al. A DNA-Fuelled Molecular Machine Made of DNA. *Nature* **2000**, *406*(6796), 605–608.
38. Balzani, V., et al. Artificial Molecular Machines. *Angew. Chem. Int. Ed.* **2000**, *39*(19), 3348–3391.
39. Prajapati, P. M.; Solanki, A. S.; Sen, D. J. Importance of Nanorobots in Health Care. *Int. Res. J. Pharmacy* **2012**, *3*(3), 122–124.
40. Sivasankar, M.; Durairaj, R. B. Brief Review on Nano Robots in Bio Medical applications. *Adv. Robot. Autom.* **2012**, *1*(101), 2.
41. Cavalcanti, A., et al. In *Nanorobotic Challenges in Biomedical Applications, Design and Control*, Electronics, Circuits and Systems, 2004. ICECS 2004. Proceedings of the 2004 11th IEEE International Conference on; IEEE, 2004.
42. Makarov, V. V., et al. "Green" Nanotechnologies: Synthesis of Metal Nanoparticles Using Plants. *Acta Naturae* **2014**, *6*(1), 35–44. (English version).
43. Lam, P.-L., et al. Recent Advances in Green Nanoparticulate Systems for Drug Delivery: Efficient Delivery and Safety Concern. *Nanomedicine* **2017**, *12*(4), 357–385.

CHAPTER 2

PLGA-BASED NANOPARTICLES FOR CANCER THERAPY

JOMON JOY[1*] and SABU THOMAS[2]

[1]*School of Chemical Sciences, Mahatma Gandhi University, P D Hills, Kottayam 686560, Kerala, India*

[2]*International and Inter University Centre for Nanoscience and Nanotechnology, Mahatma Gandhi University, Kottayam 686560, Kerala, India*

Corresponding author. E-mail: jomonjjoy87@gmail.com

ABSTRACT

The emergence of nanotechnology has made a considerable impact on clinical therapeutics in the last two decades. Progress in biocompatible nanoscale drug carriers such as liposomes and polymeric nanoparticles (NPs) has enabled more efficient and safer delivery of a numerous of drugs. Advantages in NP drug delivery, chiefly at the systemic level, include longer circulation half-lives, enhanced pharmacokinetics, and reduced side effects. In cancer treatments, NPs can additional rely on the enhanced permeability and retention effect caused by leaky tumor vasculatures for better drug accumulation at the tumor sites. These advantages have made therapeutic NPs a promising candidate to replace traditional chemotherapy, where intravenous injection of noxious agents poses a serious threat to healthy tissues and results in dose-limiting side effects. Currently, several NP-based chemotherapeutics have emerged on the market, while many are undergoing various stages of clinical or preclinical development. NP drug delivery enhances therapeutic effectiveness and reduces side effects of the drug payloads by improving their pharmacokinetics.

Poly(lactic-co-glycolic acid) (PLGA) is one of the most successfully developed biodegradable polymers. Along with the different polymers developed to formulate polymeric NPs, PLGA has attracted considerable interest due its biodegradability and biocompatibility, and it is approved by FDA and European Medicine Agency. Methods of preparation of PLGA NPs, surface modification, release mechanisms, and various applications in cancer therapy were discussed in this chapter.

2.1 INTRODUCTION

Cancer is an ever-increasing menace that needs to be curbed soon. As mortality due to cancer continues to rise, progress of research in this area requires a new paradigm in the diagnosis and therapy of the disease. Conventional therapeutic approaches such as radiation, chemotherapy, combinational chemotherapy, and surgical treatments are extensively accepted to treat or eradicate tumor(s). While chemotherapy remains a highly successful weapon to treat cancer, it is frequently associated with limitations and major side effects.[1,2] The main drawbacks of chemotherapy is the limited accessibility of drugs to the tumor tissues requiring high doses, their unbearable toxicity, development of multiple drug resistance and their nonspecific targeting. Nanoparticle (NP) drug delivery has yielded unprecedented level of control over their pharmacokinetics of chemotherapeutic agents. Recent developments in the NP-based combination therapy have shown several unique features that are untenable in traditional chemotherapy.[3,4]

For the past few decades, NP technology has been widely employed in biology and medicine, including for cancer therapy. Nanoscale particles between 10 and 200 nm in diameters have shown more favorable pharmacokinetic profiles as compared to small-molecule drugs; these drug-loaded NPs exhibit prolonged systemic circulation lifetime, constant drug release kinetics and better tumor accumulations through both passive and active mechanisms.[5-7] Recently, nanocarriers are gaining increasing attention for their ability to coencapsulate multiple therapeutic agents and to synchronize their delivery to the diseased cells. Different NP platforms such as polymeric micelles, dendrimers, liposomes, and mesoporous silica particles have been used to carry broad classes of therapeutics including chemosensitizers, cytotoxic agents, small interference RNA, and antiangiogenic agents.

Polymeric NPs advances in biomaterials research have led to the emergence of biocompatible and biodegradable polymeric NPs for drug delivery applications. Several synthetic polymers approved by the US FDA such as poly(lactic-*co*-glycolic acid) (PLGA) and polycaprolactone (PCL) and several natural polymers such as chitosan and polysaccharides have been investigated extensively for NP synthesis.[8–12] Compared with liposomes, polymeric NPs generally have higher stability, sharper size distribution, more tunable physicochemical properties, constant and more controllable drug-release profiles, and higher loading capability for poorly water soluble drugs. The polymer platform also offers higher synthetic freedom that allows particles to be tailored for specific needs. Owing to these unique characteristics, polymeric NPs have attracted tremendous interests from academia and industry.[13]

Polymer-based NPs are submicron-sized polymeric colloidal particles in which a therapeutic agent of interest can be fixed or encapsulated inside their polymeric matrix or adsorbed or conjugated onto the surface. Polymeric NPs typically consist of amphiphilic diblock copolymers that self-assemble into NPs in aqueous solutions. For in vivo drug delivery, PLGA and poly-ethylene glycol (PEG) are popular choices for the hydrophobic and the hydrophilic block, respectively, as PLGA can hydrolyze into lactic acid and PEG can significantly reduce nonspecific cellular uptake by forming a stealth layer.[14,15] Drug encapsulation is typically achieved by mixing the drugs with the polymer solutions during particle preparation process. For instance, in NP synthesis through solvent displacement technique, a water–miscible solvent such as acetonitrile is used to dissolve the hydrophobic drugs together with the diblock copolymers. The solution is subsequently mixed with water. The organic solvent diffuses into the aqueous phases and eventually evaporates, and the hydrophobic polymers self-assemble to form NPs with drugs encapsulated inside. Although polymeric NPs are most suitable for delivering hydrophobic drugs, several reports have shown success in encapsulating hydrophilic drugs through surface attachment or polymer–drug conjugation techniques.[16,17]

Many approaches have been taken to coencapsulate multiple therapeutic agents into a single polymeric NP. Presently, these approaches can be divided into three major categories, which are as follows:

- Directly encapsulating multiple drugs into the hydrophobic polymeric core;

- Incorporating an additional media compartment to the NP, usually on the NP surface, to create a separate partition for drug loading;
- Covalently conjugating several drugs to the polymer backbone before the synthesis of NP.[18–20]

PLGA is one of the most effective biodegradable polymeric NPs. It has been accepted by the US FDA to use in drug delivery systems (DDSs) due to controlled and sustained released properties, low toxicity, and biocompatibility with tissues and cells. PLGA is a copolymer synthesized via random ring opening copolymerization of two different monomers (glycolic acid and lactic acid). The forms of PLGA are usually recognized by the monomers ratio used. For example, PLGA 50:50 represents a copolymer whose composition is 50% lactic and 50% glycolic acid. PLGA is one of the most fruitfully used polymers due to its biocompatibility and long-standing track record in biomedical functions. It is a commonly used biodegradable polymer for the development of nanomedicines because it undergoes hydrolysis in the body to produce the biodegradable metabolite monomers (lactic acid and glycolic acid). Monomers obtained after hydrolysis are simply metabolized in the body via the Krebs cycle and are removed as carbon dioxide and water, thus resulting in minimal systemic toxicity.[21–23]

It is widely used in the medical field due to the following attractive properties:

i)   Biodegradability and biocompatibility

ii) European Medicine Agency and FDA approval in DDSs for parenteral administration

iii) Well described formulations and methods of fabrication adapted to various types of drugs, for example, hydrophobic or hydrophilic small molecules or macromolecules

iv) Protection of drug from degradation

v)   Possibility of sustained release

vi) Possibility to transform surface properties to provide stealthiness and/or better interaction with biological materials

vii) Possibility to target NPs to exact organs or cells.

2.2 PREPARATION AND SURFACE MODIFICATION OF PLGA NPs

2.2.1 METHODS FOR PREPARATION OF PLGA NPs

There are several methods to categorize NPs. Depending on the process of preparation, the structural organization may be different. Drug is either entrapped inside the core of a "nanocapsule" or adsorbed on the surface of a matrix nanosphere.[22–25] Dispersion of preformed polymers is the most generally used technique to prepare biodegradable NPs from poly-lactic acid (PLA); poly-DL-glycolide; poly-D-L-lactide-*co*-glycolide; and poly-cyanoacrylate. These methods usually include two important steps. The first step is the preparation of an emulsified system, and this is common to all the techniques used. The NPs are formed through the second step, which differs according to the process used. Generally, the principle of this second step gives its name to the method.[26]

(a)   Single or double-emulsion-solvent evaporation method

The most commonly used method for PLGA NP formation is the single or double-emulsion-solvent evaporation. Single-emulsion process involves oil-in-water (o/w) emulsification, whereas the double-emulsion process is a water-in-oil-in-water (w/o/w) method. The w/o/w process is best suited to encapsulate water-soluble drugs, like peptides, proteins and vaccines, whereas the o/w process is perfect for water-insoluble drugs, such as steroids.[27,28] In some cases, solid/oil/water (s/o/w) procedures have been used with PLGA-based microspheres, particularly for a higher drug loading of large water-soluble peptides, such as insulin.[28,29] In o/w method, the polymer is dissolved in an organic solvent such as chloroform, dichloromethane, or ethyl acetate. The drug is dispersed or dissolved into the preformed polymer solution, and this mixture is emulsified into an aqueous solution to make an oil (O) in water (W), that is, o/w emulsion via by means of a surfactant (emulsifying agent) like gelatin, poly(vinyl alcohol), poloxamer-188, polysorbate-80, etc. Following the formation of a stable emulsion, the organic solvent is removed either by increasing the temperature or under pressure or by nonstop stirring. The above ways use a high-speed homogenization or sonication. However, these procedures are excellent for a laboratory-scale procedure, but for a large-scale pilot production, alternative methods using low-energy emulsification are necessitated.[30,31] The size can be controlled by regulating the type and amount of dispersing agent, stir rate, viscosity of organic and aqueous phases, and temperature.[32,33] Although different types of emulsions may be used, o/w emulsions are of interest because they use water as the nonsolvent; this simplifies and thus improves process economics, because it eliminates the requirements for recycling, facilitating the washing step, and reduce agglomeration.[33] However, this technique can only be applied to liposoluble drugs, and limitations are forced by the upgrade of the high energy requirements in homogenization.[31]

(b)   Emulsification solvent diffusion method

In the method developed by Quintanar-Guerrero et al.,[36] the solvent and water are equally saturated at room temperature before use to make sure the initial thermodynamic equilibrium of both liquids. Afterwards, the organic solvent containing the dissolved polymer and the drug is emulsified in an aqueous surfactant solution (typically with PVA as a stabilizing agent) by using a high-speed homogenizer. Water is then added under regular stirring to the o/w emulsion system, as a result causing phase transformation and outward diffusion of the solvent from the internal phase, resulting to the nanoprecipitation of the polymer and the formation of colloidal NPs. Finally, the solvent can be removed by vacuum steam distillation or evaporation.[34] This method presents several advantages, for example high batch-to-batch reproducibility, high encapsulation efficiencies (generally 70%), no need for homogenization, simplicity, ease of scale-up and narrow size distribution. Disadvantages are the large volumes of water to be removed from the suspension and the leakage of water-soluble drug into the saturated-aqueous external phase through emulsification, dropping encapsulation efficiency.[33,36]

(c)   Emulsification reverse salting-out method

The emulsification reverse salting-out method involves the addition of polymer and drug solution to a water-miscible solvent, like acetone, and to an aqueous solution containing the salting-out agent, like calcium chloride, magnesium chloride, and a colloidal stabilizer, like polyvinyl pyrrolidone, under vigorous mechanical stirring. As this o/w emulsion is diluted with a plenty amount of water, it induces the formation of NPs by increasing the diffusion of acetone into the aqueous phase. The dilution produces a sudden decrease in the salt concentration in the continuous phase of the emulsion, leading the polymer solvent to migrate out of the emulsion droplets. The salting-out agent and residual solvent are removed by cross-flow filtration.[37–40] Even though the emulsification-diffusion technique is a modification of the salting-out process, it has the advantage of avoiding the use of salts and thus eliminates the requirement for severe purification steps.[41] The most important advantage of salting out is that it minimizes tension to protein encapsulants.[42] Salting out does not need a raise of temperature and, thus, may be valuable when heat sensitive substances have to be processed.[43] The greatest disadvantages are exclusive function to lipophilic drugs and the extensive NP washing steps.[33,44]

(d)   Nanoprecipitation method

The nanoprecipitation method is a one-step process, also known as the solvent displacement method.[40,45] Nanoprecipitation is performed using systems containing three basic components, namely, polymer solvent, polymer, and the nonsolvent of the polymer.[46] Usually, this method is used for hydrophobic drug entrapment, but it has been suited for hydrophilic drugs additionally. Drugs and polymers are dissolved in a polar, water-miscible solvent like acetonitrile, acetone, methanol, or ethanol. The solution is poured in a controlled way (drop-by-drop addition) into an aqueous solution with surfactant. NPs are formed instantly by fast solvent diffusion. At last, the solvent is removed under reduced pressure.[47,48]

2.2.2   SURFACE MODIFICATION OF PLGA NPs

Depending on their surface features, NPs will be taken up by the spleen, liver and other parts of the reticuloendothelial system (RES).[60–62] Surface modification of NPs is important for absconding the body's natural defense systems when transporting drugs to the bloodstream.[40,63] A long circulation time enhances the chance that the NPs will reach their target. Nanostructures with a hydrophilic surface which is smaller than 100 nm have the supreme capacity to get away from the molecular phagocytic system.[40] Hydrophobic NPs will be preferentially taken up by RES organs.

It has been shown that hydrophilic particles can remain in the circulation for a longer time and are taken up by liver to a small extent.[64] Different strategies have been used to make a hydrophilic cloud around the NPs and reduce their uptake by RES organs. These strategies consist of coating of NPs with Tween 80, PEG, polyethylene oxide, poloxamers and poloxamines,[65–67] polysorbate 80, tocopheryl PEG succinate (TPGS), and polysaccharides like dextran.[68,69] Hydrophilic polymers can be useful at the surface of NPs by adsorption of surfactants or by utilize of block copolymers or branched copolymers. Surface chemistry analysis is determined by Fourier-transform infrared spectroscopy, X-ray photoelectron spectroscopy, and nuclear magnetic resonance spectroscopy.[70–72]

The most preferred process of surface modification is the adsorption or grafting of PEG to the surface of NPs. This is a hydrophilic, nonionic polymer that has been shown to exhibit exceptional biocompatibility.[73,74] Insertion of PEG and PEG-containing copolymers to the surface of particles results in an augment in the blood circulation half-life of the NPs. The precise mechanisms by which PEG prolonged circulation time of the surface modified NPs are still not well understood. It is usually thought that the increased residency of the NPs in blood is mainly due to prevention of opsonization of NPs via a certain serum or plasma proteins (opsonins). It is assumed that PEG causes steric repulsion via creating hydrated barriers on NP surfaces that prevents coating of PEG-modified NPs by serum opsonins. Many surveys have shown that the degree to which proteins (opsonins) adsorb on to NPs surface can be reduced by means of increasing the PEG density on the particle surface. Mounting the molecular weight of the PEG chains has also been shown to reduce opsonization of NPs and improve retention in the circulation.[48,75] High-surface density and long-chain lengths of PEG are necessary for low-protein adsorption. However, surface density has a superior effect than the chain length on steric repulsion and van der Waals attraction,[31] PEG is also believed to make easy mucoadhesion and consequent transport through the Peyer's patches of the gut associated lymphoid tissue.[31,76] In addition, PEG may benefit NP's interaction with blood components. PEGylated particles showed moderately higher uptake of drug by the spleen and the brain than conventional non-PEGylated NPs.[77] Consequently, the presence of PEG on the NPs imparts additional functionality during the use of polymeric NPs.[31]

Polysorbate is another material used for the surface modification of NPs. Polysorbate 20, 40, 60, and 80 have been used to coat the surface of

PBC NPs.[74,78] Polysorbate coating leads to the alteration of surface properties of the NPs. This new surface seems to adsorb certain substances from the blood through endothelial cells.[74,79] Polysorbate-coated NPs can cross the blood–brain barrier more efficiently.

TPGS is a form of vitamin E that has been used as a solubilizer, an emulsifier, and a vehicle in drug delivery formulations. Vitamin E TPGS has been employed as an emulsifier for producing NPs enclosing hydrophobic drugs and for advancing encapsulation, drug loading, and the release profile of particles.[40,54] TPGS augments the PLGA NPs adhesion to the cells and hemodynamic properties of the NPs.[80] Random PLGA-TPGS copolymers could work as a novel and potential biocompatible polymeric matrix material proper to NP-based DDSs for cancer chemotherapy.[81]

2.3 ENCAPSULATION OF VARIOUS ANTICANCER DRUGS ON PLGA NPs

Paclitaxel (PTX) (commercially available as taxol) interferes with the normal function of microtubule breakdown via binding with β-subunit of tubulin the polymerization of tubulin causing cell death by disordering the dynamics essential for cell division. PTX has neoplastic activity against primary ovarian carcinoma breast and colon tumors. It is one of the powerful anticancer agents but less useful for clinical administration owing to its poor solubility. PLGA mix together with vitamin E, and TPGS has been used to encapsulate by solvent evaporation/extraction techniques and in vitro controlled release of this drug.[74,82] Mu and Feng used α-tocopheryl PEG 1000 succinate as well as a matrix material with other biodegradable polymers for the fabrication of a NP formulation of PTX. They concluded that vitamin E TPGS was valuable either as an emulsifier or as matrix material mixed with PLGA for the production of NPs enabling controlled release of PTX.[40,50–52]

Cisplatin is known to cross-link DNA molecule in several ways to interfere cell division via mitosis. The damaged DNA elicits DNA repair mechanism. Cisplatin is a very strong anticancer drug but the full therapeutic utilization of cisplatin is limited due to its toxicity in healthy tissues.[83] The careful delivery of cisplatin to tumor cells would considerably reduce drug toxicity and improve its therapeutic index. This drug has been encapsulated on PLGA–methoxy(PEG) (mPEG) NPs prepared by double emulsion methods.[72] Cisplatin-loaded PLGA-mPEG NPs also resulted in long-lasting cisplatin residence time in the systemic circulation when used in mice with prostate tumor.[84]

Vincristine sulfate (VCR) is a helpful chemotherapeutic agent, which has been used extensively for the treatment of various cancers. Unluckily, many tumor cells are not susceptible to VCR due to efflux from the tumor cells mediated by P-glycoprotein and associated proteins.[85] The reason behind the association of drugs with colloidal carriers against drug resistance comes from the reality that P-glycoprotein probably identifies the drug to be effluxed out of the tumoral cell just when this drug is present in the plasma membrane. As a drug-loaded NP is typically present in the endolysosomal complex after internalization by cells, it possibly escapes the P-glycoprotein pump. Based on the optimal parameters, it was found that vincristine-loaded PLGA NPs could be formulated with expectable properties by combining the o/w emulsion-solvent evaporation technique and the salting-out technique. This study also showed that two hydrophilic low-molecular-weight drugs, VCR and verapamil (VRP), a chemosensitizer, could be concurrently entrapped into PLGA NPs, with a comparatively high entrapment efficiency of 55.35 ± 4.22% for VCR and 69.47 ± 5.34% for VRP in small-sized particles of 100 nm. Moreover, their studies showed that PLGA NPs simultaneously loaded with an anticancer drug and a chemosensitizer might be the formulation with the most probable in the treatment of drug-resistant cancers in vivo.[86,87]

9-Nitrocamptothecin (9-NC) (derivative of camptothecin) and associated analogues are a promising family of anticancer agents with an exclusive mechanism of action, aiming the enzyme topoisomerase-I. All camptothecin derivatives undergo a pH-dependent rapid and reversible hydrolysis from closed lactone ring to the inactive hydroxyl carboxylated form with loss of anticancer activity. The delivery of lipophilic derivatives of 9-NC is reasonably challenging due to instability at biological pH and its low water solubility. PLGA has been employed to encapsulate 9-NC effectively by nanoprecipitation methods having more than 30% encapsulation efficiency with its complete biological activity and without disturbing lactone ring.[88,74]

Etoposide is an anticancer agent used in the treatment of a variety of malignancies, including malignant lymphomas. It acts by inhibition of topoisomerase-II and activation of oxidation–reduction reactions to create derivatives that bind directly to DNA and cause DNA damage. The successful chemotherapy of tumors depends on continual exposure to anti-cancer agents for long-lasting periods. Etoposide has a short biological half-life (3.6 h), and although intraperitoneal injection would cause initial high local tumor concentrations, long-lasting exposure of tumor cells may not be probable. It is envisaged that intraperitoneal delivery of etoposide through NPs would be a better approach for effectual treatment of peri-toneal tumors. In this perception, etoposide-loaded NPs were prepared applying nanoprecipitation and emulsion-solvent evaporation methods using PLGA in the presence of Pluronic F68 by Reddy et al. The process produced NPs with high entrapment efficiency of around 80% with con-tinuous release of the drug up to 48 h.[89]

Doxorubicin (DOX), an anthracycline antibiotic and one of the most widely used anticancer agents, shows high antitumor activity. However, its therapeutic properties are limited due to its dependent cardio toxicity and myelosuppression.[90] PLGA NPs promise to be a successful system for the targeted and controlled release of DOX with decreased systemic toxicity, patient compliance, and increased therapeutic efficiency. Moreover, mul-tifunctional PLGA NPs for combined DOX and photothermal treatments were studied by Park et al. to deliver both drug and heat simultaneously to a particular tumorigenic section.[91]

Xanthones are natural, semisynthetic, and heterocyclic compounds. Xanthone molecules having a range of substituents on the different car-bons constitute a group of compounds with a broad spectrum of biological activities.[74] These molecules inhibited the nitric oxide production from the macrophages and thus have strong inhibitory action on human cancer cell line expansion. Xanthone-loaded PLGA nanospheres have been prepared by solvent displacement techniques.[92,74]

Rapamycin and its analogues provide as promising novel drugs that use alternative mechanisms to control the growth of breast cancer cells effi-ciently.[93] Clinically, rapamycin analogues with advanced stability and phar-macological properties have been well tolerated by patients in Phase I trials, and these agents have shown a hopeful antitumor consequence in breast cancer.[94] However, despite the potency of rapamycin in preclinical studies, the clinical development of this drug. Recently, the therapeutic efficacy of rapamycin has been optimized by developing efficient delivery system for the drug, that is, nanoscale delivery vehicles (such as PLGA NPs), which are capable of controlled release of the drug, thus enhancing its regressive activity on dendritic cells through altering their maturation profile.[95]

Curcumin has been used in conventional medicine for several centuries in India and China.[96] The factor that restricts the use of free curcumin for tumor therapy is its low solubility in water, which consecutively limits its bioavailability when administered orally. Nanotechnology-based carriers (PLGA NPs) emerged as a novel hope in curcumin delivery to tumor sites. Curcumin-loaded PLGA NPs were prepared by the emulsion diffusion evaporation technique using different stabilizers such as cetyl trimethyl-ammonium bromide or PVA or PEG-5000. The present comprehensible data definitely testifies a well-established product profile, opening up new ways for the miracle molecule curcumin on PLGA owing to higher cellular uptake and increased in vitro bioactivity and better in vivo bioavailability in comparison to native curcumin.[97] Mukerjee and Vishwanatha formulated curcumin-loaded PLGA particles and recommended that a NP-based formulation of curcumin has high probable as adjuvant therapy in prostate tumor.[98]

Triptorelin is a decapeptide analogue of luteinizing releasing hormone used for the treatment of sex hormone dependent tumors. Triptorelin diminishes the production of luteinizing hormone to considerably decrease the levels of testosterone production and accumulation. This may cause contraction or slowing down the growth of the tumor. In order to optimize the treatments by means of triptorelin, maintenance of steady plasma levels of drug for prolonged time periods is required. Triptorelin-loaded PLGA nanospheres have been prepared via double emulsion solvent evaporation technique with encapsulation efficiency varying from 4% to 83%.[99,74]

A natural photosensitizer, extracted from *Hypericum perforatum*, is a tool for the treatment and detection of ovarian cancer and other cancers. Because of its hydrophobicity, administration of hypericin is problematic. Hypericin-loaded PLGA NPs suppress ovarian tumor growth efficiently.[100]

Dexamethasone is often administered previous to antibiotics in cases of bacterial meningitis. It acts to diminish the inflammatory response of the body to killed bacterial population by the antibiotics, therefore improving prognosis and outcome. Dexamethasone causes inhibitory consequence on leukocytes infiltration at the inflammatory site. Its solubility is very poor and crystalline corticoid that has been used for the treatment of diabetic macular edema administered as an implant. This drug has been incorporated into PLGA NPs via solvent evaporation technique.[101,74] The highest drug loading was obtained using 100 mg PLGA (75:25) in a mixture of acetone–dichloromethane 1:1 (v/v) and 10 mg of dexamethasone.[74]

2.4 CANCER TREATMENT WITH PLGA-BASED NANOPARTICLE

2.4.1 PASSIVE TARGETING BY EPR EFFECT AND ACTIVE TARGETING OF SPECIFIC RECEPTORS

Nanotechnology suggests a new targeted approach and could offer important benefits to cancer patients. In fact, the utilization of NPs for drug delivery and targeting is likely one of the most exciting and clinically significant applications of cancer nanotechnology.[102] Nanoparticle systems offer major improvements in therapeutics by means of site specificity, a capacity to evade multidrug resistance, and proper delivery of anticancer agents[103,40] Targeted delivery can be passively (by taking advantage of the distinct pathophysiological features of tumor tissue) or actively (by targeting the drug carrier using target-specific ligands) accomplished.[104,40]

In *passive targeting*, structural changes in vascular pathophysiology could provide probability for the use of long-circulating particulate carrier systems. Capability of vascular endothelium to present open fenestrations was described for the sinus endothelium of the liver,[105,106] when the endothelium is perturbed by inflammatory process, hypoxic areas of infracted myocardium[107] or in tumors.[27] Mainly, tumor blood vessels are usually characterized by abnormalities such as pericyte deficiency, high proportion of proliferating endothelial cells, and abnormal basement membrane formation resulting to an enhanced vascular permeability. Particles, like nanocarriers (in the size range of 20–200 nm), can extravasate and gather inside the interstitial space. Endothelial pores have sizes varying from 10 to 1000 nm.[69,106] Moreover, lymphatic vessels are absent or nonfunctional in tumor which contributes to incompetent drainage from the tumor tissue. Nanocarriers penetrated into the tumor are not removed efficiently and are thus preserved in the tumor. This passive phenomenon has been called the "enhanced permeability and retention (EPR) effect," discovered by Matsumura and Maeda.[108–110] Rapid vascularization in fast-growing cancerous tissues is known to result in leaky, imperfect architecture, and impaired lymphatic drainage. This arrangement allows an EPR effect,[108,110,102] leading to the gathering of NPs at the tumor site. To maximize circulation times and targeting capacity, the optimal size should be less than 100 nm in diameter, and the surface should be hydrophilic to circumvent clearance by macrophages.[111,112,102] The covalent linkage of amphiphilic copolymers (PCL, PLA, polycyanonacrylate chemically coupled to PEG) is commonly preferred, as it avoids aggregation and ligand desorption when in contact with blood components.[102] In vivo tumor growth inhibition by the PTX-loaded NPs was then investigated in transplantable liver tumor-bearing mice. PTX was shown to enter the tumor site through the enhanced permeation and retention effect and maintain an efficient therapeutic concentration.[106]

In *active targeting*, targeting ligands are connected at the shell of the nanocarrier for binding to appropriate receptors expressed at the target site. The ligand is selected to bind to a receptor over articulated by tumor cells or tumor vasculature and not expressed by normal cells. In addition, targeted receptors should be expressed homogeneously on all targeted cells. Targeting ligands are either monoclonal antibodies and antibody fragments or nonantibody ligands (peptidic or not). The binding affinity of the ligands controls the tumor penetration owing to the "binding-site barrier." For targets in which cells are readily reachable, usually the tumor vasculature, because of the dynamic flow environment of the bloodstream, high affinity binding shows to be preferable.[113,114] Various anticancer therapeutics, grouped under the name "ligand-targeted therapeutics," are classified into different classes based on the approach of drug delivery.[115] In the active targeting strategy, two cellular targets can be differentiated: (1) the targeting of cancer cell and (2) the targeting of tumoral endothelium. The intent of active targeting of internalization-prone cell-surface receptors, over expressed by cancer cells, is to improve the cellular uptake of the nanocarriers. Thus, the active targeting is chiefly attractive for the intracellular delivery of macromolecular drugs, such as DNA, siRNA, and proteins. The better cellular internalization rather than an increased tumor accumulation is responsible of the antitumoral efficacy of actively targeted nanocarriers. This is the basis of the design of delivery systems targeted to endocytosis-prone surface receptors.[116] The capacity of the nanocarrier to be internalized after binding to target cell is an important criterion in the selection of proper targeting ligands.[106] In this strategy, ligand-targeted nanocarriers will result in direct cell kill, as well as cytotoxicity against cells that are at the tumor periphery and are not dependent on the tumor vasculature.[106,117]

The more considered internalization-prone receptors are as follows:

i)   The transferrin receptor. Transferrin, a serum glycoprotein, transports iron through the blood and into cells by binding to the transferring receptor and then being internalized via receptor-mediated endocytosis. The transferrin receptor is a crucial protein involved in iron homeostasis and the regulation of cell growth. The high levels of expression of transferring receptor in cancer cells, which may be up to 100-fold higher than the usual expression of normal cells, its extracellular accessibility, its capacity to internalize, and its central role in the cellular pathology of human cancer, make this receptor an effective target for cancer therapy.[118,106]

ii) The folate receptor is a famous tumor marker that binds to the vitamin folic acid and folate–drug conjugates or folate-grafted nanocarriers with a high affinity and carries these bound molecules into the cells through receptor-mediated endocytosis. Folic acid is needed in one carbon metabolic reactions and, as a result, is essential for the synthesis of nucleotide bases. The alpha isoform, folate receptor-α is over expressed on 40% of human cancers. On the contrary, folate receptor-β is expressed on activated macrophages and also on the surfaces of malignant cells of hematopoietic origin.[106]

iii) Glycoproteins expressed on cell surfaces. Lectins are proteins of nonimmunological origin which are capable to recognize and bind to carbohydrate moieties attached to glycoprotein's expressed on cell surface. Cancer cells often express various glycoprotein's compared to normal cells. Lectins interaction with certain carbohydrate is very specific. Lectins can be incorporated into NPs as targeting moieties that are directed to cell-surface carbohydrates (direct lectin targeting), and carbohydrates moieties can be coupled to NPs to target lectins (reverse lectin targeting). Make use of lectins and neoglyco conjugates for direct or reverse targeting strategies is a usual approach of colon drug targeting.[119,106]

iv) Epidermal growth factor receptor (EGFR). It is a part of the ErbB family, a family of tyrosine kinase receptors. Its activation stimulates key processes involved in tumor growth and progression. EGFR is normally over expressed in a lot of cancer, principally in breast cancer. Also, it has been found to play a vital role in the progression of several human malignancies. Human epidermal receptor-2 is reported to be expressed in 14–91% of patients with breast cancer.[120,106] EGFR is expressed or over expressed in a diversity of solid tumors, including colorectal cancer, nonsmall cell lung cancer, and squamous cell carcinoma of the head and neck, as well as ovarian, pancreatic, kidney, and prostate cancer.[121,106]

2.4.1.1   TARGETING OF TUMORAL ENDOTHELIUM

Demolition of the endothelium in solid tumors can result in the death of tumor cells induced by the lack of oxygen and nutrients. In 1971, Judah Folkman suggested that the tumor growth might be inhibited by preventing tumors from recruiting new blood vessels.[122] This observation is the base of the design of nanomedicines actively targeted to tumor endothelial cells.[123] By attacking the growth of the blood supply, the size and metastatic capacities of tumors can be controlled. Therefore, in this strategy, ligand-targeted nanocarriers bind to kill angiogenic blood vessels and indirectly, the tumor cells that these vessels support, mostly in the tumor core.

The advantages of the tumoral endothelium targeting are as follows:

i)   There is no need of extravasation of nanocarriers reach to their targeted site
ii) The binding to their receptors is directly achievable after intravenous injection
iii) The possible risk of emerging resistance is reduced due to the genetically stability of endothelial cells as compared to tumor cells
iv) The majority of endothelial cells markers are expressed whatever the tumor type, involving a ubiquitous approach and an ultimate expansive application spectrum[114,106]

The major targets of the tumoral endothelium include

i)   The vascular endothelial growth factors (VEGF) and their receptors, VEGFR-1 and VEGFR-2, mediate crucial functions in tumor angiogenesis and neovascularization.[124] Tumor hypoxia and oncogenes upregulate VEGF levels in the tumor cells, resulting in an upregulation of VEGF receptors on tumor endothelial cells.
     Two major approaches to object angiogenesis via the VEGF way have been studied
     (1) targeting VEGFR-2 to reduce VEGF binding and induce an endocytotic pathway
     (2) targeting VEGF to restrain ligand binding to VEGFR-2.[125,126]

ii) The $\alpha v \beta 3$ integrin is an endothelial cell receptor for extracellular matrix proteins which includes fibrinogen (fibrin), vibronectin, thrombospondin, osteopontin, and fibronectin.[106,127] The $\alpha v \beta 3$ integrin is extremely expressed on neovascular endothelial cells but weakly expressed in resting endothelial cells and most normal organs, and is important in the calcium dependent signaling pathway leading to endothelial cell migration.[126] Cyclic or linear derivatives of RGD (Arg-Gly-Asp) oligopeptides are the mainly studied peptides which bind to endothelial $\alpha v \beta 3$ integrins. The $\alpha v \beta 3$ integrin is upregulated in both tumor cells and angiogenic endothelial cells.[106,127] Vascular cell adhesion molecule-1 (VCAM-1) is an immunoglobulin-like transmembrane glycoprotein that is expressed on the surface of endothelial tumor cells. VCAM-1 induces the cell to cell adhesion, a key step in the angiogenesis procedure. Over expression of VCAM-1 is found in various cancers, such as leukemia, lung and breast cancer, melanoma, renal cell carcinoma, gastric cancer, and nephroblastoma.[106]

iii) The matrix metalloproteinases (MMPs) are a family of zinc dependent endopeptideases. MMPs degrade the extracellular matrix, playing a significant role in angiogenesis and metastasis more particularly in endothelial cell invasion and migration, in the development of capillary tubes and in the employment of accessory cells. Membrane type 1 MMP (MT1-MMP) is expressed on endothelial tumor cells, including malignancies of lung; colon, gastric and cervical carcinomas; gliomas and melanomas.[128] Aminopeptidase N/CD13, a metalloproteinase that removes amino-acids from unblocked N-terminal segments of peptides or proteins, is an endothelial cell-surface receptor involved in tumor-cell invasion, extracellular matrix degradation by tumor cells and tumor metastasis in vitro and in vivo.[129]

2.4.2 CANCER CHEMOTHERAPY

Many small anticancer drugs have been encapsulated in PLGA-based NPs and have been evaluated in vitro and in vivo to treat various cancers. For example, long-term clinical use of DOX, a highly powerful anthracycline approved for use against a broad spectrum of tumors is compromised by toxicities, cardiomyopathies and subsequent congestive heart failures. PEGylated PLGA NPs encapsulating DOX improve DOX antitumoral efficiency compared with the free drug. Moreover, these NPs were shown to decrease drastically the cardiomyopathies compared to Doxil®, a liposomal formulation of DOX presently available on the market.[130] Chemotherapy of glioblastoma is largely ineffective as the blood–brain barrier prevents entry of most anticancer agents in the brain. Nontargeted DOX-loaded PLGA NPs coated with poloxamer 188 were found to cross the blood–brain barrier and to effectively decrease the tumor growth in rat model.[131] PTX, a mitotic inhibitor used in the treatment of various cancers, presents a low therapeutic index and a low aqueous solubility. Its current commercialized form, Taxol®, contains PTX at a concentration of 6 mg/mL solubilized in a mixture of Cremophor® EL and ethanol (1:1). However, serious side effects are associated with Cremophor® EL. The encapsulation of PTX into PLGA NPs strongly enhances the cytotoxic effect of the drug as compared to Taxol® in vitro and in vivo.[132] 9-NC is an anticancer drug which targets the Topoisomerase I nuclear enzyme. Because of instability at biological pH and low water solubility, the delivery of this drug is rather challenging. 9-NC-loaded PLGA NPs, prepared by nanoprecipitation, have shown a sustained release up to 160 h indicating the suitability of PLGA NPs in controlled 9-NC release. PLGA-based NPs targeting the tumor cells or tumor endothelium have been shown to be generally more active in preclinical studies than nontargeted NPs. A cyclic peptide, cyclo-(1,12)-penITDGEATDSGC (cLABL), has been shown to inhibit LFA-1/ICAM-1 via the binding to ICAM-1. In addition, cLABL has been shown to be internalized after binding to ICAM-1. This cyclic peptide was conjugated to PLGA NPs for the delivery of DOX. The cLABL-NPs were shown to be more rapidly taken into A549 lung epithelial cells than nontargeted NPs. The cytotoxicity study of cLABL-NPs and nontargeted NPs compared to free DOX showed similar IC50 values suggesting that the activity of DOX was maintained.[133] Folate-decorated DOX-loaded PLGA NPs induced a cellular uptake 1.5 times higher by MCF-7 cells than nontargeted NPs.[134] Recently, other folate-decorated PTX-loaded PLGA-PEG NPs showed a greater cytotoxicity against HEC-1A cancer cells both in vitro and in vivo. The accumulation of folate-decorated NPs depends on dual effects of passive and active targeting.[135] The peptide LyP-1 has been recognized for its specific binding to tumors and their lymphatics. LyP-1-PEG-PLGA NPs have been studied for their targetability to tumor lymph metastases. In vivo, the uptake of these LyP-1-targeted NPs in metastatic lymph nodes was about eight times higher than nontargeted NPs, indicating that these NPs are a capable carrier to lymphatic metastatic tumors.[136] The prostate-specific membrane antigen (PSMA) was used to functionalize PLGA-PEG NPs for the delivery of cisplatin (Pt(IV)). A comparison between Pt(IV)-encapsulated PSMA NPs and nontargeted Pt(IV)-NPs against human prostate PSMA-overexpressing LNCaP cells revealed that the effectiveness of targeted NPs is approximately one order of magnitude greater than that of free cisplatin.[137] AS1411, a DNA aptamer which binds to nucleolin highly expressed in the plasma membrane of both cancer cells and endothelial cells, was chosen to target gliomas. These NPs prolonged circulation and enhanced PTX accumulation at the tumor site.[138] The $\alpha v \beta 3$ integrin a receptor for extracellular matrix proteins is upregulated in both tumor cells and angiogenic endothelial cells.[139] The RGD peptide is the most studied peptide which binds to $\alpha v \beta 3$ integrin. The surface of PLGA NPs was linked with PEG and the RGD peptide to realize both passive and active targeting functions for the delivery of DOX. The NP targeting ability was proven through strong affinity to various integrin expressing cancer cells and much less affinity to the low integrin expressing cancer cells.[140] Other PLGA-based NPs grafted with the RGD peptide have been designed for the delivery of PTX. The targeting to the tumoral endothelium was demonstrated both in vitro and in vivo. Furthermore, therapeutic efficiency has been demonstrated by effective retardation of TLT tumor growth and long-lasting survival times of mice treated by PTX-loaded GD NPs when compared to nontargeted NPs In the past decades, increasing interest has been found in the literature toward oral administration of cytotoxic agents. PTX was encapsulated in NPs composed of PLGA with vitamin E TPGS as emulsifier for oral administration. Its oral bioavailability was 10 times higher compared with Taxol®.[141] Oral bioavailability of tamoxifen-loaded NPs was increased 11 times as compared to free tamoxifen base. In vivo oral antitumor efficacy of tamoxifen NPs was demonstrated in DMBA-induced breast tumor model with a marked reduction in hepatotoxicity when compared with free tamoxifen.[142] While numerous studies are described in the literature about PLGA-based NPs, only a few papers described active targeting using PLGA-based NPs. Moreover, majority of these papers explains the characterization of their system in vitro only. Very few research articles report actively targeted PLGA-based NPs both in vitro and in vivo.

2.4.3 CANCER DIAGNOSIS AND IMAGING

Tumor imaging plays a key role in clinical oncology by helping to identify solid tumors, determine recurrence and monitor therapeutic responses. The development of noninvasive molecular imaging systems able to detect tumors at early stages would represent a considerable improvement over the current available clinical diagnostic methods. The development of NPs for the delivery of contrast agents has emerged in recent years since the possibility of the fabrication of multifunctional NPs able to specifically target the tumor. PLGA was used to formulate NPs encapsulating super paramagnetic iron oxides for MRI. This system enhanced the imaging effects along with increasing the half-life of NPs in the blood stream, thereby reducing their side effects.[142] Another example consists in the encapsulation of a radiotracer; technetium-99m (99mTc), in PLGA NPs for sentinel lymph node detection. Biodistribution and scintigraphic imaging studies performed in Wistar rats revealed localization of 99mTc-labeled PLGA NPs.

2.4.4 THERAGNOSTICS OF CANCER

Theragnostics are defined by the combination of therapeutic and diagnosis agents on a single platform. Theragnostics may combine passive and active targeting, molecular imaging, environmentally responsive drug release, and other therapeutic functions into a single platform.[144] For example, DOX was encapsulated into magnetic NPs that were embedded in PLGA through hydrophobic interactions. DOX was released constantly without any inhibition because of the presence of magnetic nanocapsules. These NPs were tested in mice, rats, and in human.[145,146] More recently, magnetic NPs embedded in PLGA matrices were designed as dual DDS and imaging system able to encapsulate both hydrophilic (carboplatin) and hydrophobic drugs (PTX and rapamycin) in a 2:1 ratio. Magnetic resonance imaging revealed in vitro and in vivo that these NPs present an improved contrast effect than commercial contrast agents.[147]

2.4.5 PROTEINS DELIVERY FOR CANCER TREATMENT

Endostar, a 20-kDa antiangiogenic peptide, is an endostatin which shows action against solid tumors but requires multiple injections at a high dose to achieve a therapeutic effect. Endostar-loaded PEG-PLGA NPs prove longer half-life than free drug, higher amount in murine tumors, and offered higher inhibition rate and stronger antiangiogenic effect than free endostar. Recombinant human granulocyte colony-stimulating factor was fruitfully loaded in PLGA NPs and showed an in vitro sustained release with 90% of encapsulated protein released from the NPs over 1-week period.[143] PE38KDL, a model protein toxin, was loaded in PLGA NPs modified on their surface by coating of Fab′ fragments of a humanized anti-HER2 mAb (rhuMAbHER2). In vitro and in vivo data revealed that PE38KL-loaded NPs presenting rhuMAb-HER2 (PE-NPs-HER2) have a better in vitro cytotoxicity against HER2-overexpressing breast cancer lines than PE38KL-loaded NPs without rhuMAbHER2, demonstrating prominence of active targeting. Moreover, in vivo, a better antitumor activity and, above all, a higher maximum tolerated dose were observed for PENPs-HER2 than for the control, showing a decline of nonspecific toxicity owing to nanoencapsulation.[148]

2.5 PLGA RELEASE MECHANISMS AND PITFALL

2.5.1 RELEASE MECHANISM

PLGA is the most commonly used biodegradable polymer in the controlled release of encapsulated drugs. The interval of drug release can be varied from hours[149] to several months.[150–152] Moreover, pulsed drug release is also possible.[152,153] True release mechanisms is the processes in which the drug is released, and Rate-controlling release mechanisms is the processes that control the release rate.[152]

There are four true release mechanisms:

i) *Diffusion via water-filled pores:* The first stage of the release period, previous to the onset of polymer erosion.[154–156]

ii) *Diffusion via the polymer:* It is probable for small hydrophobic drugs.[157] Unlike diffusion through water-filled pores, diffusion through the polymer is not dependent on the porous structure. The drug must be dissolved in water before being released, and this process could reduce the overall release time. High porosity augments the surface area for drug dissolution and could improve drug release.[152]

iii) *Osmotic pumping:* When osmotic pressure, caused by water absorption, drives the transport of the drug, osmotic pumping occurs. This release mechanism is more frequent for DDSs using materials other than PLGA.

iv) *Erosion (no drug transport):* As a true release mechanism, that is, drug release devoid of drug transport, results in identical profiles of drug release and polymer erosion, assuming that the drug is homogeneously distributed throughout the DDS. Erosion could be the major release mechanism for low-Mw PLGA formulations, in which an important part of the polymer has a molecular weight just above the limit for water solubility.[152]

There are two major release mechanisms associated with drug release from PLGA-based DDSs. They are diffusion and degradation/erosion.

The release rate is often said to be diffusion-controlled initially and degradation/erosion controlled during the final stage of the release period.[150] The encapsulated drug may be released by more than one true release mechanism at once, and the dominating mechanism may vary with time.[157] However, numerous processes or events influence the rate of drug diffusion and the degradation kinetics, for example, dissolution of the drug (in combination with diffusion),[158] diffusion through water filled pores,[159] diffusion through the polymer matrix,[160] hydrolysis,[161] erosion,[162] osmotic pumping,[163] water absorption/swelling,[164] polymer–drug interactions,[165] drug–drug interactions,[166] polymer relaxation,[167] pore closure,[168] heterogeneous degradation,[169] formation of cracks or deformation.[170]

Controlled drug release from PLGA-based DDSs is complex, and many processes that influence drug release affect each other in many ways. The effects of different factors on drug release may differ in time and position through a polymer matrix. The complexity of drug release from PLGA-based DDSs makes it complicated to generalize results obtained with specific DDSs.[171,152]

2.5.2 PITFALLS

One of the main pitfalls of PLGA-based NPs relates to the poor loading. This low drug loading constitutes a major problem for some drugs in the design of PLGA-based NPs. A second significant pitfall consists in the high burst release of drug from NPs. This fact is described for the majority of PLGA-based NPs. As a result, the drug might not be able to reach the target tissue or cells, leading to a loss of efficacy. Because of the application of NPs in sustained release drug delivery, the drug release mechanisms are also imperative to understand. Drug release mechanisms depend on the polymer used and on the loading efficiency. Usually, the rapid initial or burst release is attributed to adsorbed drug to the NPs surface.[40,74]

The disadvantage associated with PLGA is the production of acids upon degradation. A subdiscipline of nanotechnology called "nanotoxicology" has occurred. The interactions of nanocarriers with biological systems are really complex. The size and surface properties of nanocarriers adjust the behavior of these components in the body.

2.6 CONCLUSION

PLGA is a polymer approved by the US FDA and EMA for drug delivery because of its drug biocompatibility, biodegradability, suitable biodegradation kinetics, mechanical properties, and ease of processing. PLGA-based NPs give many advantages for drug delivery. They can guard drugs from degradation and improve their stability. Additionally, due to their size, NPs can pierce specific tissues via (1) the fenestrations present in the endothelium of cancer and inflamed tissue or (2) via receptors overexpressed by target cells or in the blood–brain barrier. This permits a specific delivery of drugs, proteins, peptides, or nucleic acids to their target tissue. PLGA-based NPs can enlarge the efficiency of treatments because of the constant release of the therapeutic agent from stable NPs. They can improve pharmacodynamic and pharmacokinetic profiles. Drug delivery using PLGA or PLGA-based polymers is an attractive area with countless opportunities for biomedical research with the primary goal of increasing therapeutic (antitumor) effect while minimizing side effects. However, the opportunity remains exciting and wide open and further advances are needed to turn the idea of drug-loaded PLGA NP technology into a realistic practical application as the next generation of DDSs.

KEYWORDS

- **PLGA nanoparticles**
- **surface modification**
- **anticancer drugs**
- **cancer chemotherapy**
- **cancer diagnosis and imaging**
- **release mechanism**

REFERENCES

1. Hu, C. M.; Aryal, S.; Zhang, L. Nanoparticle-Assisted Combination Therapies for Effective Cancer Treatment. *Ther. Deliv.* **2010,** *1*(2), 323–334.
2. Davis, M. E.; Chen, Z. G.; Shin, D. M. Nanoparticle Therapeutics: An Emerging Treatment Modality for Cancer. *Nat. Rev. Drug Discov. 7*(9), 771–782.
3. Zhang, L.; Gu, F. X.; Chan, J. M.; Wang, A. Z.; Langer, R. S.; Farokhzad, O. C. Nanoparticles in Medicine: Therapeutic Applications and Developments. *Clin. Pharmacol. Ther.* **2008,** *83*(5), 761–769.
4. Peer, D.; Karp, J. M.; Hong, S.; Farokhzad, O. C.; Margalit, R.; Langer, R.; Nanocarriers as an Emerging Platform for Cancer Therapy. *Nat. Nanotechnol.* **2007,** *2*(12), 751–760.
5. Wang, A. Z.; Gu, F.; Zhang, L.; Chan, J. M.; Radovic-Moreno, A.; Shaikh, M. R., et al. Biofunctionalized Targeted Nanoparticles for Therapeutic Applications. *Expert Opin. Biol. Ther.* **2008,** *8*, 1063–1070.
6. Zhang, L.; Gu, F. X.; Chan, J. M.; Wang, A. Z.; Langer, R. S.; Farokhzad, O. C. Nanoparticles in Medicine: Therapeutic Applications and Developments. *Clin. Pharmacol. Ther.* **2008,** *83*,761–769.
7. Hu, C. M.; Kaushal, S.; Tran Cao, H. S.; Aryal, S.; Sartor, M.; Esener, S., et al. Half-Antibody Functionalized Lipid-Polymer Hybrid Nanoparticles for Targeted Drug Delivery to Carcinoembryonic Antigen Presenting Pancreatic Cancer Cells. *Mol. Pharm.* **2010,** *7*, 914–920.
8. Hu, C. M.; Zhang, L.; Therapeutic Nanoparticles to Combat Cancer Drug Resistance. *Curr. Drug Metab.* **2009,** *10*, 836–841.
9. Aryal, S.; Hu, C. M.; Zhang, L. Polymer–Cisplatin Conjugate Nanoparticles for Acid-Responsive Drug Delivery. *ACS Nano* **2010,** *4*(1), 251–258.
10. Putnam, D.; Kopecek, J. Polymer Conjugates with Anticancer Activity. *Adv. Polymer Sci.* **1995,** *122*, 55–123.
11. Takakura, Y.; Hashida, M. Macromolecular Drug Carrier Systems in Cancer Chemotherapy: Macromolecular Prodrugs. *Crit. Rev. Oncol. Hematol.* **1995,** *18*(3), 207–231.
12. Kratz, F.; Beyer, U.; Schutte, M. T. Drug–polymer Conjugates Containing Acid Cleavable Bonds. *Crit. Rev. Ther. Drug Carrier Syst.* **1999,** *16*(3), 245–288.

13. Roth, T.; Kratz, F.; Steidle, C.; Unger, C.; Fiebig, H. H. *In Vitro* and *In Vivo* Antitumor Efficacy of Acid Albumin and Poly(Ethylene)Glycol Conjugates of Doxorubicine. *Ann. Oncol.* **1998**, *9*, 102.

14. Bazile, D.; Prud'homme, C.; Bassoullet, M. T.; Marlard, M.; Spenlehauer, G.; Veillard, M.; Stealth Me. PEG–PLA Nanoparticles Avoid Uptake by the Mononuclear Phagocytes System. *J. Pharm. Sci.* **1995**, *84*(4), 493–498.

15. Allemann, E.; Brasseur, N.; Benrezzak, O., et al. PEG-Coated Poly(Lactic Acid) Nanoparticles for the Delivery of Hexadecafluoro Zinc Phthalocyanine to EMT-6 Mouse Mammary Tumours. *J. Pharm. Pharmacol.* **1995**, *47*(5), 382–387.

16. Lammers, T.; Subr, V.; Ulbrich, K., et al. Simultaneous Delivery of Doxorubicin and Gemcitabine to Tumors In Vivo Using Prototypic Polymeric Drug Carriers. *Biomaterials* **2009**, *30*(20), 3466–3475

17. Abeylath, S. C.; Turos, E.; Dickey, S.; Lim, D. V. Glyconanobiotics: Novel Carbohydrate Nanoparticle Antibiotics for MRSA and *Bacillus anthracis*. *Bioorg. Med. Chem.* **2008**, *16*(5), 2412–2418.

18. Song, X. R.; Cai, Z.; Zheng, Y., et al. Reversion of Multidrug Resistance by Co-encapsulation of Vincristine and Verapamil in PLGA Nanoparticles. *Eur. J. Pharm. Sci.* **2009**, *37*(3–4), 300–305.

19. Soma, C. E.; Dubernet, C.; Bentolila, D.; Benita, S.; Couvreur, P. Reversion of Multidrug Resistance by Co-encapsulation of Doxorubicin and Cyclosporin A in Polyalkylcyanoacrylate Nanoparticles. *Biomaterials* **2000**, *21*(1), 1–7.

20. Adel, A. L.; Dorr, R. T.; Liddil, J. D. The Effect of Anticancer Drug Sequence in Experimental Combination Chemotherapy. *Cancer Invest.* **1993**, *11*(1), 15–24.

21. Acharya, S.; Sahoo SK. PLGA Nanoparticles Containing Various Anticancer Agents and Tumour Delivery by EPR Effect. *Adv. Drug Deliv. Rev.* **2011**, *63*, 170–183.

22. Danhier, F.; Ansorena, E.; Azeitao, J., et al. PLGA-based Nanoparticles: An Overview of Biomedical Applications. *J. Control. Release* **2012**, *161*, 505–522.

23. Acharya, S.; Dilnawaz, F.; Sahoo, S. K. Targeted Epidermal Growth Factor Receptor Nanoparticle Bioconjugates for Breast Cancer Therapy. *Biomaterials* **2009**, *30*, 5737–5750.

24. Adams, G. P.; Schier, R.; McCall, A. M., et al. High Affinity Restricts the Localization and Tumor Penetration of Single-Chain fv Antibody Molecules. *Cancer Res.* **2001**, *61*, 4750–4755.

25. Agrahari, V.; Kabra, V.; Trivedi, P. *Development, Optimization and Characterization of Nanoparticle Drug Delivery System of Cisplatin*; Springer, 2009; pp 1325–1328.

26. Akbarzadeh, A.; Mikaeili, H.; Zarghami, N., et al. Preparation and In Vitro Evaluation of Doxorubicin-loaded Fe_3O_4 Magnetic Nanoparticles Modified with Biocompatible Copolymers. *Int. J. Nanomed.* **2012**, *7*, 511.

27. Jain, R. A. The Manufacturing Techniques of Various Drug Loaded Biodegradable Poly(lactide-*co*-glycolide) (PLGA) Devices. *Biomaterials* **2000**, *21*, 2475–2490.

28. Lü, J. M.; Wang, X.; Marin-Muller, C., et al. Current Advances in Research and Clinical Applications of PLGA-based Nanotechnology. *Expert Rev. Mol. Diagn.* **2009**, *9*, 325–341.

29. Zambaux, M. F.; Bonneaux, F.; Gref, R., et al. Influence of Experimental Parameters on the Characteristics of Poly(lactic acid) Nanoparticles Prepared by a Double Emulsion Method. *J. Control. Release* **1998**, *50*, 31–40.

30. Scholes, P. D.; Coombes, A. G. A.; Illum, L., et al. The Preparation of Sub-200 nm Poly(lactide-*co*-glycolide) Microspheres for Site-specific Drug Delivery. *J. Control. Release* **1993**, *25*, 145–153.

31. Soppimath, K. S.; Aminabhavi, T. M.; Kulkarni, A. R.; Rudzinski, W. E. Biodegradable Polymeric Nanoparticles as Drug Delivery Devices. *J. Control. Release* **2001**, *70*, 1–20.

32. Tice, T. R.; Gilley, R. M. Preparation of Injectable Controlled-Release Microcapsules by a Solvent-Evaporation Process. *J. Control. Release* **1985**, *2*, 343–352.

33. Pinto Reis, C.; Neufeld, R. J.; Ribeiro, A. J.; Veiga F. Nanoencapsulation. I. Methods for Preparation of Drug-Loaded Polymeric Nanoparticles. *Nanomed.: Nanotechnol. Biol. Med.* **2006**, *2*, 8–21.

34. D'Mello, S. R.; Das, S. K.; Das, N. G. *Polymeric Nanoparticles for Small-Molecule Drugs: Biodegradation of Polymers and Fabrication of Nanoparticles. Drug Delivery Nanoparticles Formulation and Characterization*; CRC Press: Boca Raton, 2006; pp 16–34.

35. Yezhelyev, M. V.; Gao, X.; Xing, Y., et al. Emerging Use of Nanoparticles in Diagnosis and Treatment of Breast Cancer. *Lancet Oncol.* **2006**, *7*, 657–667.

36. Quintanar-Guerrero, D.; Allemann, E.; Fessi, H.; Doelker E. Preparation Techniques and Mechanisms of Formation of Biodegradable Nanoparticles from Preformed Polymers. *Drug Dev. Ind. Pharm.* **1998**, *24*, 1113–1128.

37. Ibrahim, H.; Bindschaedler, C.; Doelker, E.; Buri, P.; Gurny R. Aqueous Nanodispersions Prepared by a Salting-Out Process. *Int. J. Pharm.* **1992**, *87*, 239–246.

38. Allemann, E.; Gurny, R.; Doelker E. Drug-Loaded Nanoparticles: Preparation Methods and Drug Targeting Issues. *Eur. J. Pharm. Biopharm.* **1993**, *39*, 173–191

39. Konan, Y. N.; Gurny, R.; Allémann E. Preparation and Characterization of Sterile and Freeze–Dried Sub-200 nm Nanoparticles. *Int. J. Pharm.* **2002**, *233*, 239–252.

40. Dinarvand, R.; Sepehri, N.; Manoochehri, S.; Rouhani, H.; Atyabi F. Polylactide-*co*-glycolide Nanoparticles for Controlled Delivery of Anticancer Agents. *Int. J. Nanomed.* **2011**, *6*, 877.

41. Bala, I.; Hariharan, S.; Kumar, M. N. PLGA Nanoparticles in Drug Delivery: The State of the Art. *Crit. Rev. Ther. Drug Carrier Syst.* **2004**, *21*, 387.

42. Jung, T.; Kamm, W.; Breitenbach, A., et al. Biodegradable Nanoparticles for Oral Delivery of Peptides: Is There a Role for Polymers to Affect Mucosal Uptake? *Eur. J. Pharm. Biopharm.* **2000**, *50*, 147–160.

43. Lambert, G.; Fattal, E.; Couvreur P. Nanoparticulate Systems for the Delivery of Antisense Oligonucleotides. *Adv. Drug Deliv. Rev.* **2001**, *47*, 99–112.

44. Couvreur, P.; Dubernet, C.; Puisieux, F. Controlled Drug Delivery with Nanoparticles: Current Possibilities and Future Trends. *Eur. J. Pharm. Biopharm.* **1995**, *41*, 2–13.

45. Fessi, H.; Puisieux, F.; Devissaguet, J. P.; Ammoury, N.; Benita S. Nanocapsule Formation by Interfacial Polymer Deposition Following Solvent Displacement. *Int. J. Pharm.* **1989**, *55*, R1–R4.

46. Thioune, O.; Fessi, H.; Devissaguet, J. P.; Puisieux F. Preparation of Pseudolatex by Nanoprecipitation: Influence of the Solvent Nature on Intrinsic Viscosity and Interaction Constant. *Int. J. Pharm.* **1997**, *146*, 233–238.

47. Govender, T.; Stolnik, S.; Garnett, M. C.; Illum, L.; Davis, S. S. PLGA Nanoparticles Prepared by Nanoprecipitation: Drug Loading and Release Studies of a Water Soluble Drug. *J. Control. Release* **1999,** *57,* 171–185.

48. Mahapatro, A.; Singh, D. K. Biodegradable Nanoparticles Are Excellent Vehicle for Site Directed In-Vivo Delivery of Drugs and Vaccines. *J. Nanobiotechnol.* **2011,** *9,* 55.

49. Gaumet, M.; Vargas, A.; Gurny, R.; Delie F. Nanoparticles for Drug Delivery: The Need for Precision in Reporting Particle Size Parameters. *Eur. J. Pharm. Biopharm.* **2008,** *69,* 1–9.

50. Fonseca, C.; Simoes, S.; Gaspar R. Paclitaxel-loaded PLGA Nanoparticles: Preparation, Physicochemical Characterization and In Vitro Anti-Tumoral Activity. *J. Control. Release* **2002,** *83,* 273–286.

51. Cheng, F. Y.; Wang, S. P. H.; Su, C. H., et al. Stabilizer-Free Poly(lactide-*co*-glycolide) Nanoparticles for Multimodal Biomedical Probes. *Biomaterials* **2008,** *29,* 2104–2112.

52. Mu, L.; Feng, S. S. A Novel Controlled Release Formulation for the Anticancer Drug. *J. Control. Release* **2003,** *86,* 33–48.

53. Ricci-Júnior, E.; Marchetti, J. M. Zinc (II) Phthalocyanine Loaded PLGA Nanoparticles for Photodynamic Therapy Use. *Int. J. Pharm.* **2006,** *310,* 187–195.

54. Esmaeili, F.; Ghahremani, M. H.; Ostad, S. N.; Atyabi, F.; Seyedabadi, M.; Malekshahi, M. R.; Amini, M.; Dinarvand R. Folate-Receptor-Targeted Delivery of Docetaxel Nanoparticles Prepared by PLGA–PEG–folate Conjugate. *J. Drug Target.* **2008,** *16,* 415–423.

55. Panyam, J.; Zhou, W.; Prabha, S.; Sahoo, S. K.; Labhasetwar V. Rapid Endo-lysosomal Escape of Poly(d,l-lactide-*co*-glycolide) Nanoparticles: Implications for Drug and Gene Delivery. *FASEB J.* **2002,** *16,* 1217–1226.

56. Mo, Y.; Lim, L. Y. Preparation and In Vitro Anticancer Activity of Wheat Germ Agglutinin (WGA)-Conjugated PLGA Nanoparticles Loaded with Paclitaxel and Isopropyl Myristate. *J. Control. Release* **2005,** *107,* 30–42.

57. Yang, A.; Yang, L.; Liu, W., et al. Tumor Necrosis Factor Alpha Blocking Peptide Loaded PEG-PLGA Nanoparticles: Preparation and In Vitro Evaluation. *Int. J. Pharm.* **2007,** *331,* 123–132.

58. Ravi Kumar, M. N. V.; Bakowsky, U.; Lehr, C. M. Preparation and Characterization of Cationic PLGA Nanospheres as DNA Carriers. *Biomaterials* **2004,** *25,* 1771–1777.

59. Song, K. C.; Lee, H. S.; Choung, I. Y., et al. The Effect of Type of Organic Phase Solvents on the Particle Size of Poly(d,l-lactide-*co*-glycolide) Nanoparticles. Colloids Surf., A: Physicochem. Eng. Aspects **2006,** *276,* 162–167.

60. Dong, Y.; Feng, S. S. Poly(d,l-lactide-*co*-glycolide)/Montmorillonite Nanoparticles for Oral Delivery of Anticancer Drugs. *Biomaterials* **2005,** *26,* 6068–6076.

61. Nie, S.; Xing, Y.; Kim, G. J.; Simons, J. W. Nanotechnology Applications in Cancer. *Annu. Rev. Biomed. Eng.* **2007,** *9,* 257–288.

62. Mozafari, M. R.; Pardakhty, A.; Azarmi, S., et al. Role of Nanocarrier Systems in Cancer Nanotherapy. *J. Liposome Res.* **2009,** *19,* 310–321.

63. Storm, G.; Belliot, S. O.; Daemen, T.; Lasic, D. D. Surface Modification of Nanoparticles to Oppose Uptake by the Mononuclear Phagocyte System. *Adv. Drug Deliv. Rev.* **1995,** *17,* 31–48.

64. Araujo, L.; Löbenberg, R.; Kreuter J. Influence of the Surfactant Concentration on the Body Distribution of Nanoparticles. *J. Drug Target.* **1999,** *6,* 373–385.

65. Moghimi, S. M.; Hunter, A. C.; Murray, J. C. Nanomedicine: Current Status and Future Prospects. *FASEB J.* **2005,** *19,* 311–330.

66. Gelperina, S. E.; Khalansky, A. S.; Skidan, I. N., et al. Toxicological Studies of Doxorubicin Bound to Polysorbate 80-Coated Poly(butyl cyanoacrylate) Nanoparticles in Healthy Rats and Rats with Intracranial Glioblastoma. *Toxicol. Lett.* **2002,** *126,* 131–141.

67. Tang, N.; Du, G.; Wang, N., et al. Improving Penetration in Tumors with Nanoassemblies of Phospholipids and Doxorubicin. *J. Natl. Cancer Inst.* **2007,** *99,* 1004–1015.

68. Stolnik, S.; Illum, L.; Davis, S. S. Long Circulating Microparticulate Drug Carriers. *Adv. Drug Deliv. Rev.* **1995,** *16,* 195–214.

69. Torchilin, V. P.; Trubetskoy, V. S. Which Polymers Can Make Nanoparticulate Drug Carriers Long-Circulating? *Adv. Drug Deliv. Rev.* **1995,** *16,* 141–155.

70. Si-Shen, F.; Li, M.; Win, K. Y.; Guofeng H. Nanoparticles of Biodegradable Polymers for Clinical Administration of Paclitaxel. *Curr. Med. Chem.* **2004,** *11,* 413–424.

71. Choi, S. W.; Kim, J. H. Design of Surface-Modified Poly(d,l-lactide-*co*-glycolide) Nanoparticles for Targeted Drug Delivery to Bone. *J. Control. Release* **2007,** *122,* 24–30.

72. Avgoustakis, K.; Beletsi, A.; Panagi, Z., et al. PLGA–mPEG Nanoparticles of Cisplatin: In Vitro Nanoparticle Degradation, In Vitro Drug Release and In Vivo Drug Residence in Blood Properties. *J. Control. Release* **2002,** *79,* 123–135.

73. Tobıo, M.; Sanchez, A.; Vila, A., et al. The Role of PEG on the Stability in Digestive Fluids and In Vivo Fate of PEG–PLA Nanoparticles Following Oral Administration. *Colloids Surf. B: Biointerfaces* **2000,** *18,* 315–323.

74. Kumari, A.; Yadav, S. K.; Yadav, S. C. Biodegradable Polymeric Nanoparticles Based Drug Delivery Systems. *Colloids Surf. B: Biointerfaces* **2010,** *75,* 1–18.

75. Gref, R.; Lück, M.; Quellec, P., et al. 'Stealth' Corona-Core Nanoparticles Surface Modified by Polyethylene glycol (PEG): Influences of the Corona (PEG Chain Length and Surface Density) and of the Core Composition on Phagocytic Uptake and Plasma Protein Adsorption. *Colloids Surf. B: Biointerfaces* **2000,** *18,* 301.

76. Vila, A.; Sanchez, A.; Tobıo, M.; Calvo, P.; Alonso, M. J. Design of Biodegradable Particles for Protein Delivery. *J. Control. Release* **2002,** *78,* 15–24.

77. Calvo, P.; Gouritin, B.; Brigger, I., et al. PEGylated Polycyanoacrylate Nanoparticles as Vector for Drug Delivery in Prion Diseases. *J. Neurosci. Methods* **2001,** *111,* 151–155.

78. Kreuter, J.; Petrov, V. E.; Kharkevich, D. A.; Alyautdin, R. N. Influence of the Type of Surfactant on the Analgesic Effects Induced by the Peptide Dalargin after Its Delivery Across The Blood–Brain Barrier Using Surfactant-Coated Nanoparticles. *J. Control. Release* **1997,** *49,* 81–87.

79. Alyautdin, R. N.; Tezikov, E. B.; Ramge, P., et al. Significant Entry of Tubocurarine into the Brain of Rats by Adsorption to Polysorbate 80-Coated Polybutylcyanoacry-

late Nanoparticles: An In Situ Brain Perfusion Study. *J. Microencapsul.* **1998**, *15*, 67–74.

80. Zhang, Z.; Feng, S. S. The Drug Encapsulation Efficiency, In Vitro Drug Release, Cellular Uptake and Cytotoxicity of Paclitaxel-Loaded Poly(Lactide)–tocopheryl Polyethylene Glycol Succinate Nanoparticles. *Biomaterials* **2006**, *27*, 4025–4033.

81. Ma, Y.; Zheng, Y.; Liu, K., et al. Nanoparticles of Poly(lactide-*co*-glycolide)-da-tocopheryl Polyethylene Glycol 1000 Succinate Random Copolymer for Cancer Treatment. *Nanoscale Res. Lett.* **2010**, *5*, 1161–1169.

82. Wang, Y. M.; Sato, H.; Adachi, I.; Horikoshi I. Preparation and Characterization of Poly(Lactic-*co*-glycolic Acid) Microspheres for Targeted Delivery of a Novel Anticancer Agent, Taxol. *Chem. Pharm. Bull.* **1996**, *44*, 1935.

83. Rosenberg B.; Charles, F. Kettring Prize. Fundamental Studies with Cisplatin. *Cancer* **2006**, *55*, 2303–2316.

84. Gryparis, E. C.; Hatziapostolou, M.; Papadimitriou, E.; Avgoustakis K. Anticancer Activity of Cisplatin-loaded PLGA-mPEG Nanoparticles on LNCaP Prostate Cancer Cells. *Eur. J. Pharm. Biopharm.* **2007**, *67*, 1–8.

85. Ambudkar, S. V.; Kimchi-Sarfaty, C.; Sauna, Z. E.; Gottesman, M. M. *P*-Glycoprotein: From Genomics to Mechanism. *Oncogene* **2003**, *22*, 7468–7485.

86. Song, X.; Zhao, Y.; Wu, W., et al. PLGA Nanoparticles Simultaneously Loaded with Vincristine Sulfate and Verapamil Hydrochloride: Systematic Study of Particle Size and Drug Entrapment Efficiency. *Int. J. Pharm.* **2008**, *350*, 320–329.

87. Song, X. R.; Cai, Z.; Zheng, Y., et al. Reversion of Multidrug Resistance by Co-encapsulation of Vincristine and Verapamil in PLGA Nanoparticles. *Eur. J. Pharm. Sci.* **2009**, *37*, 300–305.

88. Derakhshandeh, K.; Erfan, M.; Dadashzadeh S. Encapsulation of 9-Nitrocamptothecin, A Novel Anticancer Drug, in Biodegradable Nanoparticles: Factorial Design, Characterization and Release Kinetics. *Eur. J. Pharm. Biopharm.* **2007**, *66*, 34–41.

89. Reddy, L. H.; Sharma, R. K.; Chuttani, K.; Mishra, A. K.; Murthy, R. R. Etoposide-incorporated Tripalmitin Nanoparticles with Different Surface Charge: Formulation, Characterization, Radiolabeling, and Biodistribution Studies. *AAPS J.* **2004**, *6*, 55–64.

90. Misra, R.; Sahoo, S. K. Intracellular Trafficking of Nuclear Localization Signal Conjugated Nanoparticles for Cancer Therapy. *Eur. J. Pharm. Sci.* **2010**, *39*, 152–163.

91. Park, H.; Yang, J.; Lee, J., et al. Multifunctional Nanoparticles for Combined Doxorubicin and Photothermal Treatments. *ACS Nano* **2009**, *3*, 2919–2926.

92. Teixeira, M.; Alonso, M. J.; Pinto, M. M. M.; Barbosa, C. M. Development and Characterization of PLGA Nanospheres and Nanocapsules Containing Xanthone and 3-Methoxyxanthone. *Eur. J. Pharm. Biopharm.* **2005**, *59*, 491–500.

93. Noh, W. C.; Mondesire, W. H.; Peng, J., et al. Determinants of Rapamycin Sensitivity in Breast Cancer Cells. *Clin. Cancer Res.* **2004**, *10*, 1013–1023.

94. Hidalgo, M.; Rowinsky, E. K. The Rapamycin-sensitive Signal Transduction Pathway as a Target for Cancer Therapy. *Oncogene* **2000**, *19*, 6680.

95. Garber K. Rapamycin May Prevent Post-Transplant Lymphoma. *J. Natl. Cancer Inst.* **2001**, *93*, 1519.

96. Shishodia, S.; Sethi, G.; Aggarwal, B. B. Curcumin: Getting Back to the Roots. *Ann. N.Y. Acad. Sci.* **2005**, *1056*, 206–217.

97. Shaikh, J.; Ankola, D. D.; Beniwal, V.; Singh, D.; Kumar, M. N. V. Nanoparticle Encapsulation Improves Oral Bioavailability of Curcumin by At Least 9-Fold When Compared to Curcumin Administered with Piperine as Absorption Enhancer. *Eur. J. Pharm. Sci.* **2009**, *37*, 223–230.

98. Mukerjee, A.; Vishwanatha, J. K. Formulation, Characterization and Evaluation of Curcumin-loaded PLGA Nanospheres for Cancer Therapy. *Anticancer Res.* **2009**, *29*, 3867–3875.

99. Nicoli, S.; Santi, P.; Couvreur, P., et al. Design of Triptorelin Loaded Nanospheres for Transdermal Iontophoretic Administration. *Int. J. Pharm.* **2001**, *214*, 31–35.

100. Zeisser-Labouèbe, M.; Lange, N.; Gurny, R.; Delie F. Hypericin-loaded Nanoparticles for the Photodynamic Treatment of Ovarian Cancer. *Int. J. Pharm.* **2006**, *326*, 174–181

101. Gómez-Gaete, C.; Tsapis, N.; Besnard, M.; Bochot, A.; Fattal E. Encapsulation of Dexamethasone into Biodegradable Polymeric Nanoparticles. *Int. J. Pharm.* **2007**, *331*, 153–159.

102. Nie, S.; Xing, Y.; Kim, G. J.; Simons, J. W. Nanotechnology Applications in Cancer. *Annu. Rev. Biomed. Eng.* **2007**, *9*, 257–288.

103. Parveen, S.; Sahoo, S. K. Polymeric Nanoparticles for Cancer Therapy. *J. Drug Target.* **2008**, *16*, 108–123.

104. Lamprecht, A.; Ubrich, N.; Yamamoto, H., et al. Biodegradable Nanoparticles for Targeted Drug Delivery in Treatment of Inflammatory Bowel Disease. *J. Pharm. Exp. Therapeutics* **2001**, *299*, 775–781.

105. Roerdink, F. H.; Dijkstra, J.; Spanjer, H. H.; Scherphof, G. L. Interaction of Liposomes with Hepatocytes and Kupffer Cells In Vivo and In Vitro. *Biochem. Soc. Trans.* **1984**, *12*, 335.

106. Danhier, F.; Feron, O.; Préat V. To Exploit the Tumor Microenvironment: Passive and Active Tumor Targeting of Nanocarriers for Anti-Cancer Drug Delivery. *J. Control. Release* **2010**, *148*, 135–146.

107. Palmer, T. N.; Caride, V. J.; Caldecourt, M. A.; Twickler, J.; Abdullah V. The Mechanism of Liposome Accumulation in Infarction. *BBA Gen. Subj.* **1984**, *797*, 363–368.

108. Matsumura, Y.; Maeda H. A New Concept for Macromolecular Therapeutics in Cancer Chemotherapy: Mechanism of Tumoritropic Accumulation of Proteins and the Antitumor Agent Smancs. *Cancer Res.* **1986**, *46*, 6387–6392.

109. Maeda, H.; Sawa, T.; Konno T. Mechanism of Tumor-Targeted Delivery of Macromolecular Drugs, Including the EPR Effect in Solid Tumor and Clinical Overview of the Prototype Polymeric Drug SMANCS. *J. Control. Release* **2001**, *74*, 47–61.

110. Maeda, H.; Bharate, G. Y.; Daruwalla J. Polymeric Drugs for Efficient Tumor-Targeted Drug Delivery Based on EPR-Effect. *Eur. J. Pharm. Biopharm.* **2009**, *71*, 409–419.

111. Ringsdorf, H. Structure and Properties of Pharmacologically Active Polymers. *J. Polym Sci.* **51**, 135–153.

112. Moghimi, S. M.; Hunter, A. C. Poloxamers and Poloxamines in Nanoparticle Engineering and Experimental Medicine. *Trends Biotechnol.* **2000**, *18*, 412–420.

113. Adams, G. P.; Schier, R.; McCall, A. M., et al. High Affinity Restricts the Localization and Tumor Penetration of Single-Chain fv Antibody Molecules. *Cancer Res.* **2001**, *61*, 4750–4755.

114. Gosk, S.; Moos, T.; Gottstein, C.; Bendas G. VCAM-1 Directed Immunoliposomes Selectively Target Tumor Vasculature *In Vivo*. *BBA,Biomembr.* **2008**, *1778*, 854–863.

115. Allen, T. M. Ligand-Targeted Therapeutics in Anticancer Therapy. *Nat. Rev. Cancer* **2002**, *2*, 750–763.

116. Kirpotin, D. B.; Drummond, D. C.; Shao Y, et al. Antibody Targeting of Long-Circulating Lipidic Nanoparticles Does not Increase Tumor Localization but Does Increase Internalization in Animal Models. *Cancer Res.* **2006**, *66*, 6732–6740

117. Pastorino, F.; Brignole, C.; Di Paolo, D., et al. Targeting Liposomal Chemotherapy via Both Tumor Cell-Specific and Tumor Vasculature-Specific Ligands Potentiates Therapeutic Efficacy. *Cancer Res.* **2006**, *66*, 10073–10082.

118. Daniels, T. R.; Delgado, T.; Helguera, G.; Penichet, M. L. The Transferrin Receptor Part II: Targeted Delivery of Therapeutic Agents into Cancer Cells. *Clin. Immunol.* **2006**, *121*, 159–176.

119. Minko T. Drug Targeting to the Colon with Lectins and Neoglycoconjugates. *Adv. Drug Deliv. Rev.* **2004**, *56*, 491–509.

120. Scaltriti, M.; Baselga J. The Epidermal Growth Factor Receptor Pathway: A Model for Targeted Therapy. *Clin. Cancer Res.* **2006**, *12*, 5268–5272.

121. Lurje, G.; Lenz, H. J. EGFR Signaling and Drug Discovery. *Oncology* **2009**, *77*, 400–410.

122. Folkman J. Transplacental Carcinogenesis by Stilbestrol. *N. Engl. J. Med.* **1971**, *285*, 404.

123. Lammers, T.; Hennink, W. E.; Storm G. Tumour-Targeted Nanomedicines: Principles and Practice. *Br. J. Cancer* **2008**, *99*, 392–397.

124. Shadidi, M.; Sioud M. Selective Targeting of Cancer Cells Using Synthetic Peptides. *Drug Resist. Updat.* **2003**, *6*, 363–371.

125. Carmeliet P. VEGF as a Key Mediator of Angiogenesis in Cancer. *Oncology* **2005**, *69*, 4–10.

126. Byrne, J. D.; Betancourt, T.; Brannon-Peppas L. Active Targeting Schemes for Nanoparticle Systems in Cancer Therapeutics. *Adv. Drug Deliv. Rev.* **2008**, *60*, 1615–1626.

127. Desgrosellier, J. S.; Cheresh, D. A. Integrins in Cancer: Biological Implications and Therapeutic Opportunities. *Nat. Rev. Cancer* **2010**, *10*, 9–22.

128. Genís, L.; Gálvez, B. G.; Gonzalo, P.; Arroyo, A. G. MT1-MMP: Universal or Particular Player in Angiogenesis? *Cancer Metastasis Rev.* **2006**, *25*, 77–86.

129. Saiki, I.; Yoneda, J.; Azuma, I., et al. Role of Aminopeptidase N (CD13) in Tumor-Cell Invasion and Extracellular Matrix Degradation. *Int. J. Cancer* **1993**, *54*, 137–143.

130. Park, J.; Fong, P. M., Lu, J.; Russell, K. S.; Booth, C. J.; Saltzman, W. M.; Fahmy, T. M. PEGylated PLGA Nanoparticles for the Improved Delivery of Doxorubicin. *Nanomedicine* **2009**, *5*, 410–418.

131. Wohlfart, S.; Khalansky, A. S.; Gelperina, S.; Maksimenko, O.; Bernreuther, C.; Glatzel, M.; Kreuter, J. Efficient Chemotherapy of Rat Glioblastoma Using Doxorubicin-loaded PLGA Nanoparticles with Different Stabilizers. *PLoS One* **2011**, *6*, e19121.

132. Danhier, F.; Lecouturier, N.; Vroman, B.; Jerome, C.; Marchand-Brynaert, J.; Feron, O.; Preat, V. Paclitaxel-loaded PEGylated PLGA-based Nanoparticles: In Vitro and In Vivo Evaluation. *J. Control. Release* **2009,** *133,* 11–17.

133. Chittasupho, C.; Xie, S. X.; Baoum, A.; Yakovleva, T.; Siahaan, T. J.; Berkland, C. J. ICAM-1 Targeting of Doxorubicin-loaded PLGA Nanoparticles to Lung Epithelial Cells. *Eur. J. Pharm. Sci.* **2009,** *37,* 141–150.

134. Zhang, Z.; Huey, L. S.; Feng, S. S. Folate-decorated Poly(Lactide-*co*-glycolide)-vitamin ETPGS Nanoparticles for Targeted Drug Delivery. *Biomaterials* **2007,** *28,* 1889–1899.

135. Liang, C.; Yang, Y.; Ling, Y.; Huang, Y.; Li, T.; Li, X. Improved Therapeutic Effect of Folate-decorated PLGA-PEG Nanoparticles for Endometrial Carcinoma. *Bioorg. Med. Chem.* **2011,** *19,* 4057–4066.

136. Luo, G.; Yu, X.; Jin, C.; Yang, F.; Fu, D.; Long, J.; Xu, J.; Zhan, C.; Lu, W. LyP-1-conjugated Nanoparticles for Targeting Drug Delivery to Lymphatic Metastatic Tumors. *Int. J. Pharm.* **2010,** *385,* 150–156.

137. Dhar, S.; Gu, F. X.; Langer, R.; Farokhzad, O. C.; Lippard, S. J. Targeted Delivery of Cisplatin to Prostate Cancer Cells by Aptamer Functionalized Pt(IV) Prodrug-PLGAPEG Nanoparticles. *Proc. Natl. Acad. Sci. U.S.A.* **2008,** *105,* 17356–17361.

138. Guo, J.; Gao, X.; Su, L.; Xia, H.; Gu, G.; Pang, Z.; Jiang, X.; Yao, L.; Chen, J.; Chen, H. Aptamer-functionalized PEG-PLGA Nanoparticles for Enhanced Anti-glioma Drug Delivery. *Biomaterials* **2011,** *31,* 8010–8020.

139. Desgrosellier, J. S.; Cheresh, D. A. Integrins in Cancer: Biological Implications and Therapeutic Opportunities. *Nat. Rev. Cancer* **2010,** *10,* 9–22.

140. Wang, Z.; Chui, W. K.; Ho, P. C. Design of a Multifunctional PLGA Nanoparticulate Drug Delivery System: Evaluation of Its Physicochemical Properties and Anticancer Activity to Malignant Cancer Cells. *Pharm. Res.* **2009,** *26,* 1162–1171.

141. Danhier, F.; Vroman, B.; Lecouturier, N.; Crokart, N.; Pourcelle, V.; Freichels, H.; Jerome, C.; Marchand-Brynaert, J.; Feron, O.; Preat, V. Targeting of Tumor Endo-thelium by RGD-grafted PLGA-nanoparticles Loaded with Paclitaxel. *J. Control. Release* **2009,** *140,* 166–173.

142. Wang, Y.; Ng, Y. G.; Chen, Y.; Shuter, B.; Yi, J.; Ding, J.; Wang, S.; Feng, S. G. Formulation of Superparamagnetic Iron Oxides by Nanoparticles of Biodegradable Polymers for Magnetic Resonance Imaging. *Adv. Funct. Mater.* **2008,** *18,* 308–318.

143. Choi, S. H.; Park, T. G. G-CSF Loaded Biodegradable PLGA Nanoparticles Prepared by a Single Oil-in-Water Emulsion Method. *Int. J. Pharm.* **2006,** *311,* 223–228.

144. Janib, S. M.; Moses, A. S.; Mackay, J. A. Imaging and Drug Delivery Using Ther-anostic Nanoparticles. *Adv. Drug Deliv. Rev.* **2010,** *62,* 1052–1063.

145. Lubbe, A. S.; Bergemann, C.; Huhnt, W.; Fricke, T.; Riess, H.; Brock, J. W.; Huhn, D. Preclinical Experiences with Magnetic Drug Targeting: Tolerance and Efficacy. *Cancer Res.* **1996,** *56,* 4694–4701.

146. Lubbe, A. S.; Bergemann, C.; Riess, H.; Schriever, F.; Reichardt, P.; Possinger, K.; Matthias, M.; Dorken, B.; Herrmann, F.; Gurtler, R.; Hohenberger, P.; Haas, N.; Sohr, R.; Sander, B.; Lemke, A. J.; Ohlendorf, D.; Huhnt, W.; Huhn, D. Clinical Experi-ences with Magnetic Drug Targeting: A Phase I Study with 4′-Epidoxorubicin in 14 Patients with Advanced Solid Tumors. *Cancer Res.* **1996,** *56,* 4686–4693.

147. Singh, A.; Dilnawaz, F.; Mewar, S.; Sharma, U.; Jagannathan, N. R.; Sahoo, S. K. Composite Polymeric Magnetic Nanoparticles for Co-delivery of Hydrophobic and Hydrophilic Anticancer Drugs and MRI Imaging for Cancer Therapy. *ACS Appl. Mater. Interfaces* **2011**, *3*, 842–856.

148. Chen, H.; Gao, J.; Lu, Y.; Kou, G.; Zhang, H.; Fan, L.; Sun, Z.; Guo, Y.; Zhong, Y. Preparation and Characterization of PE38KDEL-loaded Anti-HER2 Nanoparticles for Targeted Cancer Therapy. *J. Control. Release* **2008**, *128*, 209–216

149. Ratajczak-Enselme, M.; Estebe, J. P.; Dollo, G., et al. Epidural, Intrathecal and Plasma Pharmacokinetic Study of Epidural Ropivacaine in PLGA-Microspheres in Sheep Model. *Eur. J. Pharm. Biopharm.* **2009**, *7*, 54–61.

150. D'Souza, S. S.; Selmin, F.; Murty, S. B., et al. Assessment of Fertility in Male Rats after Extended Chemical Castration with a GnRH Antagonist. *AAPS J.* **2004**, *6*, 94–99.

151. Lagarce, F.; Renaud, P.; Faisant, N., et al. Baclofen-loaded Microspheres: Preparation and Efficacy Testing in a New Rabbit Model. *Eur. J. Pharm. Biopharm.* **2005**, *59*, 449–459.

152. Fredenberg, S.; Wahlgren, M.; Reslow, M.; Axelsson, A. The Mechanisms of Drug Release in Poly(Lactic-*co*-glycolic Acid)-based Drug Delivery Systems—A Review. *Int. J. Pharm.* **2011**, *415*, 34–52.

153. Dorta, M. J.; Santoveña, A.; Llabrés, M.; Fariña, J. B. Potential Applications of PLGA Film-implants in Modulating In Vitro Drugs Release. *Int. J. Pharm.* **2002**, *248*, 149–156.

154. Alexis, F.; Venkatraman, S. S.; Rath, S. K.; Boey, F. In Vitro Study of Release Mechanisms of Paclitaxel and Rapamycin from Drug-Incorporated Biodegradable Stent Matrices. *J. Control. Release* **2004**, *98*, 67–74.

155. Johnson, O. F. L.; Jaworowicz, W.; Cleland, J. L., et al. The Stabilization and Encapsulation of Human Growth Hormone into Biodegradable Microspheres. *Pharm. Res.* **1997**, *14*, 730–735.

156. Lam, X. M.; Duenas, E. T.; Daugherty, A. L.; Levin, N.; Cleland, J. L. Sustained Release of Recombinant Human Insulin-like Growth Factor-I for Treatment of Diabetes. *J. Control. Release* **2000**, *67*, 281–292.

157. Wischke, C.; Schwendeman, SP. Principles of Encapsulating Hydrophobic Drugs in PLA/PLGA Microparticles. *Int. J. Pharm.* **2008**, *364*, 298–327.

158. Wong, H. M.; Wang, J. J.; Wang, C. H. In Vitro Sustained Release of Human Immunoglobulin G from Biodegradable Microspheres. *Ind. Eng. Chem. Res.* **2001**, *40*, 933–948.

159. Kim, H. K.; Chung, H. J.; Park, T. G. Biodegradable Polymeric Microspheres with "Open/Closed" Pores for Sustained Release of Human Growth Hormone. *J. Control. Release* **2006**, *112*, 167–174.

160. Sun, Y.; Wang, J. C.; Zhang, X., et al. Synchronic Release of Two Hormonal Contraceptives for About One Month from the PLGA Microspheres: In Vitro and In Vivo Studies. *J. Control. Release* **2008**, *129*, 192–199.

161. Bishara, A.; Domb, A. J. PLA Stereocomplexes for Controlled Release of Somatostatin Analogue. *J. Control. Release* **2005**, *107*, 474–483.

162. Shah, S. S.; Cha, Y.; Pitt, C. G. Poly(Glycolic Acid-*co*-dl-lactic Acid): Diffusion or Degradation Controlled Drug Delivery? *J. Control. Release* **1992**, *18*, 261–270.

163. Jonnalagadda, S.; Robinson, D. H. A Bioresorbable, Polylactide Reservoir for Diffusional and Osmotically Controlled Drug Delivery. *AAPS Pharm. Sci. Technol.* **2000,** *1*, 26–34.

164. Mochizuki, A.; Niikawa, T.; Omura, I.; Yamashita S. Controlled Release of Argatroban from PLA Film—Effect of Hydroxylesters as Additives on Enhancement of Drug Release. *J. Appl. Polymer Sci.* **2008,** *108*, 3353–3360.

165. Manuela Gaspar, M.; Blanco, D.; Cruz, M. E. M.; José Alonso M. Formulation of l-Asparaginase-loaded Poly(Lactide-*co*-glycolide) Nanoparticles: Influence of Polymer Properties on Enzyme Loading, Activity and In Vitro Release. *J. Control. Release* **1998,** *52*, 53–62.

166. Zhu, G.; Schwendeman, S. P. Stabilization of Proteins Encapsulated in Cylindrical Poly(lactide-*co*-glycolide) Implants: Mechanism of Stabilization by Basic Additives. *Pharm. Res.* **2000,** *17*, 351–357.

167. Gagliardi, M.; Silvestri, D.; Cristallini, C., et al. Combined Drug Release from Biodegradable Bilayer Coating for Endovascular Stents. *J. Biomed. Mater. Res., B: Appl. Biomater.* **2010,** *93*, 375–385.

168. Kang, J.; Schwendeman, S. P. Pore Closing and Opening in Biodegradable Polymers and Their Effect on the Controlled Release of Proteins. *Mol. Pharm.* **2007,** *4*, 104–118.

169. Park, T. G. Degradation of Poly(Lactic-*co*-glycolic Acid) Microspheres: Effect of Copolymer Composition. *Biomaterials* **1995,** *16*, 1123–1130.

170. Matsumoto, A.; Matsukawa, Y.; Horikiri, Y.; Suzuki T. Rupture and Drug Release Characteristics of Multi-reservoir Type Microspheres with Poly(dl-lactide-*co*-glycolide) and Poly(dl-lactide). *Int. J. Pharm.* **2006,** *327*, 110–116.

171. Friess, W.; Schlapp M. Release Mechanisms from Gentamicin Loaded Poly(Lactic-*co*-glycolic Acid) (PLGA) Microparticles. *J. Pharm. Sci.* **2002,** *91*, 845–855.

CHAPTER 3

MAGNETIC NANOPARTICLES: ADVANCES IN DESIGN AND DEVELOPMENT FOR BIO-INSPIRED APPLICATIONS

ANN ROSE ABRAHAM[1], SABU THOMAS[2], and NANDAKUMAR KALARIKKAL[1,2*]

[1]*School of Pure and Applied Physics, Centre for Nanoscience and Nanotechnology, Mahatma Gandhi University, Kottayam 686560, Kerala, India*

[2]*International and Inter University Centre for Nanoscience and Nanotechnology, Mahatma Gandhi University, Kottayam 686560, Kerala, India*

[*]Corresponding author. E-mail: nkkalarikkal@mgu.ac.in

ABSTRACT

Magnetic nanoparticles (MNPs) are extensively used for applications in nanobiomedicine like gene therapy, chemotherapy, magnetic resonance imaging, contrast enhancement, protein therapy, tissue repair, magnetic hyperthermia therapy, and anticancer drug delivery. These applications necessitate MNPs to be colloidally stable in biological media. The goal of this chapter is to study the current advances in designing and surface engineering of MNPs for biomedical applications. The superparamagnetic nanoparticles stabilized by coating with biologically compatible polymers are analyzed using X-ray diffraction, transmission electron microscopy, zeta potential measurements, dynamic light scattering, superconducting quantum interference device magnetometry and inductively coupled plasma optical emission spectrometry, and cell viability measurements that indicate the toxicity of the developed system to typical cell lines. Here, we report the accelerated growth in the demand for surface modified MNPs as theranostic tools for application in the field of oncology and bionanomedicine.

3.1 INTRODUCTION

Magnetic nanoparticles (MNPs), particularly/principally ferrites (MFe_2O_4, M = Fe, Co, Ni, Mn) with spinel structure,[1,2] are extensively exploited for biomedical applications because of their amazing functionalities. MNPs are attractive for a broad array of applications such as drug delivery, magnetic hyperthermia, magnetic resonance imaging (MRI), and biosensors. MNPs are functionalized with drugs[3] via targeting ligands to release the required dosage of drug at the targeted site. In addition, distinctive magnetic properties exhibited by MNPs facilitate effective evaluation of efficiency of drug delivery, image guidance during delivery process, magnetic field guided localization, and noninvasive scrutiny of the treatment responses. Hence, MNPs, because of their imaging attributes and therapeutic applications as drug and gene carriers, are admirable candidates in the field of nanobiomedicine and image-guided therapeutics.[4]

Superparamagnetic[5] nanoparticles are highly promising and crucial for cancer therapy, since they do not retain magnetism after removal of external field.[6,7] The Mössbauer spectrum at room temperature of superparamagnetic $ZnFe_2O_4$ is given in Figure 3.1. The occurrence of single doublet is the characteristic-distinguishing feature of superparamagnetic zinc ferrite. The biocompatibility and the low cytotoxicity of MNPs in human body illustrated by numerous studies make them attractive for biomedical applications. Superparamagnetic iron oxide nanoparticles (SPIONs) are exclusively used for magnetic hyperthermia,[8] because of their low toxicity. MNPs get degraded into Fe ions at the lysosomes,[9] and the iron accumulates in the body's iron stores to contribute to the production of hemoglobin.[10]

Polymer-coated magnetic particles have aroused meticulous interest for biomedical applications. Theranostic[11] MNPs coated with polymers display interesting imaging properties. In the recent years, MNPs are simultaneously used for therapy and diagnostics and hence proved to be an auspicious theranostic agent[12,13] for biomedical applications.

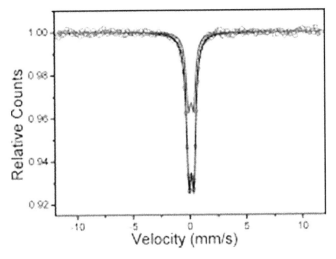

FIGURE 3.1   The Mössbauer spectrum of superparamagnetic $ZnFe_2O_4$ at room temperature. Reprinted from Reference [1] with permission from Elsevier.

3.2   MECHANISM OF CONJUGATION OF MAGNETIC NANOPARTICLES WITH THERAPEUTIC AGENTS

The utilization of MNPs for chemotherapy reduces the detrimental side-effects, without hindering its efficacy. The superparamagnetic nanoparticles[14] are linked to a therapeutic agent and injected into a tumor-supporting artery. The MNP-drug system[4] is then guided by an external magnetic field to the tumor site. Surface modification of the MNPs with the amphiphilic polymer coating like polyethylene glycol (PEG),[15] chitosan,[16] or protein (bovine serum albumin)[17] is indispensable for appropriate loading of drugs and its successful release at the diseased sites. Drugs are incorporated by physical or chemical (covalent bond) interactions between the functional groups attached to MNPs and the drug molecules. The hydrophobic drug is dispersed in an organic solvent like dimethylformamide, dimethyl sulfoxide, methanol, etc., with appropriate polarity. The surface modified ferrite[18] NPs are then drenched in the drug solution, and the hydrophobic coating on MNP surface facilitates the uptake or diffusion of drugs onto the MNP surface. MNPs are engineered with proper coatings to enhance the drug loading capacity. The nature and potential of functional group attached to the MNP surface highly influence the drug loading efficiency of MNP. Increased thickness or extra coating layers also facilitate effective encapsulation of hydrophobic drugs on engineered MNP surface and prevent premature release of drugs.[19]

3.3 SURFACE MODIFICATION OF FERRITE NANOPARTICLES

The MNPs[20] are surface engineered[21] to facilitate various biomedical tasks such as drug carriers, MRI contrast agents, and therapeutic agents for local heat induction (hyperthermia). The molecular imaging technique gives an in vivo image of molecular events at the cellular level. Bioconjugation of SPIONs with polymeric biomaterials, which are biodegradable,[22] are highly promising for various clinical applications like spleen and liver imaging, cardiovascular diseases, and apoptosis.[6] Chemically adsorbed coating agents show more stability than physically adsorbed (by hydrogen binding or electrostatic interactions) ones. MNPs surface modified with alkoxysilanes are attractive for magnetic bioseparations.[18] MNPs are capped with oleic acids to enhance the colloidal stability in polar solvents like ethanol.[23] Here, we briefly describe the surface engineering of ferrite nanoparticles with various systems like biocompatible polymers, polymers and drugs, graphene oxide (GO), and traditional anticancer drugs.

3.3.1 NANOFERRITES FOR BIOMEDICAL APPLICATIONS

The $MgFe_2O_4$[24] nanoparticles exhibit toxicity toward MCF-7 human breast cancer cells.[25] The cell death was observed by apoptosis and necrosis, and the cytotoxicity results show dose-dependent toxicity of $MgFe_2O_4$ nanoparticles toward cancer cells. These findings reveal the potential utility of $MgFe_2O_4$ nanoparticles in cancer therapy due to their toxicity toward MCF-7 cells. Hence, the toxicity of $MgFe_2O_4$ nanoparticles toward cancer cells remains an area of potential interest.

The $ZnFe_2O_4$[26,27] ferrite nanoparticle system doped with cobalt is interesting for biomedical applications.[28] The $Zn_xCo_{1-x}Fe_2O_4$ ferrite nanoparticles synthesized by coprecipitation method were characterized by X-ray diffraction (XRD) and transmission electron microscopy (TEM), and the crystallinity and spherical shape of the nanoparticles were confirmed. Magnetic characterizations of the nanoparticles at room temperature were carried out. The specific absorption rate (SAR) was analyzed by calorimetric method. The MNP system was coated with oppositely charged polyelectrolites to enable specific drug binding. The magnetization versus applied magnetic field is studied using a vibrating sample magnetometer. The superparamagnetic-to-ferrimagnetic transition and the decrease in saturation magnetization with Zn substitution can be observed in Figure 3.2.

In magnetic hyperthermia, the magnetic particles are exposed to an alternating magnetic field (AMF), and the cancer cells are heated. The nanoparticles induce heat into the cancerous cells, and the temperature of cells increases. The heating curves of $Zn_xCo_{1-x}Fe_2O_4$ ferrite solution on ac field application is shown in Figure 3.3. It is observed that the Zn content highly influence the maximum temperature attained. The temperature is found to increase as Zn substitution is decreased.

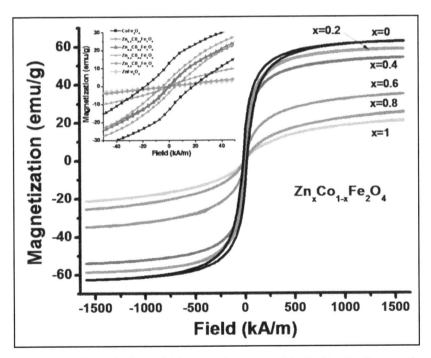

FIGURE 3.2 **(See color insert.)** Magnetization curves of the $Zn_xCo_{1-x}Fe_2O_4$ magnetic nanoparticles. Reprinted from Reference [28] with permission from AIP Publishing.

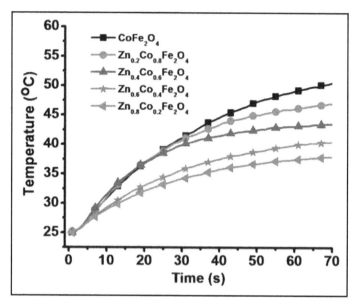

FIGURE 3.3 **(See color insert.)** Heating curves of $Zn_xCo_{1-x}Fe_2O_4$ magnetic nanoparticles at 1.95 MHz frequency and 4.44 kA/m field. Reprinted from Reference [28] with permission from AIP Publishing.

The slope $\dfrac{dT}{dt}$ of the temperature–time plots is used to evaluate the SAR,

$$SAR = C\frac{dT}{dt}\frac{1}{m_{NP}} \tag{3.1}$$

where C indicates the heat capacity of the sample, m_{NP} is the mass of nanoparticle, and dT is the temperature increase in the sample for field application time interval dt. The SAR is found to decrease with increasing zinc substitution level (Fig. 3.4). The maximum value of SAR was achieved for $CoFe_2O_4$ (52.2 W/g) sample at a 4.44-kA/m field. SAR values are observed to decrease with decrease in the magnetic field. The nanoparticles, dispersed in agar phantoms, are found to be promising contrast agents for MRI. The $Zn_xCo_{1-x}Fe_2O_4$ nanoparticles are confirmed to be potential candidates for drug delivery, magnetic hyperthermia, and other biomedical applications.

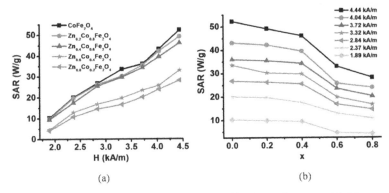

FIGURE 3.4 **(See color insert.)** Variation of specific absorption rate with (a) ac magnetic field intensity of aqueous $Zn_xCo_{1-x}Fe_2O_4$ ferrite solution (b) with Zn content at different ac magnetic field. Reprinted from Reference [28] with permission from AIP Publishing.

3.3.2 FERRITE NANOPARTICLES COATED WITH BIOCOMPATIBLE POLYMERS

The polymers mostly used for surface engineering of MNPs are derivatives of dextran, arabinogalactan, starch, siloxane, glycosaminoglycan, PEG, sulfonated styrene-divinylbenzene, biodegradable poly-ε-caprolactone,[29] polylactic acid, or polyalkylcyanoacrylate. Polysaccharides are extensively used for the stabilization of SPIONs because of their biocompatibility and solubility in water.[22]

Multipurpose theranostics agents are designed and developed by covering the ferrite nanoparticles with biocompatible polymers.[30] Polymer-coated MNPs were widely used for bioimaging purposes.[31] Magnetic nanoassemblies of PEGylated mesoporous $MgFe_2O_4$[32,33] are exciting for applications in chemothermal therapy. The successful development of hollow microcapsules made of $CoFe_2O_4$ nanoparticles coated with silica studied are reported.[34] $MnFe_2O_4$ nanoparticles coated with chitosan[35] are reported to be effective for magnetic hyperthermia applications.[36] Nanoclusters of SPIONS coated with poly(dopamine) are amazing for biomedical applications[2,37] like MRI, photothermal cancer therapy and field-targeting.[38] Coating with PEG and dextran helps to reduce the cytotoxicity of nanoparticle.[15] Yu et al. have reported the reduction in cytotoxicity of the iron-oxide nanoparticles on dextran and PEG coating in 2D and 3D cell culture and explained by the reactive oxygen species (ROS) formation.[15]

Thermotherapeutic applications of $NiFe_2O_4$ nanoparticles coated with chitosan and PEG have been reported.[39] $NiFe_2O_4$ nanoparticles were synthesized by coprecipitation method, and their particle sizes were tuned by proper heat treatment in the temperature range varying from 200°C to 800°C. Crystallinity of the samples and particle size of 2–58 nm was established from the XRD and TEM results. The superparamagnetic–ferromagnetic transition that occurs with increase in particle size was established by VSM measurements and Mössbauer spectroscopic studies. The nanoparticles of nickel ferrite with most favorable particle size of 10 nm were coated with chitosan and PEG for biomedical applications. Using dynamic light scattering, the polydispersity index and hydrodynamic diameter were observed to be 0.21 and 187 nm for nanoparticles coated with chitosan and 0.32 and 285 nm for PEG-coated ones. It is verified that by modulating the shape and particle size and thereby modulating the magnetization of ferrites, the temperature required for hyperthermia heating might be tuned. The higher magnitude of T_2 relaxivity (89 mM^{-1}/s) than T_1 (0.348 mM^{-1}/s) signifies the potential of the system for MRI as an efficient T_2 contrast agent.

3.3.3   FERRITE NANOPARTICLES FUNCTIONALIZED WITH BIOCOMPATIBILE NANOCONJUGATES AND ANTICANCER DRUGS

Silanized SPIONs coated with PEG and conjugated with herceptin (HER) to target receptors expressed over HER2+ human metastatic breast cancer cells are amazing for magnetic hyperthermia.[8] Monodisperse nickel-ferrite nanoparticles surface coated with PMAO polymer are promising for enhanced magnetic hyperthermia treatment response and other biomedical applications.[2]

Zinc ferrite nanoparticles functionalized with PEG-cyclodextrin are loaded with anticancer drug curcumin[40] as pH-responsive drug delivery system for antiinflammation and anticancer application on human cervical cancer HeLa cell lines.[41] Biocompatible nanoconjugates of cobalt ferrite with traditional anticancer drug curcumin encapsulated with PEG are excellent for drug delivery applications.[42] Zinc ferrite core–shell[43] nanoassembly loaded with curcumin and encapsulated with chitosan is promising for biocompatible drug delivery on chicken embryonic stem cells.[44]

Biofunctional MNPs conjugated with anticancer antigen 19-9 (Anti-CA 19-9) monoclonal antibody (a pancreatic cancer-targeting antibody) show potential for MRI for detection of pancreatic cancer as investigated by atomic force microscopy and TEM techniques.[10]

3.3.4 DECORATION OF FERRITE NANOPARTICLES WITH GO FOR CANCER THERAPY

Multifunctionalized GO targeted for prostate cancer has been reported to be promising for MRI and drug delivery applications.[45] Fe_3O_4 nanoparticles of 6 nm coated with graphite were developed by a chemical vapor deposition process.[46] Peng et al. have reported the decoration of $MnFe_2O_4$ nanoparticles with GO nanocomposites (M-GO NCs).[47] The $MnFe_2O_4$ nanoparticles embedded on GO (M-GO NCs) are again conjugated with PEG for biomedical applications.[47] Heating by exposure of M-GO NCs systems to an AMF demonstrates efficiency of the system as hyperthermia agents. The excellent biocompatibility of the PEGylated M-GO NCs is illustrated by the cytotoxicity results.

3.3.5 FERRITE NANOPARTICLES CONJUGATED WITH GO AND DRUG FOR CANCER THERAPY

Yang et al. have reported the exciting applications of nanohybrids of GO and manganese ferrite for photothermal therapy, MRI, and drug delivery.[48] Manganese ferrite ($MnFe_2O_4$) is a better choice to be employed for MRI as a T_2 contrast enhancement agent, due to its large magnetic spin magnitude and excellent biocompatibility. Graphene and GO of nanosize range are biocompatible and efficient nanocarriers for gene and drug delivery owing to their large surface area and desirable surface properties. The integration of $MnFe_2O_4$ NPs with GO will merge the desirable attributes of the two functional materials into a single system. Hence, the developed GO/$MnFe_2O_4$ magnetic composite system is highly preferred for MRI and chemophotothermal cancer treatment. Superparamagnetic manganese ferrite ($MnFe_2O_4$) nanoparticles were decorated on GO, by the thermal decomposition of iron(III) acetylacetonate ($Fe(acac)_3$) and manganese(II) acetylacetonate ($Mn(acac)_2$) in triethylene glycol in presence of GO synthesized by a modified Hummers method. The hysteresis loops in the magnetization curves of GO/MnFe$_2$O$_4$ hybrids prove the superparamagnetic nature of the hybrid system for MRI applications. The GO sheets reduce the value of saturation magnetization (M$_s$) of GO/MnFe$_2$O$_4$ system to 19.0 emu/g. HeLa cells were treated with GO/MnFe$_2$O$_4$/DOX to determine the chemotherapy and PTT performance. For chemotherapeutic application, the nanohybrids of GO/MnFe$_2$O$_4$ are loaded with the drug doxorubicin (DOX) under influence of π–π conjugate effect. To carry out the drug loading, the GO/MnFe$_2$O$_4$ was incubated with DOX_HCl in phosphate buffer solution for 24 h under-dark conditions. The DOX-loaded GO/MnFe$_2$O$_4$ (GO/MnFe$_2$O$_4$/DOX) was centrifuged and excess drug was removed by washing with PBS. The drug-loading capacity was estimated by the absorbance maximum at 490 nm. Near-infrared (NIR) laser light and pH have a great impact and control the release of DOX from GO/MnFe$_2$O$_4$. On irradiation with near-infrared light, GO/MnFe$_2$O$_4$/DOX composites demonstrated enhanced cancer cell killing effect, suggestive of the synergistic effect of nanohybrids for DOX chemotherapy and photothermal therapy. The cytotoxic effect of free DOX and GO/MnFe$_2$O$_4$/DOX on HeLa cells was analyzed by MTT assays. In order to analyze the in vivo toxic effect, GO/MnFe$_2$O$_4$ was injected intravenously into mouse via the tail vein. After administration, no abnormal behaviors or no difference in the number or morphology of blood cells was exhibited by the treated mice. Thus, the insignificant toxicity of the material is verified. To further investigate the toxicity of GO/MnFe$_2$O$_4$ after drug administration, the main organs of the mice were collected, sectioned into thin slices, and H&E stained for histological analysis. In the organs of the GO/MnFe$_2$O$_4$-exposed mice, no tissue damage, swelling, or lesions were observed.

The GO/MnFe$_2$O$_4$ composite was dispersed in water at different Fe concentration, and T_1 and T_2 relaxation time was measured on a 0.5-T MRI scanner, to confirm the diagnostic potential of GO/MnFe$_2$O$_4$ as an MRI contrast agent. With increasing Fe concentration, both T_1- and T_2-weighted images vary astonishingly in signal intensity. The nanohybrids are observed to be effective T_2 contrast agent for tumor imaging, from the T_2-weighted MRI studies. The nanohybrids of GO/MnFe$_2$O$_4$ demonstrated very low cytotoxicity and negligible hemolytic activity. The intensive optical absorbance in the NIR region and superior photothermal stability exhibited by the GO/MnFe$_2$O$_4$ nanohybrids prove the potential of the nanohybrid system for photothermal ablation of cancerous cells.

Multifunctional nanohybrid structures of manganese ferrite and GO were reported to show superior heating efficiency and magnetic anisotropy.[49] Nanoparticles of $MnFe_2O_4$ (MFO) were deposited onto GO nanosheets by modified coprecipitation method using $FeCl_3 \cdot 6H_2O$ and $MnCl_2 \cdot 4H_2O$ precursors and characterized by XRD, TEM, and selected area electron diffraction patterns. The anchoring of MFO nanoparticles on the GO surface was confirmed by the Fourier-transform infrared measurements. The superparamagnetic nature of MFO and the MFO–GO system and a significant increase of the effective anisotropy for the MFO–GO system are verified by the magnetic measurements and the modified Langevin model analysis. Magnetic hyperthermia experiments carried out by ac magnetometry and calorimetric methods denote the higher effective anisotropy of the MFO–GO nanocomposite and enhanced SAR for ac fields greater than 300 Oe. The GO minimizes the MFO nanoparticle aggregation and enhances the potential of the system for advanced hyperthermia. The employability of the MFO–GO nanocomposite as a magnetic biomarker for biosensing applications are verified by the magnetoimpedance-based biodetection studies. Thus, the significance of MFO–GO nanocomposite for various biomedical applications including targeted drug delivery, magnetic hyperthermia, and biomolecular detection are elucidated.

In summary, though other nanoparticles with various modifications, say, multifunctional hydroxyapatite (HAp) nanoparticles[50] or gold (Au) nanoparticles[51] with silica[52,53] coating, are employed for drug delivery and imaging applications, nanomagnetic ferrites like $ZnFe_2O_4$, $MnFe_2O_4$ with various surface modifications are the most preferred for biomedical applications because of their amazing properties like high magnetic moment, chemical stability, superparamagnetic attributes, the reactive surface, cost effectiveness, biocompatibility, and advances over conventional chemotherapy.[37] Like conventional magnetic nanoparticles, magneto-electric or multiferroic nanoparticles $(CoFe_2O_4@BaTiO_3)$[54,55] with non-zero magnetic moment are also used to treat human ovarian cancer cells (SKOV-3), making use of the electric field interactions between the magneto-electric nanoparticles (MENs) and drug and body cells.[56,57]

3.4 CONCLUSION

Magnetic-based theranostic nanoparticle systems are of particular interest for detection and treatment of diseases like cancer. The versatile characteristics of superparamagnetic nanoparticles promise new harbors for multifunctional applications. Biofunctionality of the nanomagnetic particles make them eye catching for biomedical applications such as MRI, anticancer drug delivery, phototherapy, magnetic hyperthermia, chemotherapy, and biosensors. Surface-modified nanomagnetic ferrites are highly preferred for drug delivery, imaging and other biomedical applications because of their wonderful magnetic attributes, reactive surface and cost effectiveness. MENs are also promising for high specificity and targeted anticancer drug delivery applications to eliminate cancer cells sparing normal cells. Theranostic MNPs encrusted with polymers display amazing imaging properties in reaction to stimuli, and also capably deliver the conjugated drugs or therapeutic genes at the targeted site, for desired application. The multifunctionality imparted by surface modification of superparamamgnetic nanoparticles provide multiple imaging platforms for diagnosis, exact localization of the affected cells and the rapid delivery in conjugation with anticancer drugs of therapeutic agents to the target. The potential of MNPs and the strategies for surface engineering with various moieties like PEG, chitosan, GO and along with anticancer drugs like DOX, herceptin, curcumin for interesting applications in nanobiomedicine are briefly accomplished in this paper. Recent interest and curiosity in utilization of MNPs for biomedical applications ranging from therapeutics and diagnosis particularly in the field of oncology reflect the revolutionary demand and futuristic applications of the MNPs.

KEYWORDS

- magnetic nanoparticles
- superparamagnetism
- surface modification
- theranostic tools
- magnetic hyperthermia
- magneto-electric nanoparticles

REFERENCES

1. Thomas, J. J.; Krishnan, S.; Sridharan, K.; Philip, R.; Kalarikkal, N. A Comparative Study on the Optical Limiting Properties of Different Nano Spinel Ferrites with Z-Scan Technique. *Mater. Res. Bull.* **2012,** *47,* 1855–1860.

2. Abraham, A. R., et al. *An Overview of Prospects of Spinel Ferrites and Their Varied Applications, Theoretical Models and Experimental Approaches in Physical Chemistry, Series: Innovations in Physical Chemistry: Monograph Series*; Haghi, A. K., Thomas, S., Praveen, K. M., Pai, A. R., Eds.; Apple Academic Press, 2017 (Hard ISBN: 9781771886321) (in press) (http://www.appleacademicpress.com/theoretical-models-and-experimental-approaches-in-physical-chemistry-research-methodology-and-practical-methods/9781771886321).

3. Singh, D., et al. Formulation Design Facilitates Magnetic Nanoparticle Delivery to Diseased Cells and Tissues. *Nanomedicine (Lond)* **2014**, *9*, 469–485.

4. Dumitrescu, A. M., et al. Advanced Composite Materials Based on Hydrogels and Ferrites for Potential Biomedical Applications. *Colloids Surf. A Physicochem. Eng. Asp.* **2014**, *455*, 185–194.

5. Abraham, A. R., et al. Magnetic Response of Superparamagnetic Multiferroic Core-Shell Nanostructures, *AIP Conf. Proc.* **2016**, *1731*(1), 050151. DOI:10.1063/1.4947805.

6. Laurent, S.; Mahmoudi, M. Superparamagnetic Iron Oxide Nanoparticles: Promises for Diagnosis and Treatment of Cancer. *Int. J. Mol. Epidemiol. Genet.* **2011**, *2*, 367–390.

7. Gomes, D. A., et al. Synthesis of Core–Shell Ferrites Nanoparticles for Ferrofluids: Chemical and Magnetic Analysis. **2008**, *112*(16), 6220–6227.

8. Almaki, J. H., et al. Synthesis, Characterization and *In Vitro* Evaluation of Exquisite Targeting SPIONs–PEG–HER in HER2+ Human Breast Cancer Cells. *Nanotechnology* **2016**, *27*, 105601.

9. Yu, M. K.; Park, J.; Jon, S. Targeting Strategies for Multifunctional Nanoparticles in Cancer Imaging and Therapy. *Theranostics* **2012**, *2*, 3–44.

10. Huang, M.; Qiao, Z.; Miao, F.; Jia, N.; Shen, H. Biofunctional Magnetic Nanoparticles as Contrast Agents for Magnetic Resonance Imaging of Pancreas Cancer. *Microchim. Acta* **2009**, *167*, 27–34.

11. Nandwana, V., et al. Engineered Theranostic Magnetic Nanostructures: Role of Composition and Surface Coating on Magnetic Resonance Imaging Contrast and Thermal Activation. *ACS Appl. Mater. Interfaces* **2016**, *8*, 6953–6961.

12. Fan, Z.; Fu, P. P.; Yu, H.; Ray, P. C. Theranostic Nanomedicine for Cancer Detection and Treatment. *J. Food Drug Anal.* **2014**, *22*, 3–17.

13. Ahmed, N.; Fessi, H.; Elaissari, A. Theranostic Applications of Nanoparticles in Cancer. *Drug Discov. Today* **2012**, *17*, 928–934.

14. Xiao, D.; Lu, T.; Zeng, R.; Bi, Y. Preparation and Highlighted Applications of Magnetic Microparticles and Nanoparticles: A Review on Recent Advances. *Microchim. Acta* **2016**, 1–21. DOI:10.1007/s00604-016-1928-y.

15. Yu, M.; Huang, S.; Yu, K. J.; Clyne, A. M. Dextran and Polymer Polyethylene Glycol (PEG) Coating Reduce Both 5 and 30 nm Iron Oxide Nanoparticle Cytotoxicity in 2D and 3D Cell Culture. *Int. J. Mol. Sci.* **2012**, *13*, 5554–5570.

16. Liu, L.; Xiao, L.; Zhu, H. Y.; Shi, X. W. Preparation of Magnetic and Fluorescent Bifunctional Chitosan Nanoparticles for Optical Determination of Copper Ion. *Microchim. Acta* **2012**, *178*, 413–419.

17. Tiwari, A. P.; Rohiwal, S. S.; Suryavanshi, M. V.; Ghosh, S. J.; Pawar, S. H. Detection of the Genomic DNA of Pathogenic α-Proteobacterium *Ochrobactrum anthropi* via Magnetic DNA Enrichment Using pH Responsive BSA@Fe$_3$O$_4$ Nanoparticles

Prior To In-Situ PCR and Electrophoretic Separation. *Microchim. Acta* **2016**, *183*, 675–681.

18. Bruce, I. J.; Sen, T. Surface Modification of Magnetic Nanoparticles with Alkoxysilanes and Their Application in Magnetic Bioseparations. **2005**, *21*(15), 7029–7035.

19. Huang, J., et al. Magnetic Nanoparticle Facilitated Drug Delivery for Cancer Therapy with Targeted and Image-Guided Approaches. *Adv. Funct. Mater.* **2016**, n/a–n/a. DOI:10.1002/adfm.201504185.

20. Augustine, R.; Abraham, A. R.; Kalarikkal, N.; Thomas, S. *Monitoring and Separation of Foodborne Pathogens Using Magnetic Nanoparticles*; Academic Press: Oxford, 2016; Vol. 1, pp 271–312.

21. Veiseh, O.; Gunn, J.; Zhang, M. Design and Fabrication of Magnetic Nanoparticles for Targeted Drug Delivery and Imaging. *Adv. Drug Deliv. Rev.* **2011**, *62*, 284–304.

22. Nishida, H.; Tokiwa, Y. Effects of Higher-Order Structure of Poly(3-Hydrolxybutyrate) on Biodegradation. II. Effects of Crystal Structure on Microbial Degradation. *J. Environ. Polym. Degrad.* **1993**, *1*, 65–80.

23. Lee, S. Y.; Harris, M. T. Surface Modification of Magnetic Nanoparticles Capped by Oleic Acids: Characterization and Colloidal Stability in Polar Solvents. *J. Colloid Interface Sci.* **2006**, *293*, 401–408.

24. Sreekanth, P., et al. Nonlinear Transmittance and Optical Power Limiting in Magnesium Ferrite Nanoparticles: Effects of Laser Pulsewidth and Particle Size. *RSC Adv.* **2016**. DOI:10.1039/C6RA15788B

25. Kanagesan, S., et al. Cytotoxic Effect of Nanocrystalline $MgFe_2O_4$ Particles for Cancer Cure. *J. Nanomater.* **2013**, *2013*, 165.

26. Thomas, J. J.; Shinde, A. B.; Krishna, P. S. R.; Kalarikkal, N. Temperature Dependent Neutron Diffraction and Mössbauer Studies in Zinc Ferrite Nanoparticles. *Mater. Res. Bull.* **2013**, *48*, 1506–1511.

27. Thankachan, R. M., et al. Cr^{3+}-Substitution Induced Structural Reconfigurations in the Nanocrystalline Spinel Compound $ZnFe_2O_4$ as Revealed From X-Ray Diffraction, Positron Annihilation and Mössbauer Spectroscopic Studies. *RSC Adv.* **2015**, *5*, 64966–64975.

28. Doaga, A.; Cojocariu, A. M.; Constantin, C. P.; Hempelmann, R.; Caltun, O. F. Magnetic Nanoparticles for Medical Applications: Progress and Challenges. *AIP Conf. Proc.* **2013**, *1564*, 123–131.

29. Abedalwafa, M.; Wang, F.; Wang, L.; Li, C. Biodegradable Poly-Epsilon-Caprolactone (PCL) for Tissue Engineering Applications : A Review. **2013**, *34*, 123–140.

30. Zahraei, M., et al. Versatile Theranostics Agents Designed by Coating Ferrite Nanoparticles with Biocompatible Polymers. *Nanotechnology* **2016**, *27*, 255702.

31. Uthaman, S.; Lee, S. J.; Cherukula, K.; Cho, C.; Park, I. Polysaccharide-Coated Magnetic Nanoparticles for Imaging and Gene Therapy. **2015**, *2015*.

32. Abraham A. R., et al. Realization of Enhanced Magneto-electric Coupling and Raman Spectroscopic   Signatures in 0−0 Type Hybrid Multiferroic Core−Shell Geometric Nanostructures. *J. Phys. Chem.* **2017**, 121, 4352−4362.

33. Abraham, A.R., et al. Interface Engineered Ferrite@Ferroelectric Core-Shell Nanostructures: A Facile Approach to Impart Superior Magneto-Electric Coupling. *AIP Conf. Proc.* **1942**, 020002 (2018); DOI: 10.1063/1.5028581.

34. Kumar, S.; Daverey, A.; Sahu, N. K.; Bahadur, D. In Vitro Evaluation of PEGylated Mesoporous MgFe$_2$O$_4$ Magnetic Nanoassemblies (MMNs) for Chemo-Thermal Therapy. *J. Mater. Chem. B* **2013**, *1*, 3652–3660.

35. Patel, M. P.; Patel, R. R.; Patel, J. K. Chitosan Mediated Targeted Drug Delivery System: A Review. *J. Pharm. Pharm. Sci.* **2010**, *13*, 536–57.

36. Oh, Y.; Lee, N.; Kang, H. W.; Oh, J. In Vitro Study on Apoptotic Cell Death by Effective Magnetic Hyperthermia with Chitosan-Coated MnFe$_2$O$_4$. *Nanotechnology* **2016**, *27*, 115101.

37. Cheng, K., et al. Magnetic Nanoparticles: Synthesis, Functionalization, and Applications in Bioimaging and Magnetic Energy Storage. *Chem. Soc. Rev.* **2009**, *38*, 2532–2542.

38. Wu, M., et al. Nanocluster of Superparamagnetic Iron Oxide Nanoparticles Coated with Poly (Dopamine) for Magnetic Field-Targeting, Highly Sensitive MRI and Photothermal Cancer Therapy. *Nanotechnology* **2015**, *26*, 115102.

39. Hoque, S. M., et al. Thermo-Therapeutic Applications of Chitosan- and PEG-Coated NiFe$_2$O$_4$ Nanoparticles. *Nanotechnology* **2016**, *27*, 285702.

40. Modasiya, M. K.; Patel, V. M. Studies on Solubility of Curcumin. *Int. J. Pharmacy Life Sci.* **2012**, *3*, 1490–1497.

41. Sawant, V. J.; Bamane, S. R.; Patil, S. V.; Kanase, D. G. PEG-Cyclodextrin Coated Curcumin Loaded Zinc Ferrite Core Nanocomposites as pH-Responsive Drug Delivery System for Anti Inflammation and Anticancer Application. *Arch. Appl. Sci. Res.* **2014**, *6*, 44–54.

42. Sawant, V. J.; Bamane, S. R. PEG Encapsulated Curcumin Conjugated Cobalt Ferrite Core Shell Nanoassembly for Biocompatibility Testing and Drug Delivery. *Int. J. Pharm. Sci. Rev. Res.* **2013**, *20*, 159–164.

43. Thomas, J. J.; Kalarikkal, N.; Garg, A. B.; Mittal, R.; Mukhopadhyay, R. Mossbauer Study of Ni, Ni–Co, and Co Ferrite Nanoparticles. **2011**, *1176*, 1175–1176.

44. Sinica, D. C.; Sawant, V. J.; Bamane, S. R.; Pachchapurkar, S. M. Chitosan Encapsulated Curcumin Loaded Zinc Ferrite Core Shell Nanoassembly for Biocompatible Drug Delivery on Chicken Embryonic Stem Cells. **2013**, *4*, 67–78.

45. Guo, L., et al. Prostate Cancer Targeted Multifunctionalized Graphene Oxide for Magnetic Resonance Imaging and Drug Delivery. *Carbon N. Y.* **2016**, *107*, 87–99.

46. Espinoza-Rivas, A.; Pérez-Guzmán, M. Synthesis and Magnetic Characterization of Graphite-Coated Iron Nanoparticles. *J. Nanotechnol.* **2016**, *2016*.

47. Peng, E., et al. Synthesis of Manganese Ferrite/Graphene Oxide Nanocomposites for Biomedical Applications. *Small* **2012**, *8*, 3620–3630.

48. Yang, Y., et al. Graphene Oxide/Manganese Ferrite Nanohybrids for Magnetic Resonance Imaging, Photothermal Therapy and Drug Delivery. *J. Biomater. Appl.* **2016**, *30*, 810–822.

49. Le, A. T., et al. Enhanced Magnetic Anisotropy and Heating Efficiency in Multi-Functional Manganese Ferrite/Graphene Oxide Nanostructures. *Nanotechnology* **2016**, *27*, 155707.

50. Bao, G.; Mitragotri, S.; Tong, S. Multifunctional Nanoparticles for Drug Delivery and Molecular Imaging. *Annu. Rev. Biomed. Eng.* **2013**, *15*, 253–282.

51. Alula, M. T.; Yang, J. Photochemical Decoration of Gold Nanoparticles on Polymer Stabilized Magnetic Microspheres for Determination of Adenine by Surface-Enhanced Raman Spectroscopy. *Microchim. Acta* **2015,** *182,* 1017–1024.

52. Shore, A.; Mazzochette, Z.; Mugweru, A. Mixed Valence Mn,La,Sr-Oxide Based Magnetic Nanoparticles Coated with Silica Nanoparticles for Use in an Electrochemical Immunosensor for IgG. *Microchim. Acta* **2016,** *183,* 475–483.

53. Jiang, X.; Huang, L.; Liu, X. Silica-Coated Magnetic Nanoparticles Loaded with Cu(II): Preparation and Application to the Purification of Bovine Serum Albumin. *Microchim. Acta* **2013,** *180,* 219–226.

54. Abraham, A. R.; Thomas, S.; Kalarikkal, N. Control of Magnetism by Voltage in Multiferroics: Theory and Prospects, Innovations in Physical Chemistry: Monograph Series, Modern Physical Chemistry, Engineering Models, Materials, and Methods with Applications; Haghi, E., Besalú, M., Jaroszewski, S., Thomas, K. M., Eds.; Apple Academic Press; 2018 (Hard ISBN: 9781771886437) (in press).

55. Abraham A. R.; Thomas, S.; Kalarikkal, N. Novel Applications of Multiferroic Heterostructures. In *Chemical Technology and Informatics in Chemistry with Applications*; Vakhrushev, A. V., Mukbaniani, O. V., Susanto, H., EDs.; Innovations in Physical Chemistry: Monograph Series; Apple Academic Press, 2018 (Hard ISBN: 9781771886666) (in press).

56. Rodzinski, A.; et al. Targeted and Controlled Anticancer Drug Delivery and Release with Magnetoelectric Nanoparticles. *Sci. Rep.* **2016,** *6,* 20867.

57. Demirer, G. S. Synthesis and Design of Biologically Inspired Biocompatible Iron Oxide Nanoparticles for Biomedical Applications. *J. Mater. Chem. B* **2015,** *3*(40), 7831–7849.

CHAPTER 4

BIOMEDICAL APPLICATIONS OF 3D PRINTING

NEELAKANDAN M. S.[1], YADU NATH V. K.[2], BILAHARI ARYAT[1], PARVATHY PRASAD[1], SUNIJA SUKUMARAN[1], JIYA JOSE[1], SABU THOMAS[2*], and NANDAKUMAR KALARIKKAL[3]

[1]*International and Inter University Centre for Nanoscience and Nanotechnology, Mahatma Gandhi University, Kottayam 686560, Kerala, India*

[2]*School of Chemical Sciences, Mahatma Gandhi University, Kottayam 686560, Kerala, India*

[3]*School of Pure and Applied Physics, Centre for Nanoscience and Nanotechnology, Mahatma Gandhi University, Kottayam 686560, Kerala, India*

Corresponding author. E-mail: sabuchathukulam@yahoo.co.uk

ABSTRACT

Three-dimensional (3D) printing has a very important role in biomedical applications. 3D printing has most definitely entered the mainstream. Entrepreneurial types are using the technique to build practically everything: protein models, fabrics, jewellery, dental implants, and lots of others. Medical uses for 3D printing, both actual and potential, can be organized into several broad categories, including tissue and organ fabrication; creation of customized prosthetics, implants, and anatomical models; and pharmaceutical research regarding drug dosage forms, delivery, and discovery. Here, we are discussing the principles, mechanism, and applications of 3D printing in the biomedical field.

4.1   INTRODUCTION

Three-dimensional printing or 3D printing is a fabrication technique in which objects were prepared by the fusion or deposition of materials such as plastic, metal, ceramic composites, powders, liquids, or even living cells in a layered form to produce a 3D object.[1] Additive manufacturing (AM), rapid prototyping (RP), or solid free-form technologies (SFFs) are the other commonly using names of 3D printing. Some 3D printers are similar to traditional inkjet printers; however, the end product differs in that a 3D object is produced. 3D printing is expected to revolutionize medicine and other fields, not unlike the way the printing press transformed publishing. Currently, two dozen 3D printing processes are reported, which use varying printer technologies, speeds, and resolutions, and hundreds of materials.[2] These technologies can build a 3D object in almost any shape imaginable as defined in a computer-aided design (CAD) file. In a basic setup, the 3D printer first follows the instructions in the CAD file to build the foundation for the object, moving the print head along the x–y plane. The printer then continues to follow the instructions, moving the print head along the z-axis to build the object vertically layer by layer. It is important to note that two-dimensional (2D) radiographic images, such as X-rays, magnetic resonance imaging (MRI), or computerized tomography (CT) scans, can be converted to digital 3D print files, allowing the creation of complex, customized anatomical, and medical structures.[3,4]

Applications of 3D printing in biomedical fields are expanding rapidly and are expected to revolutionize the health care system. Medical uses for 3D printing, both actual and potential, can be categorized into several broad categories, which includes tissue and organ fabrication; creation of customized prosthetics, implants, and anatomical models; and pharmaceutical research regarding drug dosage forms, delivery, and discovery.[5,6] Benefits of 3D printing over other techniques can be illustrated as the customization and personalization of medical products, drugs, and equipment; cost-effectiveness; increased productivity; the democratization of design and manufacturing; and enhanced collaboration. However, it is important that notable scientific and regulatory challenges remain in recent advancement in all fields especially in medical fields involving 3D printing, and the most transformative applications for this technology will need time to evolve.[7]

"3D printing just opens up a number of possibilities that we didn't have before, from a number of standpoints," says David Zopf, Assistant Professor of Otolaryngology, University of Michigan, Ann Arbor, MI. Biomedical applications of 3D printing are termed as bioprinting, a subset of 3D printing. In this case, the "ink" can be either cell aggregates or cells in the hydrogel, and they are positioned in a matrix (predesigned 3D locations in the computer design for whatever is being constructed) by the printer.[8]

In this chapter, we have discussed the history, principle, mechanism, and biomedical applications of 3D printing and stereolithography (SLA).

4.2   3D PRINTING—AN OVERVIEW

3D printing uses software which slices the 3D model into layers (0.01 mm thick or less in most cases). Each layer is then traced onto the build plate by the printer; once the pattern is finished, the build plate is lowered and the next layer is added on top of the previous one. Typical producing techniques are known as "subtractive manufacturing" because with the help of this process, the materials are removed from a preformed block. This method is also referred to as AM, RP, or SFF.[9] Some 3D printers are similar to traditional inkjet printers; however, the end product differs; in that, a 3D object is produced.[10] 3D printing is expected to revolutionize medicine and other fields, not unlike the way the printing press transformed publishing. This type of process produces a lot of waste as the material that is cut off generally cannot be used for anything else and is simply sent out as scrap. 3D printing discharges such waste since the material is placed in the location that is needed only, the rest will be left out as empty space. 3D printing technologies for biomedicine are being made due to the replacement of the more commonly used resins and plastics with nontoxic, biocompatible, and bioresorbable bioplastics such as polylactic acid (PLA) and poly(ε-caprolactone) (PCL). These materials permit the production of heterogeneous physical objects or structures of high complexity without loss of resolution. These materials also permit the fabrication of heterogeneous physical objects with control over many design characteristics including stack orientation, dopant, and polymer composition. 3D printing also provides the benefit of affordable customization and manufacture of more efficient designs that are lighter and stronger; requires less assembly; and reduces waste.[10]

Layer by layer manufacturing allows for much greater flexibility and creativity in the designing process. 3D printing significantly speeds up the design and prototyping process. There is no problem with making one part at a time and changing the design each time it is produced. Parts can be produced within hours, bringing the design cycle down to a matter of days or weeks compared to months. Also, since the price of 3D printers has decreased over the years, some 3D printers are now within financial reach of the ordinary consumers or minor companies.[11] Expensive hardware and expensive materials can be described as the disadvantages of 3D printing. This leads to expensive parts, thus making it hard if you were to compete with mass manufacture. It also requires a CAD designer to create what the customer has in mind and can be expensive if the part is very intricate. 3D printing is not the answer to every type of synthesis method; however, its advancement is helping to accelerate the designing and engineering more than ever before. Through the use of 3D printers, designers are able to make one of a kind piece of art, intricate building and product designs and also make parts while in space.

Printing physical 3D objects from digital data was first developed by Charles Hull in 1984. He named the technique as SLA and obtained a patent for the technique in 1986. While SLA systems had become popular by the end of the 1980s, other similar technologies like fused deposition modelling (FDM) and selective laser sintering (SLS) were introduced. In 1993, Massachusetts Institute of Technology (MIT) patented another technology, named "3 Dimensional Printing techniques," which is similar to the inkjet technology used in 2D printers. In 1996, three major products, "Genisys" from Stratasys, "Actua 2100" from 3D Systems, and "Z402" from Z Corporation were introduced. In 2005, Z Corp. launched a breakthrough product, named Spectrum Z510, which was the first high-definition color 3D printer in the market. Another breakthrough in 3D printing occurred in 2006 with the initiation of an open source project, named Reprap, which was aimed at developing a self-replicating 3D printer.[11]

4.2.1 TYPES OF 3D PRINTING MECHANISM

4.2.1.1 FUSED DEPOSITION MODELING

FDM is an AM technology commonly used for modelling, prototyping, and production applications. FDM works on an "additive" principle by laying down the materials in layers. FDM, a prominent form of RP, is used for prototyping and rapid production. RP enables iterative testing, and for very short runs, rapid manufacturing can be a relatively inexpensive alternative. Its main advantage is that it is cheaper than others since it uses plastic; more expensive models use a different (water soluble) material to remove supports completely. Even cheap 3D printers have enough resolution for many applications. Its disadvantage is that the supports leave marks that require removing and sanding. Warping limited testing allowed due to thermoplastic material.[12,13]

4.2.1.2   STEREOLITHOGRAPHY

SLA is an additive engineering process which employs a vat of liquid ultraviolet curable photopolymer "resin" and an ultraviolet laser to build parts' layers one at a time. For each layer, the laser beam traces a cross-section of the part pattern on the surface of the liquid resin. Exposure to the ultraviolet laser light cures and hardens the pattern traced on the resin and joins it to the layer below. After the pattern has been traced, the SLA's elevator platform inclines by a distance equal to the thickness of a single layer, typically 0.05–0.15 mm (0.002–0.006″). Then, a resin-filled blade bends across the cross-section of the part, recoating it with fresh material. On this new liquid surface, the subsequent sheet pattern is traced, joining the previous layer. A complete 3D part is formed by this process. After being built, parts are immersed in a chemical bath in order to be cleaned of excess resin and are subsequently cured in an ultraviolet oven. One of the benefits of SLA is its speed; functional parts can be produced within a day. The length of time it takes to synthesis one particular part depends on the size and complexity of the project and can last from a few hours to more than a day. Prototypes made by SLA are strong enough to be machined and can be used as master patterns for injection moulding, thermoforming, blow moulding, and various metal casting processes.

4.2.1.3   SELECTIVE LASER SINTERING

It is an AM method that uses a high power laser (e.g., a carbon dioxide laser) to fuse small particles of plastic, metal (direct metal laser sintering), ceramic, or glass powders into a mass that has a desired 3D shape. The laser selectively fuses powdered material by scanning cross-sections generated from a 3D digital description of the part (e.g., from a CAD file or scan data) on the surface of a powder bed. After each cross-section is scanned, the powder bed is lowered by one layer thickness, a new layer of material is applied on top, and the process is repeated until the part is completed. Some SLS machines use single-component powder, such as direct metal laser sintering. However, most SLS machines use two-component powders, typically either coated powder or a powder mixture. In single-component powders, the laser melts only the outer surface of the particles (surface melting), fusing the solid nonmelted cores to each other and to the previous layer. Compared with other types of AM, SLS can synthesize parts from a relatively wide range of commercially available powder materials, which includes polymers such as nylon (neat, glass-filled, or with other fillers) or polystyrene, metals including steel, titanium, alloy mixtures, and composites and green sand. SLS is performed by machines called SLS systems. SLS technology is in extensive use around the world due to its ability to easily make very complex geometries directly from digital CAD data. SLS has many advantages over traditional manufacturing techniques. Speed is the most obvious because no special tooling is required and parts can be built in a matter of hours. Additionally, SLS shows more rigorous testing of prototypes. Since SLS can use most alloys; prototypes can now be functional hardware made out of the same material as production components. SLS is also one of the less AM technologies being used in production. Since the components are built layer by layer, it is also possible to design internal features and passages that could not be cast or otherwise machined.[13,14]

4.3   MECHANISM AND PRINCIPLE OF 3D PRINTING

3D printing technology has developed as a promising instrument to manufacture platforms with high exactness and precision, making complicatedly point by point biomimetic 3D structures.[15] The procedures as of now being utilized to accomplish 3D printing of scaffold include a layer-by-layer process, which incorporates, yet is not restricted to, direct 3D printing, intertwined testimony demonstrating, SLA, and SLS. These systems have been utilized to create scaffolds ranging from millimetre to nanometer estimated scaffolds. It is likewise essential to note the solid freeform fabrication. The added substance assembling and 3D printing have turned out to be synonymous over the previous decade and are currently utilized conversely. Points of interest of utilizing 3D printing incorporate the capacity to manufacture flexible scaffolds with complex shapes equipped with homogeneous cell distribution and the capacity to mimic the extracellular matrix (ECM). However, the accessibility of biomaterials with the security and wanted properties for 3D printing of platforms is confined relying upon the printing innovation utilized. Another disadvantage is the creation time that it takes to manufacture scaffolds, which significantly increments as the scaffolds configuration turns out to be more exact and intricate.[16] This is particularly the case for ordinary techniques which include a considerable measure of physical work contrasted with a computerized process.[17] With expanded research and comprehension of 3D printing, the utilization of cross-breed materials and numerous scaffolds equipped for overcoming current limitations. Advancing from customary systems, 3D printing of tissue builds an approach to configure the scaffolds, fit for emulating the intricate structures of the ECM and along these lines giving a microenvironment to cell connection, multiplication, appropriation, and separation with the possibility to shape useful tissue.

4.3.1 MATERIALS USED FOR 3D PRINTING

Imperative criteria to consider while creating reasonable platforms are biocompatibility, biodegradability, pore interconnectivity, pore size, porosity, and mechanical properties. Biocompatibility and biodegradability are imperative properties for scaffold materials to have, guaranteeing they are degraded into nontoxic items while deserting just the desirable living tissue. Likewise, the material ought to instigate negligible inflammatory responses, along these lines decreasing the probability of dismissal by the host's resistant system. It would likewise be advantageous if scaffold materials could carry on as substrates that advance cellular attachment, proliferation, and differentiation. Moreover, as cells multiply and separate, the scaffold must have the capacity to withstand the strengths being connected by the cells; generally, its fall would bring about poor dissemination of oxygen, supplements and waste, prompting wasteful tissue arrangement. At long last, the mechanical soundness of the platform must be basically solid in order to withstand everyday activities and ordinary body movements.[18] Naturally inferred materials, for example, alginate, chitosan, collagen, fibronectin, and hyaluronic corrosive have favorable position over engineered materials as they give all the more intrinsically organic capacities. Utilizing actually inferred materials, that typically constitute or occupy the ECM, brings about a superior emulating of certifiable ECM, and this, therefore, improves cell connection and directs cell multiplication more effectively than manufactured polymers.[19] Although common materials are gainful for cell forms, the utilization of engineered polymers, for example, PCL and poly(d,l-lactic-*co*-glycolic corrosive) (PLGA) for scaffolds, has yielded higher mechanical qualities, higher processability, and controllable debasement rates.[19,20] However, these engineered polymer platforms have generally low organic movement, as far as advancing tissue recovery, contrasted with normally determined ECM polymers. Notwithstanding being less organically dynamic, the characteristic hydrophobicity of engineered polymers, for example, polyesters, by and large outcomes in poor cell adhesion[21] brings about imperfect expansion and separation, at last prompting substandard tissue formation.

For 3D printing scaffolds using powder beds, grain size and grain measure circulations must be considered to deliver permeable scaffolds,[22] as these variables impact microporosity which has been believed to impact cell appropriation, connection, multiplication, and differentiation.[23,24] To accomplish biomimicry of the ECM, platforms should be naturally dynamic, have high mechanical qualities, be anything but difficult to prepare, and have controllable debasement rates. To make these intricate platforms, half breed scaffold involving both engineered and characteristic polymers have been utilized and are probably going to be utilized as a part of the future.[25–27] It is imperative to remember that diverse powdered blends, materials, and structure measure affect the platform printability, just like the case for most materials in 3D printing. To be a practical alternative for tissue recovery, the materials utilized for 3D printing of platforms for tissue building ought to be printable with a high level of reproducibility. Such materials ought to likewise be financially savvy and flexible to frame the coveted morphology of the planned scaffolds.

Polymers speak to a noteworthy classification of materials with potential for use in the 3D printing of scaffolds for tissue building. Such polymers incorporate manufactured poly (ethylene glycol) diacrylate and regular gelatin methacrylate (GelMA),[28] which are both utilized as a part of the arrangement of hydrogels. Hydrogels are alluring biomaterials for tissue designing since they have customizable mechanical properties, are biocompatible, and can be hydrated while staying insoluble and keeping up their 3D structure. What is more, the hydrating properties of hydrogels enable them to copy those of organic tissue.[29,30] Both PCL and PLGA platforms have been made utilizing fast prototyping. The utilization of these scaffolds in the imperfections of rabbit tibias exhibited their wellbeing and also their ability to advance the era of bone tissue.[31] Preference of utilizing engineered polymer materials, for example, PLGA and PCL, is that both these manufactured polymers have been affirmed by the FDA for clinical use.[32] Another preferred standpoint of utilizing PCL polymers and PLGA copolymers is the low danger of their corruption items, which bolsters into metabolic pathways. A disservice of utilizing PLGA is that it can bring about fiery reactions when there is a development of acidic oligomers.[33] Inflammation assumes a key part in tissue regeneration,[34] yet it is vital to control or farthest point this reaction, as elevated amounts of irritation can prompt fibrosis bringing about poor tissue work, or even dismissal of the embedded platform. Hence, it is essential to comprehend the provocative impact of the specific biomaterial(s) utilized and the scaffolds structure used to create the coveted tissue. The fiery reaction, for the most part mounted by the inborn insusceptible scaffolds, advances the enlistment of cells (principally neutrophils and monocytes) to the region of tissue harm so as to help with tissue repair and regeneration.[35] In one review, 3D printed polylactic corrosive (PLA) and chitosan platforms were contrasted for their capacity with prompt irritation and therefore affect tissue regeneration.[36] Through the examination of macrophage morphology and human monocyte cytokine profiles (tumor putrefaction consider tumor necrosis factor (TNF)-α, interleukin (IL)-6, IL-10, IL-12/23(p40), and changing development calculate transforming growth factor (TGF)-β1), it was presumed that the provocative properties of the scaffolds were dictated by both platform geometry and piece. PLA-based scaffolds advanced a higher generation of Il-6, IL-12/23, and IL-10 contrasted with chitosan platforms. While chitosan platforms just advanced a higher discharge of TNF-α when contrasted with PLA-based scaffolds. It was likewise reasoned that orthogonal platforms initiated more grounded fiery responses contrasted with that of inclining scaffolds because of an expanded nearness of multinucleated goliath cells in the orthogonal scaffolds.[37] However, the perfect levels of irritation to create practical tissue were not tried, and additionally, including different cells, sorts and the arrangement of useful tissue, may be required.

Another intriguing issue for these polymers is their rate of biodegradation, which is regularly too quick using PLGA and too moderate while using PCL. In a polymer, a corruption test was performed utilizing scaffolds; it was demonstrated that with practically identical convergences of PCL and PLGA, PLGA debased by 18% at 14 days and 56% by 28 days contrasted with PCL whose debasement was 33% at 21 days and 39% at 28 days.[38] Nevertheless, both PLGA and PCL may even now have valuable regenerative characteristics relying upon the kind of injury.[39] Long-term recuperating might be important in open-bone fractures.[40,41] PCL is conceivably an ideal decision for open cracks because of its slower debasement rate. A more extended mending period is frequently required in open breaks on the grounds that the bone has infiltrated the skin, which can as a rule prompt contaminations that expand the ideal opportunity for recuperating. The moderate corruption rate of PCL enables the platform to offer help for developing cells for a more drawn-out timeframe empowering more thick tissues to form. For cases, for example, shut breaks (a broken bone that has not entered the skin), PLGA could be a potential possibility for bone recovery.

PCL-based copolymers, for example, PLGA–PCL–PLGA[42] and PCL–PEG–PCL,[43] have been integrated to control the debasement of PCL for controlled medication discharge applications. Be that as it may, these copolymers can possibly be utilized also for tissue building applications. Other manufactured polymers utilized for platform incorporate polyglycolic corrosive, poly(propylene fumarate) (PPF), and poly(hydroxybutyrate). Characteristic polymers, for example, proteins and polysaccharides, have additionally been utilized for scaffolds manufacture, and among these polymers, the most mainstream contender for tissue building has been collagen sort 1.[44] Collagen platforms stacked with cationic PEI-pDNA edifices were as of late used to make a fix equipped for bone recovery in calvarial abandons in rats (Fig. 4.1).[44]

4.3.2 TECHNIQUES

In the most recent decades, various procedures have been utilized to shape permeable 3D biomimetic scaffold and have included stage detachment, self-gathering, electrospinning, solidify drying, dissolvable throwing/particulate filtering, gas frothing, and liquefy forming. By utilizing platforms, the design of local extracellular grids can be emulated at the

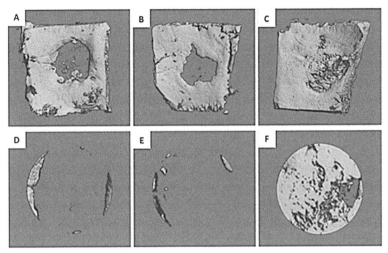

FIGURE 4.1   In vivo bone formation in rat calvarial defects implanted with collagen scaffolds: representative micro-CT scans showing the level of regenerated bone tissue in calvarial defects 4 weeks after implantation with (A and D) empty defects (no collagen scaffolds), (B and E) empty scaffolds (collagen scaffolds alone), and (C and F) PEI-pPDGF-B complex-loaded scaffolds (collagen scaffold). (Reprinted with permission from Reference [44]. © 2014 Elsevier.)

nanoscale level and in this manner gives the essential base to the recovery of new tissue.[46] Originally, a "beat-down" approach was utilized as a tissue-building technique for scaffold manufacture. In this strategy, cells are seeded onto a biodegradable and biocompatible scaffold and are anticipated to move and fill the platform, henceforth making their own particular network. By utilizing this system, a few vascular tissue, for example, bladder[47] and skin,[48] have been built viable. In any case, because of the restricted dispersion properties of these scaffolds, this procedure confronts a few difficulties for creation of more intricate tissues, for example, heart and liver.[49] Therefore, "base-up" strategies have been produced to defeat this problem.[50] Bottom-up methodologies incorporate cell-exemplification with microscale hydrogels, cell collection without anyone else's input gathering, era of cell sheets, and direct printing of cells.[51] These mind-boggling tissue pieces can be amassed utilizing different techniques, including microfluidics,[52] attractive fields,[53] acoustic fields,[54] and surface tension.[55] These strategies are moderately simple and have given a strong establishment to the manufacture of platforms. Be that as it may, as specified beforehand, these regular strategies experience the ill effects of a few confinements, including deficient control over platform properties, for example, pore measure, pore geometry, dissemination of elevated amounts of interconnectivity, and mechanical quality. Accordingly, it is important to create innovations with adequate control to outline more multifaceted tissue-particular platforms. Also, scaffolds can be covered by utilizing surface alteration procedures (e.g., presenting utilitarian gatherings) to improve cell movement, connection, and expansion. 3D printing enables platforms to wind up plainly more unequivocally manufactured [like that of the PC helped outline (CAD)] with higher adaptability in the kind of materials used to make such scaffolds. 3D printing utilizes an added substance producing process where a structure is manufactured by utilizing a layer-by-layer preparation. Materials stored for the arrangement of the scaffold might be cross-connected or polymerized through warmth, bright light, or fastener arrangements. Utilizing this innovation, 3D printed scaffold can be set up for streamlined tissue designing.

For the suitable arrangement of tissue engineering, the seeding cells (regularly undifferentiated cells) require a 3D situation/scaffold-like that of the ECM. The ECM goes about as a medium to give proteins and proteoglycans among different supplements for cell development. The ECM additionally gives auxiliary support to permit to cell usefulness, for example, directing cell correspondence, development, and assembly.[56] With this as the main priority, researchers and architects initially endeavored to reproduce the ECM through regular systems, which thus settled a structure for utilizing more propelled methods, for example, 3D printing, to yield higher quality scaffold. The 3D printing procedure can make characterized platform structures with controlled pore size and interconnectivity and the capacity to bolster cell development and tissue formation.[57,58] The present techniques for 3D printing include a CAD, which is then transferred to every 3D printing scaffold to "print" the coveted platform structure. Through different 3D printing innovations, examined underneath, analysts are attempting to create a biocompatible scaffold that effectively bolsters tissue arrangement.

4.3.3   COMPUTER-AIDED DESIGN AND DIGITAL IMAGING

The beginning of numerous 3D printing forms includes a CAD that must be drawn or taken from known organ structures. By and large 2D cuts procured from imaging, instruments are arranged and stacked on top of each other to frame a 3D structure.[59,60] In tissue building, it is basic to develop tissue like that of the local tissue, and keeping in mind the end goal to finish this, imaging methods can be utilized to deliver scaffolds that firmly emulate the structure of local tissues.[61] These pictures illuminate platform outlines by giving morphology and size parameters to which platforms need to adjust so as to fit into unpredictably molded imperfections/breaks where tissue arrangement is fancied. The scaffolds shaped likewise coordinates the development of cells and give shape to the last tissue.[62] It is additionally important that platform shape can influence the kind of tissue recovered as can be found in dentin tissue recovery with differentially molded platforms utilizing dental mash determined cells.[63] The complexities in morphology and engineering of tissues can be depicted with imaging advancements, for example, attractive reverberation imaging (MRI) and PC tomography (CT). These imaging innovations help to take cross-sectional cuts of organs and arrange them into a 3D picture, therefore enabling the plan of platforms to be a nearby portrayal of local organs.[64]

X-ray works by utilizing attractive fields and throbbing radio waves to yield definite pictures of organs and delicate tissues. Utilizing inclination loops to translate vitality signals delivered by water atoms inside the tissue, 2D pictures are generated.[65] These 2D pictures are then stacked to make a 3D picture of the checked region. Since MRI requires hydrogen particles by and large in water, they are best utilized for delicate tissue imaging, for example, ligaments and organs of the trunk and stomach area (heart, spleen, pancreas, liver, and kidneys). They are additionally used to picture pelvic organs, for example, the bladder and conceptive organs.

CT, otherwise called electronic pivotal tomography, is an innovation that utilizes X-beams to deliver pictures from a filtered range. In a CT scanner, X-beam tubes are turned around a patient's body, creating signals that are taken up by computerized X-beam indicators and sent to be handled by a PC to produce cross-sectional pictures of the body. These cross-areas are then stacked to make a 3D picture of the examined organ(s). CT sweeps are by and large utilized for imaging bone because of its thickness, while delicate tissues can be risky as they have different capacities to hinder X-beam infiltration bringing about swoon or unclear pictures. With a specific end goal to picture these delicate tissues, differentiating specialists, for example, iodine or barium-based mixes, might be utilized to encourage complexity and increment visibility.[66]

X-ray is better than CT while endeavoring to picture delicate tissues and different organs other than bones as the differentiation of firmly set organs can all the more be promptly observed when changes to radio waves and attractive fields are connected. The radio waves and attractive field empower the capacity of the instrument to highlight the coveted tissue in firmly weaved regions. Notwithstanding, CT filters make better quality pictures of bone structures than MRI because of the low grouping of water in bones bringing about less hydrogen iota's emanating vitality to briefly make a cross-sectional picture. Making a scaffold straightforwardly from the pictures is not generally attainable because of the likelihood of checking unhealthy or harmed organs. For this situation, PC displaying might be important to reproduce the missing parts of the organ or tissue. With MRI and CT imaging procedures, the remaking of both 2D and 3D pictures is a capable apparatus to reproduce the many-sided quality of tissue structures. These apparatuses enable specialists to be one bit nearer to manufacturing an exact imitation of the required extracellular lattice to improve utilitarian tissue arrangement.

4.3.4   DIRECT 3D PRINTING

3D printing includes the manufacture of structures through progressive layer testimony by utilizing an automated procedure. The initial "3D printer" was produced by specialists at the MIT in the 1990s and depended on the innovation of a standard inkjet printer. This printing strategy can now and again be alluded to as "folio streaming" or "drop-on-powder." In a standard 2D inkjet printer, the ink spout moves in a side-to-side movement incrementally along one plane with the end goal that the literature has the two measurements of length and width. A 3D printer utilizes a similar innovation and moving side to side along one plane, the printer has a stage fit for moving up and down (90 degrees to the side-to-side movement), thus including the measurement of stature and in this way imprinting in three measurements. The 3D printer composed by MIT has comparative qualities to the 2D inkjet printer, notwithstanding, rather than ink, the 3D printer utilizes a fluid cover arrangement that is specifically saved on a powder bed rather than paper. The procedure starts with a powder bed, which could change contingent upon materials utilized, that is spread onto the fabricate stage and leveled utilizing a roller framework.

The printer spout then administers cover arrangement in the assigned powdered ranges coordinated by the CAD. Once the cover arrangement and powder are joined, the overabundance powder is evacuated (passed over). The fabricate stage is then brought down, and another powder layer is kept and leveled. This procedure is then rehashed until the last structure is made (Fig. 4.2). This procedure likewise has the adaptability to change the organization of cover and powder on the off chance that it is regarded fundamental where certain parts of the platform may require a material with higher mechanical qualities or potentially littler pore sizes. A case of this might construct a platform with bigger pore sizes that profound inside the framework, while having littler surface pores. With the expansion in pore sizes, cells profound inside the framework will have the capacity to keep up their cell forms as indispensable assets, for example, supplements, oxygen, and waste can diffuse without trouble contrasted with little pore sizes that may bring about the supplement hardship of cells prompting cell demise. Cell demise on an expansive scale eventually prompts the crumple of the framework and the failure to shape utilitarian tissue. The subsequent coveted impact of the created platform will be to give a medium that ensures high multiplication and separation of cells to produce utilitarian tissues. The upsides of this technique are the far-reaching rundown of powder–cover arrangements accessible to yield the coveted platform. The utilization of folios, in any case, can prompt danger on the off chance that they are not totally expelled once the framework is prepared to be embedded, as on account of natural solvents that are utilized as fasteners for some powdered polymer materials. Another impediment for this printing procedure is the post handling required, where warm treatment might be important to guarantee durability.[67]

4.4 BIOMEDICAL APPLICATIONS OF 3D PRINTING

4.4.1 ORTHOPEDIC IMPLANTS

Orthopaedics is a surgical discipline that is commonly tied to biomedical engineering, which has been applied to different orthopaedic disciplines, ranging from trauma surgery, joint arthroplasty, and tumor surgery to deformity correction. From preoperative planning to training and education to implant designing, the use of 3D printing is rising and has become more prevalent in medical applications over the last decade as surgeons and researchers are increasingly utilizing the technology's flexibility in manufacturing objects. 3D printing is a type of manufacturing process in which materials such as plastic or metal are deposited in layers to create a 3D object from a digital model. (1) The process is distinct from traditional manufacturing methods; in that, it is an additive rather than a subtractive process. Specifically, in surgical applications, the 3D printing techniques can not only generate models that give a better understanding of the complex anatomy and pathology of the patients (2) but can also produce patient-specific instruments (PSIs)[68–76] or even custom implants[77,78] that are tailor-made to the surgical requirements.

4.4.1.1 APPLICATIONS OF 3D PRINTING IN PATIENT-SPECIFIC ORTHOPEDICS

In the last decades, 3D printing has undergone tremendous development and now has important applications in various fields of medicine. However, reports on orthopaedic applications are limited and mainly consist of case reports or small series. These include anatomic models for surgical planning, education and training, PSI for assisting surgical procedures, and fabrication of complex custom metal implants.

4.4.1.2 ANATOMIC MODELS FOR SURGICAL PLANNING

In traditional preoperative planning, orthopaedic surgeons utilized 2D plain X-ray and CT images to assess the bony anatomy and pathology of the patient. As a more advanced image, post-processing software can generate 3D nonaxial reformatted images and 3D images, patient-specific treatment choice(s) can be selected by studying all the digital images. Despite these advances in image processing technology, the 3D anatomy is still being viewed as a flat, 3D image.

Patient-specific physical bone models can be recreated from patients' CT image data by 3D printing. The models not only allow surgeons to have a tactile and visual understanding of the patient-specific anatomy and pathology but also anticipate the operative challenges that will aid in the planning of orthopaedic procedures.

4.4.2   3D-PRINTED CUSTOM IMPLANTS

Off-the-shelf, standard-sized bone implants are mainly fabricated using traditional subtractive manufacturing method. They are readily available to meet the surgical requirement for most patients. In contrast to an off-the-shelf implant in which the patient's anatomy may need to be modified for a proper fit, a custom implant is a perfect match for the unique anatomy of that patient as it is based on the patient's own medical images. Custom implants may be indicated when (1) patients 'bony geometries fall outside the range of standard implants with respect to implant size- or disease-specific requirements and (2) improved surgical results are anticipated due to a better fit between implants and patients' anatomical needs.[79]

4.4.2.1   POTENTIAL ADVANTAGES AND LIMITATIONS OF 3D-PRINTED CUSTOM IMPLANTS

The nature of additive layer manufacturing in 3D printing allows fabrication of custom implants with any complex shape or geometric feature; this is impossible using traditional subtractive manufacturing techniques. In addition to the advantage of being anatomically conformed to an individual patient, a 3D-printed implant can be generated with a complex free-form surface such as scaffold lattice in a metal monoblock. This interconnected pore surface structure has the potential to facilitate osteointegration and therefore the possibility to reduce the stiffness mismatch at bone-implant junctions. The stress-shielding problem may be further minimized as the porosity, and pore size in the implants can be modified, and the elastic modulus of porous titanium can be tailored to be comparable to that of the patient's bone with the use of modern 3D printing techniques. Therefore, the modern design of a 3D-printed custom implant not only has the structural geometry that can match the surgical requirements of an individual patient but also allows biomechanical evaluation under patient-specific loading conditions for design modification prior to the actual implant fabrication.[2] The use of 3D-printed custom implants has recently been reported in difficult revision hip arthroplasty with severe acetabular bone loss.[12] Sixteen patients with large acetabular defects underwent revision hip arthroplasty 3D-printed custom trabecular titanium implants. The implants were designed from a detailed analysis of patients' CT images with special reference to the bone quality and the bone geometry of deficient acetabulum. The implant consisted of a porous augment and cage that recreated the artificial acetabulum. Flanges were added to the design for optimum screw fixation to the remaining pelvic bone. Patient-specific aids and instruments such as 3D-printed physical models of the hemipelvis, trial implants, and drill guides were provided for preoperative rehearsal and intraoperative reference and guidance. However, the difficulty was reported in placing the custom implants accurately as planned in 7 out of 16 implants, which were found to be malpositioned in one or more parameters of measurement. This was the only study that addressed the accuracy of placement of custom acetabular implants by comparing the preoperative CT planning with CT data on the postoperative position. Further follow-up is needed to evaluate the clinical outcome of this new 3D-printed custom implant.

4.4.3 DENTAL IMPLANTS

The first industrial applications of 3D printing were for rapid tooling and RP. Its use in dentistry, where single, personalized objects were manufactured, was an obvious next step. Today, by combining oral scanning, computer-aided designing and 3D printing, dental labs can accurately and rapidly produce crowns, bridges, plaster/stone models, and a range of orthodontic appliances such as surgical guides and aligners. Instead of uncomfortable impressions, which were in use until recent times, a 3D scan can be taken, which can be later transformed into a 3D model and sent to be 3D printed. The printed model can be used to create a full range of orthodontic appliances, delivery and positioning trays, clear aligners, and retainers. 3D printing allows one to digitize the whole workflow, dramatically accelerating production times and significantly increasing production capacity. In addition, they allow one to eliminate physical impressions and the storage of models.

By 2015, printing complete dental prostheses and restorations with the push of a button have come closer to reality. With the dental printer that can print layers of different material over one another[80] and FDA approved denture material for 3DP,[81] 3D printing was considered the future.

Currently, in the dental clinic, the applications of 3D printers are fabricating 3D printer models for crowns,[82] bridges, and prostheses,[83] dental, orthodontic brace models,[84] and surgical guides. 3D printers enhance lab economics by increasing productivity and improving patient satisfaction as they allow self-production. With clear aligner systems, teeth are easy, convenient, and safely straightened with wireless orthodontics, precise 3D images of teeth are captured by dental 3D scanner, tooth movement is elaborately predicted and designed by computer clear aligner program, and it is rapidly produced by directly sending the designed 3D image to its own 3D printer.[85,86] Figure 4.2 shows clear aligner manufactured in person by using a 3D printer in a dental clinic nowadays.

4.4.4 TISSUE AND ORGAN IMPLANTS

Organ and tissue failures from birth complications, ageing, diseases, and accidents are common medical problems.[87] Nowadays, treatments for organ failure rely mostly on organ transplants from both living and deceased donors. The problem is that there is a chronic shortage viable donor available for transplant. In 2009, 154,324 patients in America were waiting for an organ. Only 18% of them received an organ transplant, and an average of 25 persons died per day, while they were on the waiting list. Also, the cost of organ transplant surgery and follow-up is also expensive. The additional problem is that organ transplantation involves the often

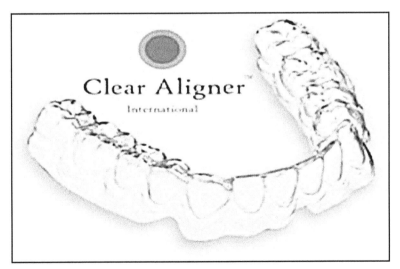

FIGURE 4.2   Clear aligner manufactured in person by using a 3D printer. (Reprinted from *Indian Journal of Science and Technology*, Vol. 9(20), DOI: 10.17485/ijst/2016/ v9i20/94696, May 2016. Open access.)

difficult task of finding a viable donor.[1] This problem can be eliminated by using cells taken from the patient's own body to build a replacement organ,[88] minimizing the risk of tissue rejection, as well as the need to take lifelong immunosuppressant drugs. Lately, tissue engineering and regenerative medicine are being pursued as a solution for the organ donor shortage for the near future. Adding to this, 3D bioprinting offers additional advantages beyond the traditional regenerative methods (which provide scaffold support alone), such as the precise cell placement and control over speed, resolution, cell concentration, drop volume, and diameter of printed cells. Organ printing is used to produce cells, biomaterials, and cell-laden biomaterials separately or in tandem, layer by layer, creating 3D tissue-like structures. Although hydrogels are usually considered to be most suitable for producing soft tissues, various other materials are also used to build the scaffolds, depending on desired strength, porosity, and type of tissue.[92]

Although 3D bioprinting systems are usually laser-based, inkjet-based, or extrusion-based, inkjet-based bioprinting is commonly preferred. In this method, droplets of living cells or biomaterials are dropped onto a substrate using digital instructions to reproduce human tissues or organs. The process includes the creation of a blueprint of an organ with its vascular architecture, generation of a bioprinting process plan, isolation stem cells, differentiation of the stem cells into organ-specific cells, preparation of biomaterial reservoir with organ-specific cells, blood vessel cells, and support medium and loading them into the printer, bioprinting; and placement the bioprinted organ in a bioreactor prior to transplantation. Laser printers can be employed in the cell printing process, using the laser energy to excite the cells in a particular pattern, providing spatial control of the cellular environment.

Although tissue and organ bioprinting is still in its early stages of development, many studies have provided its proof of concept. Researchers have used 3D printers to create knee meniscus, heart valve, spinal disk, other types of cartilage and bone, and an artificial ear. Cui et al. applied inkjet 3D printing in repairing human articular cartilage.[91] Wang et al. used 3D bioprinting technology for depositing different cells within various biocompatible hydrogels for producing an artificial liver.

A number of biotech companies have focused on creating tissues and organs for medical research, so that it may be possible to rapidly screen new potential therapeutic drugs on patient tissue, cutting research costs and time. Scientists at Organovo are developing strips of printed liver tissue for this purpose; soon, they expect the material will be advanced enough to use in screening new drug treatments.[89] Other researchers are working on techniques to grow complete human organs that can be used for screening purposes during drug discovery.[90]

4.4.5   PROSTHETICS

Another biomedical application for 3D printing is the development of prosthetics. Medical prosthetic devices which are manufactured using traditional practices are typically either too expensive or too unsophisticated. Almost 185,000 amputations take place in America every year, about 54% of which are attributed to vascular problems and 45% to accidents, according to the Amputee Coalition, a nonprofit supported by the Centers for Disease Control and Prevention (CDC). A majority of these amputations are of extremities, which are subject to high levels of variance from individual to the next. Although a number of effective, low-cost prosthetics for extremities are available (such as the Jaipur foot), these are often uncomfortable to the amputee, especially on the point of contact between the stump and the prosthetic.

Using a patient's MRI or CT scan, a prosthetic catering to that specific person can be made using 3D printing.[93] 3D printed prostheses are a good example of the influence this technique can exert since, on the one hand, they are inexpensive, and, on the other, they are fully customized to the wearer. In addition, they are more comfortable than the traditional prostheses and can be manufactured in a day.[94] Low costs of the 3D printed limb prostheses are especially important in prosthetics for children since they are at a growing stage. Moreover, expandable 3D prosthetics may soon be available for children that could "grow" with the child taking into account the high cost of the traditional prostheses, this leads to a revolution disrupting the prosthesis market.

4.5   SLA AND APPLICATIONS IN TISSUE ENGINEERING

In recent years, tissue engineering has been an area of immense research interests because of its vast potential in the repair or replacement of damaged tissues and organs.[95,96] The prospects for using SLA fabrication methods for biomedical applications are various. A number of researchers in the biomedical engineering field are employing 3D printing as a transformative tool for biomedical applications, especially for tissue engineering and regenerative medicine. Nowadays, researchers have greatly increased awareness of the dramatic differences in cell behavior between 2D and 3D culture systems. Culturing cells in 3D delivers a more physiologically applicable environment to guide cell behaviors and enhance their functions. Therefore, excessive efforts have been made to develop 3D bio fabrication techniques that can generate complex, functional 3D architectures with proper biomaterials and cell types to mimic the native microenvironment and biological components.[97–99]

The idea of tissue engineering was formalized in 1993 when Langer and Vacanti published a historical milestone paper in *Science*, in which the characteristics and applications of biodegradable 3D scaffolds were first exhaustive.[100] Preferably, 3D scaffolds should be highly porous, have well-interconnected pore networks, and have consistent and adequate pore size for cell migration and infiltration.[101] A number of conventional manufacturing techniques were applied to fabricating porous 3D scaffolds, such as fiber bonding, phase separation, solvent casting, particulate leaching, membrane lamination, molding, and foaming. Nevertheless, all these methods share a major drawback, that is, they do not permit enough control of scaffold architecture, pore network, and pore size, giving rise to inconsistent and less-than-ideal 3D scaffolds. To overwhelm this problem, researchers proposed the use of 3D-printing methods (also known as RP, solid free-form fabrication or AM) in order to fabricate customized scaffolds with controlled pore size and pore structure.[102–104] Out of more than 40 different 3D-printing techniques in development, FDM, SLA, inkjet printing, SLS, and color jet printing seemed to be the most popular, due to their capacity to process plastics. As a result, in the second decade of this field (2003–2012), the number of studies in the area of 3D printing for tissue engineering increased rapidly. These studies covered scaffold design, process modeling and optimization, comparisons of 3D-printing methods, postprocessing and characterization of 3D printed scaffolds, in vitro and in vivo applications of 3D printed scaffolds, new scaffold materials for 3D printing, new 3D-printing methods for scaffold fabrication, and even the branching out of an entirely new field—3D bioprinting, or organ printing.[105–107]

The prospects for using SLA fabrication methods for biomedical applications are numerous. But, despite the technique has been commercially available for more than 20 years. But still it is not extensively used in the medical field. So here below, an overview is given of the possibility of biomedical applications of SLA.

4.5.1 PATIENT-SPECIFIC MODELS AND FUNCTIONAL PARTS

The ability to use data from (clinical) imaging techniques like MRI or CT makes SLA particularly useful for biomedical applications. Initially, these images were only used for diagnosis and preoperative planning. Now with the advancement in CAD and manufacturing technologies, it is now possible to make use of this information in directing the surgery itself. By making use of patient-specific models of parts of the body fabricated by SLA, the time in the operating theatre and the risks convoluted can significantly be reduced. In implantations using drill guides built by SLA, the accuracy of the implantation was much improved when compared to using conventional surgical guides, particularly in oral surgery. SLA has also been used to formulate moulds for the preparation of implants in cranial surgery, using data from CT scans. The function of the structures prepared by SLA is helping the surgeons. Additionally, the combination of medical imaging and SLA has been used to create models or moulds for making anatomically shaped implants. In this way, Societal et al. prepared customized heart valves that could be placed without the need for suturing. Other studies on such use of computer-aided fabrication are ear-shaped implants and aortas. Other tailored parts, such as hearing aids, have specific functionalities as well and are applied in contact with the body. They are now regularly manufactured using SLA. These applications demonstrate the profits of SLA in the manufacturing of tailored parts for use in the clinical practice. [108,109]

4.5.2 IMPLANTABLE DEVICES

Besides for the fabrication of surgical models, SLA can be used to fabricate patient-specific implantable devices. In trauma surgery, for example, a tailored implant for facial reconstruction can be obtained by imaging the undamaged side of the head and reproducing its mirror image by SLA. Also, a CT-image of bone with a defect due to the removal of a tumor could be used to prepare a custom-fit, biodegradable implant that supports the regeneration of bone. For prototyping, the appearance of the built parts, its mechanical properties and biocompatibility are not very important, and most commercially available resins suffice for this purpose. Nevertheless, for the direct fabrication of implantable devices, these properties are of highest importance, and special resins need to be developed and used. The implantation of devices prepared by SLA has only been reported in limited cases.

4.5.3   TISSUE ENGINEERING

Tissue engineering is a fast-developing field in biomedical engineering. In tissue engineering, a resorbable scaffolding structure is used in amalgamation with cells and or biologically active compounds to make the regeneration of tissues in vitro or in vivo. The scaffold is a porous implant, envisioned as a temporary support structure for seeded cells and formed tissues. Upon implantation, it should not produce severe inflammatory responses, and it should degrade into nontoxic compounds as newly formed tissue formed.

The scaffold is a template that permits cell adhesion and directs the development of tissue. Therefore, the architecture of its inner pore network and its outer geometry need to be well defined. For this reason, solid freeform fabrication techniques show significant role in the manufacturing of advanced tissue engineering scaffolds. The manufacture of precisely defined tissue engineering scaffolds by SLA is becoming a new standard. A lot of work has been done in developing biodegradable macromeres and resins. In 2000, the first report on the preparation of biodegradable structures by SLA was performed. Liquid low-molecular-weight copolymers of ε-caprolactone (ε-CL) and trimethylene carbonate (TMC) are prepared by ring opening polymerization using a polyol as an initiator, which subsequently derivate at the hydroxyl term with coumarin.[110] Later, the same group published the use of similar oligomers, but now end-functionalized with methacrylate groups. Figure 4.3 illustrates the structures fabricated using these macromers. The biodegradable macromers that was applied in SLA are based on functionalized oligomers with hydrolysable ester or

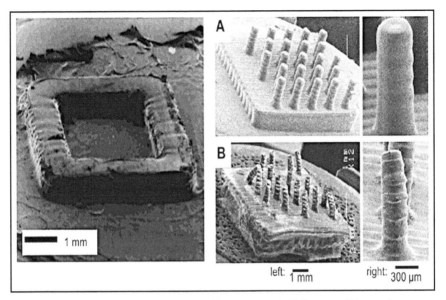

FIGURE 4.3 SEM images of biodegradable structures built by stereolithography envisioned for application in tissue engineering. Left: structures prepared from p(TMC-*co*-CL)-coumarin macromers. Right: structure prepared from p(TMC-*co*-CL)-acrylate (A) before and (B) after 1 month of implantation. (Reprinted with permission from Reference [25]. © 2010 Elsevier.)

carbonate linkages in the main chain. Among these, macromers are those based on PPF, TMC, and ε-CL, or d,l-lactide (DLLA) shown in Figure 4.4.

The viscosity of low-molecular-weight macromers that have been a low glass transition temperature such as those based on TMC and CL can be adequately low to allow direct processing in the SLA apparatus. It should be noted that this can result in a cross-linked structure with relatively poor mechanical properties. In the case of macromers of higher molecular weights with relatively high glass transition temperatures such as DLLA or PPF diluents is suffers the decrease of the viscosity of the resin. Typically, the highest resin viscosities that can be engaged in SLA are approximately 5 Pa s (pascal second). Macromers that contain fumarate (end)groups require reactive diluents such as diethyl fumarate or *N*-vinyl-2-pyrrolidone to reach proper reaction rates. Solid or highly viscous macromers that are functionalized with more reactive groups such as (meth)acrylates can also dilute with nonreactive diluents. In such case, the cured material shrinks upon extraction of the diluents. As this shrinkage

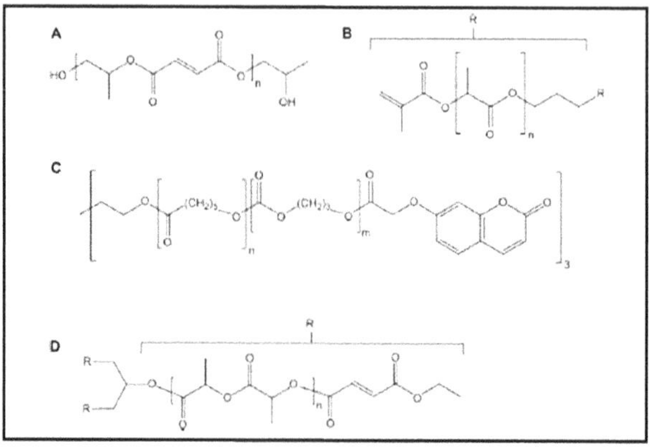

FIGURE 4.4   Examples of biodegradable macromeres used in stereolithography. (A) Polypropylenefumarate (PPF), (B) linear poly(d,l-lactide)-methacrylate, (C) poly(TMC-*co*-CL)-coumarin, (D) star-shaped poly(d,l-lactide)-fumarate.

is isotropic, it can be taken into interpretation when designing the scaffold structures. The SLA resins summarized above have enabled the creation of biodegradable tissue engineering scaffolds with a range of properties. The size scales are significant for tissue engineering. Scaffolds with hexagonal and cubic pores both prepared from a PPF/DEF resin and scaffolds with gyroid pore network architecture (PDLLA-fumarate/NVPresin) are represented in Figure 4.5.

The scaffold materials accessible above are shown to support cell adhesion and growth. In general nevertheless, the interaction between cells and synthetic polymers is not very strong. Therefore, adhesion of the cells is facilitated by the adsorption of proteins that is from the culture medium on the surface of the pores of the scaffold. Instead, bioactive compounds can be included in the scaffold fabrication method. To improve the bioactivity of scaffolds in bone tissue engineering by using composite structures containing hydroxyl apatite have been fabricated using resins containing dispersed hydroxylapatite particles. By mixing PPF and hydroxyapatite particles in diethyl fumarate as reactive diluents, a photopolymerizable composite resin can be acquired. Moreover, specific protein-binding molecules have been included in the resins, and proteins have been grafted on to the surface of the scaffold network. Recently, most promising is the developments in the functionalization of natural polymers. For the first time, using methacrylate-functionalized gelatin, an object has made by SLA from

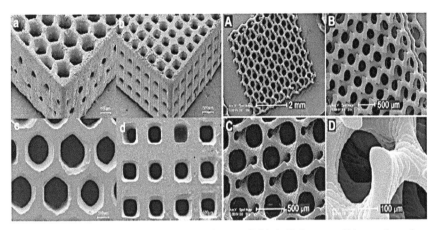

FIGURE 4.5   Examples of tissue engineering scaffolds built by stereolithography using PPF or PDLLA-fumarate and a reactive diluent. Adapted with permission from Reference [30]. Left: structures were prepared from PPF and DEF as a reactive diluent. Right: structures prepared from PDLLA-fumarate and NVP as a reactive diluent.

a natural polymer. Before, porous structures prepared by photocuring modified chitosan exhibited very good endochondral ossification upon implantation. Other natural polymers that have been modified to enable photocuring are (meth)acrylate oligopeptides and methacrylated hyaluronic acid.[111–112] It is probable that these materials will be used in SLA in the near future.

4.6   RECENT ADVANCES IN 3D PRINTING OF BIOMATERIALS

The expression "3D printing" has turned out to be universal; it can be portrayed as a solitary procedure. Similarly, as there are numerous approaches to print on paper, 3D printing envelops a few assortments of advances. A portion of the 3D printers can be work like ink-fly printers, while in some others, light was utilized to instigate polymerization. Some utilization warmth to liquefy and expel plastic materials as though they were toothpaste, while others utilized a laser to combine a fine material into a strong question, a procedure called laser sintering. "3D printing is an iterative, added substance innovation. As opposed to beginning with a square of material and evacuating what is undesirable in a subtractive procedure (as in chiseling or processing), added substance fabricating begins from nothing and specifically manufactures, one layer at any given moment, a question of enthusiasm as per PC directions," composes Jordan Miller, colleague teacher of bioengineering at Rice University, in an article. The noticeable preferred standpoint of 3D printers that it can be utilized as a part of restorative applications is the opportunity to deliver uniquely crafted medicinal items and hardware. For instance, the utilization of 3D printing to redo prosthetics and inserts can give incredible incentive to both patients and doctors. Also, 3D printing can create made-to-request dances and apparatuses for use in working rooms. Handcrafted inserts, installations, and surgical apparatuses can have a positive effect regarding the time required for surgery, understanding recuperation time, and the achievement of the surgery or embed. It is additionally expected that 3D printing advancements will in the end permit tranquilize measurement shapes, discharge profiles, and administering to be modified for every patient. Another essential advantage offered by 3D printing is the capacity to create things inexpensively. Customary assembling techniques stay more affordable for substantial scale creation; in any case, the cost of 3D printing is ending up noticeably more focused for little generation runs. This is particularly valid for little measured standard inserts or prosthetics, for example, those utilized for spinal, dental, or craniofacial issue. The cost to custom-print a 3D question is insignificant, with the primary thing being as reasonable as the last. This is particularly invaluable for organizations that have low generation volumes or that create parts or items that are exceptionally intricate or require visit changes. 3D printing can likewise diminish producing costs by diminishing the utilization of superfluous assets. For instance, a pharmaceutical tablet measuring 10 mg could possibly be specially created on request as a 1-mg tablet. A few medications may likewise be imprinted in measurements frames that are simpler and more financially savvy to convey to patients. "Quick" in 3D printing implies that an item can be made inside a few hours. This makes 3D printing innovation considerably quicker than conventional techniques for making things, for example, prosthetics and inserts, which require processing, producing, and a long conveyance time. Notwithstanding speed, different qualities, for example, the determination, precision, dependability, and repeatability of 3D printing advances, are additionally progressing.

3D printing guarantees to deliver complex biomedical gadgets as indicated by PC configuration utilizing quiet particular anatomical information. Since its underlying use as presurgical perception models and tooling moulds, 3D printing has gradually developed to make unique gadgets, inserts, frameworks for tissue building, analytic stages, and medication conveyance frameworks. Energized by the current blast out in the open intrigue and access to reasonable printers, there is reestablished enthusiasm to join undifferentiated cells with custom 3D platforms for customized regenerative medication. Before 3D printing can be utilized routinely for the recovery of complex tissues (e.g., bone, ligament, muscles, vessels, and nerves in the craniomaxillofacial complex) and complex organs with mind-boggling 3D microarchitecture (e.g., liver and lymphoid organs), a few mechanical impediments must be tended. 3DP materials incorporate calcium polyphosphate and PVA, HA and TCP, TCP, TCP with SrO and MgO doping, HA and apatite-wollastonite glass artistic with a water-based fastener, calcium phosphate with collagen in folio, PLGA, and Farringtonite powder ($Mg_3(PO_4)_2$). Materials utilized as a part of roundabout 3DP gelatin preforms supplanted with PCL and chitosan. In vitro thinks about with cow-like chondrocytes for articular ligament tissue building, bone tissue designing, monocytic cells from the RAW 264.7 cell line, human osteoblasts, C2C12 premyoblastic cell line, and bone marrow stromal cells.[113–120] In vivo contemplates have been performed with calvarial rabbit bone, rabbit tibia bone and porcine maxillary bone, rodent femoral deformities mouse femoral imperfections and rabbit femoral bone. Utilizing combined statement demonstrating (FDM innovation), platforms with biocompatible materials have been made with various pore morphology and channel sizes by controlling the x-y development of the expulsion head. Materials can likewise be joined in this innovation, for example, poly (ethylene glycol) terephthalate/poly(butylene terephthalate) or polypropylene/TCP.[121,122] Different composites, for example, PCL/HA or PCL/TCP are utilized with FDM because of good mechanical and biochemical properties for bone recovery. FDM has normally utilized biocompatible polymers with low dissolving temperatures. Materials utilized as a part of FDM to make platforms are PCL and bioactive glass composites,[123] L-lactide/ε-CL,[46] PLGA with collagen penetration, PCL-TCP with gentamicin,[48] PCL-TCP,[49] PLGA-TCP and covered with HA, PCL-PLGA-TCP,[50] PLGA-PCL,[51] PCL covered with gelatin, PCL, PMMA,[55] and PLA. In vitro examines have been performed with porcine chondrocytes, mouse preosteoblasts, and bone-marrow-inferred mesenchymal foundational microorganism. In vivo thinks about with murine creature models for wound mending human patient for craniofacial deformity, and rabbit bone imperfection. Applications incorporate ligament tissue designing, antitoxin conveyance system, bony craniofacial surrenders in people, and bone tissue building.[123–130]

Additional advance for 3D printing advances is required for expanding determination without yielding shape, quality, and hand ability of platforms. Anatomical elements and tissue engineering may have subtle elements on the size of several microns (e.g., villi of the small digestive tract with ~500 μm distances across). Dissemination utilization displaying has demonstrated a 200-μm restrain in frameworks for oxygen transport to cells, bringing about a greatest of 400 μm distance across components for cell survival.[128] For both SLS and 3DP, there is a test making more grounded structures without expanding measurements. To make little components which survives the creation procedure, powder particles much be bound together firmly. By expanding the quality of the laser for SLS or measure of folio for 3DP, extra powder particles would tie and in this way increment the measurements. Extra work is expected to move SLS and 3DP to resolutions beneath 400–500 μm. Also, unbound caught powder is hard to expel from little channels. Future work is expected to make powder that is effectively removable with conventional strategies for high-constrained air. One methodology is to make circular powder particles which would encourage evacuation in tight spaces. While SLA can achieve amazingly high resolutions, there are a set number of biodegradable, biocompatible gums. Propels have been made to integrate new macromers with biodegradable moieties, be that as it may, these materials have not been FDA affirmed. FDM, SLS, and 3DP can utilize polymers, for example, PLGA, PLLA, and PCL without concoction alteration which will help speed up future FDA endorsement for biomedical gadgets. Albeit large scale and microarchitecture has made incredible walks in the previous 5 years, extra work ought to concentrate on the nanoarchitecture (e.g., biochemical atoms). Because of cruel preparing states of SFF strategies (e.g., warm, natural dissolvable), biochemical particles are not by and large fused specifically into the framework. While biochemical atoms can be covered onto structures in post preparing, there is a requirement for supported development figure discharge after some time. Along these lines, techniques to fuse biochemical particles straightforwardly into frameworks for delayed discharge will be required.

4.7   CONCLUSION

3D printing has a long history of applications in biomedical engineering. The development and expansion of traditional biomedical applications are advanced and enriched by new printing technologies. Bioprinting is a new emerging technique which is highly attractive and trendy. 3D printing is a solid freeform fabrication technique that especially varies with the design of the structures to build and the scales used. It has proven to facilitate, speed up, and improve the quality of surgical procedures such as implant placements and complex surgeries. Also, anatomically shaped implants and tailor-made biomedical devices, such as hearing aids, have been prepared using SLA. The development of new resins has enabled to fabricate implantable devices like biodegradable tissue engineering scaffolds directly. With the introduction of hydroxyapatite composites, peptide-grafted structures, cell-containing hydrogels and modified natural polymers, these techniques have developed into a broadly applicable technique for biomedical engineering purposes.

KEYWORDS

- **3D printing**
- **lithography**
- **stereolithography**
- **prosthetics**
- **laser sintering**
- **bioprinting**

REFERENCES

1. Schubert, C.; van Langeveld, M. C.; Donoso, L.A. Innovations in 3D Printing: A 3D Overview from Optics to Organs. *Br. J. Ophthalmol.* **2014,** *98*(2), 159–161.
2. Gross, B.C.; Erkal, J.L.; Lockwood, S. Y., et al. Evaluation of 3D Printing and Its Potential Impact on Biotechnology and the Chemical Sciences. *Anal. Chem.* **2014,** *86*(7), 3240–3253.
3. Ursan, I.; Chiu, L.; Pierce, A. Three-Dimensional Drug Printing: A Structured Review. *J. Am. Pharm. Assoc.* **2013,** *53*(2), 136–144.
4. Lipson, H. New World of 3-D Printing Offers "Completely New Ways of Thinking:" Q&A with Author, Engineer, and 3-D Printing Expert Hod Lipson. *IEEE Pulse* **2013,** *4*(6), 12–14.

5. Klein, G. T.; Lu, Y.; Wang, M. Y. 3D Printing and Neurosurgery—Ready for Prime Time? *World Neurosurg.* **2013,** *80*(3–4), 233–235.

6. Banks J. Adding Value in Additive Manufacturing: Researchers in the United Kingdom and Europe Look to 3D Printing for Customization. *IEEE Pulse* **2013,** *4*(6), 22–26.

7. Mertz, L. Dream It, Design It, Print It in 3-D: What Can 3-D Printing Do for You? *IEEE Pulse* **2013,** *4*(6), 15–21.

8. 3D Print Exchange. National Institutes of Health; Available at http://3dprint.nih.gov (accessed July 9, 2014).

9. Gross, B. C.; Erkal, J. L.; Lockwood, S. Y., et al. Evaluation of 3D Printing and Its Potential Impact on Biotechnology and the Chemical Sciences. *Anal. Chem.* **2014,** *86*(7), 3240–3253.

10. Schubert, C.; van Langeveld, M. C.; Donoso, L. A. Innovations in 3D Printing: A 3D Overview from Optics to Organs. *Br. J. Ophthalmol.* **2014,** *98*(2), 159–161.

11. Costello, J. P.; Olivieri, L. J.; Krieger, A.; Thabit, O.; Marshall, M. B., et al. Utilizing Three Dimensional Printing Technology to Assess the Feasibility of High-Fidelity Synthetic Ventricular Septal Defect Models for Simulation in Medical Education. *World J. Pediatr. Congenit. Heart Surg.* **2014,** *5*, 421–426.

12. Berman, B. 3-D Printing: The New Industrial Revolution. *Bus. Horiz.* **2012.**

13. Hoy, M. B. 3D Printing: Making Things at the Library. *Med. Ref. Serv. Q.* **2013,** *32*(1), 94–99.

14. Klein, G. T.; Lu, Y.; Wang, M. Y. 3D Printing and Neurosurgery—Ready for Prime Time? *World Neurosurg.* **2013,** *80*(3–4), 233–235.

15. Leong, K. F.; Cheah, C. M.; Chua, C. K. *Biomaterials* **2003,** *24*, 2363.

16. Dhandayuthapan, B.; Yoshida, Y.; Maekawa, T.; Kumar, D. S. Polymeric Scaffolds in Tissue Engineering Application: A Review, *Int. J. Polym. Sci.* **2011,** *19*.

17. Chen, M.; Le, D. Q.; Baatrup, A.; Nygaard, J. V.; Hein, S.; Bjerre, L.; Kassem, M.; Zou, X.; Bunger, C.; *Acta Biomater.* **2011,** *7*, 2244.

18. Peter, J.; Miller, M. J.; Yasko, A. W.; Yaszemski, M. J.; Mikos, A. G. *J. Biomed. Mater. Res.* **1998,** *43*, 422.

19. Zhu.Y , C. Gao, and J. Shen, "Surface Modification of Polycaprolactone with Poly(Methacrylic Acid) and Gelatin Covalent Immobilization for Promoting Its Cytocompatibility," *Biomaterials*, 2002.

20. Spath, S.; Seitz, H. *Int. J. Adv. Manuf. Technol.* **2014,** *70*, 135.

21. Cook, R.; Crute, B. E.; Patrone, L. M.; Gabriels, J.; Lane, M. E.; Van Buskirk, R. G. In Vitro Cell. *Dev. Biol.* **1989,** *25*, 914.

22. Lawrence, J.; Madihally, S. V. *Cell Adhes. Migration* **2008,** *2*, 9.

23. Lee, H.; Ahn, S.; Bonassar, L. J.; Kim, G. *Macromol. Rapid Commun.* **2013,** *34*, 142.

24. Shim, H.; Kim, J. Y.; Park, M.; Park, J.; Cho, D. W. *Biofabrication* **2011,** *3*, 034102.

25. Melchels, F. B. W.; Feijen, J.; Grijpma, D. W., A Review on Stereolithography and Its Applications in Biomedical Engineering. *Biomaterials* **2010,** *31*, 24.

26. Bose.S; Vahabzadeh, S.; Bandyopadhyay, A. *Mater. Today* **2013,** *16*, 496.

27. Liu, W. Y.; Li, Y.; Liu, J. Y.; Niu, X. F.; Wang, Y.; Li, D. Y. *J. Nanomaterials* **2013,** *7*.

28. Billiet, T.; Gevaert, E.; De Schryver, T.; Cornelissen, M.; Dubruel, P. *Biomaterials* **2014,** *35*, 49.

29. Shapiro, J.; Oyen, M. *JOM* **2013,** *65*, 505.

30. Zhu, J.; Marchant, R. E. *Expert Rev. Med. Devices* **2011**, *8*, 607.

31. Park, S. H.; Park, D. S.; Shin, J. W.; Kang, Y. G.; Kim, H. K.; Yoon, T. R.; Shin, J. W. *J. Mater. Sci. Mater. Med.* **2012**, *23*, 2671.

32. Griffith, G. *Acta Mater.* **2000**, *48*, 263.

33. Intra, J.; Glasgow, J. M.; Mai, H. Q.; Salem, A. K. *J. Controlled Release* **2008**, *127*, 280.

34. Mountziaris, P. M.; Spicer, P. P.; Kasper, F. K.; Mikos, A. G. *Tissue Eng., B: Rev.* **2011**, *17*, 393.

35. Rajan, V.; Murray, R. *Wound Pract. Res.* **2008**, *16*, 122.

36. Almeida, R.; Serra, T.; Oliveira, M. I.; Planell, J. A.; Barbosa, M. A.; Navarro, M. *Acta Biomater.* **2014**, *10*, 613.

37. Sung, J.; Meredith, C.; Johnson, C.; Galis, Z. S. *Biomaterials* **2004**, *25*, 5735.

38. Ma, X. Scaffolds for tissue fabrication, *Mater. Today* **2004**, *7*, 30.

39. Oryan, A.; Moshiri, A. *Hard Tissue* **2013**, *2*.

40. Aydin.A; Memisoglu, K.; Cengiz, A.; Atmaca, H.; Muezzinoglu, B.; Muezzinoglu, U. S. *J. Orthop. Sci.* 2012, *17, 796.*

41. Choi, H.; Park, T. G. *J. Biomater. Sci. Polym. Ed.* **2002**, *13*, 1163.

42. Gong.C.Y.; Shi, S.; Dong, P.; Kan, B.; Gou, M.; Wang, X.; Li, X.; Luo, F.; Zhao, X.; Wei, Y.; Qian, Z. *Int. J. Pharm.* **2009**, *365*, 89.

43. Inzana; Olvera, D.; Fuller, S. M.; Kelly, J. P.; Graeve, O. A.; Schwarz, E. M.; Kates, S. L.; Awad, H. A. *Biomaterials* **2014**, *35*, 4026.

44. Elangovan.S; D'Mello, S. R.; Hong, L.; Ross, R. D.; Allamargot, C.; Dawson, D. V.; Stanford, C. M.; Johnson, G. K.; Sumner, D. R.; Salem, A. K. The Enhancement of Bone Regeneration by Gene Activated Matrix Encoding for Platelet Derived Growth Factor *Biomaterials* **2014**, *35*, 737.

45. Lam, X. F.; Mo, X. M.; Teoh, S. H.; Hutmacher, D. W. Scaffold development using 3D printing with a starch-based polymer. *Mater. Sci. Eng.: C* **2002**, *20*, 49.

46. Wei.G; Ma, P. X. *Adv. Funct. Mater.* **2008**, *18*, 3568.

47. Korossis, F.; Southgate, S. J.; Ingham, E. Fisher, J. *Biomaterials* **2009**, *30*, 266.

48. Groeber, F.; Holeiter, M.; Hampel, M.; S. Hinderer, Schenke-Layland, K. *Clin. Plast. Surg.* **2012**, *39*, 33.

49. Guoyou, H.; Lin, W.; ShuQi, W.; Yulong, H.; Jinhui, W.; Qiancheng, Z.; Feng, X.; Tian Jian, L. *Biofabrication* **2012**, *4*, 042001.

50. Nichol, J. W.; Khademhosseini, A. *Soft Matter* **2009**, *5*, 1312.

51. Napolitano, P.; Chai, P.; Dean, D. M.; Morgan, J. R. *Tissue Eng.* **2007**, 13, 2087.

52. Chung, S. E.; Park, W.; Shin, S.; Lee, S. A.; Kwon, S. *Nat. Mater.* **2008**, *7*, 581.

53. Xu, F.; Wu, C. A.; Rengarajan, V.; Finley, T. D.; Keles, H. O.; Sung, Y.; Li, B.; Gurkan, U. A.; Demirci, U. *Adv. Mater.* **2011**, *23*, 4254.

54. Xu, F.; Finley, T. D.; Turkaydin, M.; Sung, Y.; Gurkan, U. A.; Yavuz, A. S.; Guldiken, R. O.; Demirci, U. *Biomaterials* **2011**, *32*, 7847.

55. Kachouie, N. N.; Du, Y.; Bae, H.; Khabiry, M.; Ahari, A. F.; Zamanian, B.; Fukuda, J.; Khademhosseini, A. *Organogenesis* **2010**, *6*, 234.

56. Geckil, H.; Xu, F.; Zhang, X.; Moon, S.; Demirci, U. *Nanomedicine* **2010**, *5*, 469.

57. Lee, K. W.; Wang, S.; Lu, L.; Jabbari, E.; Currier, B. L.; Yaszemski, M. J. *Tissue Eng.* **2006**, *12*, 2801.

58. Li.X; Cui, R.; Sun, L.; Aifantis, K. E.; Fan, Y.; Feng, Q.; Cui, F.; Watari, F. *Int. J. Polym. Sci.* **2014**, *2014*, 13.

59. Murphy, S. V.; Atala, A. *Nat. Biotechnol.* **2014**, *32*, 773.

60. Wu.B; Klatzky, R. L.; Stetten, G. *J. Exp. Psychol. Appl.* **2010**, *16*, 45.

61. Geckil, H.; Xu, F.; Zhang, X.; Moon, S.; Demirci, U. *Nanomedicine* **2010**, 5, 469.

62. Talib, M.; Covington, J. A.; Bolarinwa, A. *AIP Conf. Proc.* **2014**, *1584*, 129.

63. Tonomura.A; Mizuno, D.; Hisada, A.; Kuno, N.; Ando, Y.; Sumita, Y.; Honda, M. J.; Satomura, K.; Sakurai, H.; Ueda, M.; Kagami, H. *Ann. Biomed. Eng.* **2010**, *38*, 1664.

64. Berger.A, How does it work?: Magnetic resonance imaging, *BMJ* **2002**, *324*, 35.

65. N. I. o. B. I. a. Bioengineering, Vol. 2014; U.S. Department of Health & Human Services, National Institute of Health, 2014.

66. Sachs, E. M.; Haggerty, J. S.; Cima, M. J.; Williams, P. A. U.S. 5,204,055, 1993.

67. Cima, M.; Sachs, E.; Fan, T.; Bredt, J. F.; Michaels, S. P.; Khanuja, S.; Lauder, A.; Lee, S. J. J.; Brancazio, D.; Curodeau, A.; Tuerck, H. U.S. 5,387,380, 1995.

68. Mac-Thiong, J. M.; Labelle, H.; Rooze, M.; Feipel, V.; Aubin, C. E. Evaluation of a Transpedicular Drill Guide for Pedicle Screw Placement in the Thoracic Spine. *Eur. Spine J.* **2003**, *12*, 542–547.

69. Lu, S.; Xu, Y. Q.; Lu, W. W., et al. A Novel Patient-Specific Navigational Template for Cervical Pedicle Screw Placement. *Spine* **2009**, *34*(26), E959–E966.

70. Hananouchi, T.; Saito, M.; Koyama, T.; Sugano, N.; Yoshikawa, H. Tailor Based Surgical Guide Reduces Incidence of Outliers of Cup Placement. *Clin. Orthop. Relat. Res.* **2010**, *468*, 1088–1095.

71. Miyake, J.; Murase, T.; Moritomo, H.; Sugamoto, K.; Yoshikawa, H. Distal Radius Osteotomy with Volar Locking Plates Based on Computer Simulation. *Clin. Orthop. Relat. Res.* **2011**, *469*(6), 1766–1773.

72. Ng, V. Y.; DeClaire, J. H.; Berend, K. R.; Gulick, B. C.; Lombardi, A. V. Jr. Improved Accuracy of Alignment with Patient-Specific Positioning Guides Compared with Manual Instrumentation in TKA. *Clin. Orthop. Relat. Res.* **2012**, *470*(1), 99–107.

73. Wong, K. C.; Kumta, S. M.; Sze, K. Y.; Wong, C. M. Use of a Patient-Specific CAD/CAM Surgical Jig in Extremity Bone Tumor Resection and Custom Prosthetic Reconstruction. *Comput. Aided Surg.* **2012**, *17*(6), 284–293.

74. Buller, L.; Smith, T.; Bryan, J.; Klika, A.; Barsoum, W.; Innotti, J. P. The Use of Patient-Specific Instrumentation Improves the Accuracy of Acetabular Component Placement. *J. Arthroplasty* **2013**, *28*, 631–636.

75. Takeyasu, Y.; Oka, K.; Miyake, J.; Kataoka, T.; Moritomo, H.; Murase, T. Preoperative, Computer Simulation-Based, Three-Dimensional Corrective Osteotomy for Cubitus Varus Deformity with Use of a Custom-Designed Surgical Device. *J. Bone Joint Surg. Am.* **2013**, *95*(22), e173.

76. F. Gouin, L. Paul, G. A. Odri, and O. Cartiaux, "Computer-Assisted Planning and Patient-Specific Instruments for Bone Tumour Resection Within the Pelvis: A Series of 11 Patients," *Sarcoma*, 2014.

77. Baauw, M.; van Hellemondt, G. G.; van Hooff, M. L.; Spruit, M. The Accuracy of Positioning of a Custom-Made Implant Within a Large Acetabular Defect at Revision Arthroplasty of the Hip. *Bone Joint J.* **2015**, *97-B*(6), 780–785.

78. Fan, H.; Fu, J.; Li, X., et al. Implantation of Customized 3-D Printed Titanium Prosthesis in Limb Salvage Surgery: A Case Series and Review of the Literature. *World J. Surg. Oncol.* **2015,** *13,* 308.

79. Rengier, F.; Mehndiratta, A.; von Tengg-Kobligk, H., et al. 3D Printing Based on Imaging Data: Review of Medical Applications. *Int. J. Comput. Assist. Radiol. Surg.* **2010,** *5,* 335–341.

80. Stratasys Debuts New 3D Printer for Larger Labs. https://www.dentalproductsreport.com/lab/article/stratasys-debuts-new-3d-printer-larger-labs (accessed Aug 11, 2016).

81. FDA Approves 3D Printable Denture Base Material. http://www.dentalproductsreport.com/dental/article/fda-approves-3d-printable-denture-base-material (accessed Aug 11, 2016).

82. Hoang, L. N.; Thompson, G. A.; Cho, S. H. Die Spacer Thickness Reproduction for Central Incisor Crown Fabrication with Combined Computer-Aided Design and 3D Printing Technology: An In Vitro Study. *J. Prosthet. Dent.* **2015,** *113*(5), 398–404.

83. Bharath Reddy, R.; Sai Gopi Nihal, S.; Taneja, A. S.; Kalyana Rama, J. S. Comparative Study on the Lateral Load Resistance of Multi-Storied Structure with Bracing Systems. *Indian J. Sci. Technol.* **2015,** 8(Dec (36)), 1–9.

84. Miresmaeili, A.; Sajedi, A.; Moghimbeigi, A.; Farhadian. Three-Dimensional Analysis of the Distal Movement of Maxillary 1st Molars in Patients Fitted with Mini-Implant-Aided Transpalatal Arches. *Korean J Orthod.* **2015,** 45(Sep (5)), 236–244.

85. P. X. Ma and J. W. Choi, "Biodegradable polymer scaffolds with well-defined interconnected spherical pore network.," *Tissue Eng.,* 2001.

86. Byun, C.; Kim, C.; Cho, S.; Baek, S. H.; Kim, G.; Kim, S. G.; Kim, S. Y. Endodontic Treatment of an Anomalous Anterior Tooth with the Aid of a 3-Dimensional Printed Physical Tooth Model. *J Endod.* **2015,** *41*(6), 961–965.

87. Schubert, C.; van Langeveld, M. C.; Donoso, L. A. Innovations in 3D Printing: A 3D Overview from Optics to Organs. *Br. J. Ophthalmol.* **2014,** *98*(2), 159–161.

88. D'Aveni, R. The 3-D Printing Revolution. *Harvard Business Review.* 2015.

89. Gross, B. C.; Erkal, J. L.; Lockwood, S. Y., et al. Evaluation of 3D Printing and Its Potential Impact on Biotechnology and the Chemical Sciences. *Anal. Chem.* **2014,** *86*(7), 3240–3253.

90. Bartlett, S. Printing Organs on Demand. *Lancet Respir. Med.* **2013,** *1*(9), 684.

91. Cui, X.; Boland, T.; D'Lima, D. D.; Lotz, M. K. Thermal Inkjet Printing in Tissue Engineering and Regenerative Medicine. *Recent Pat. Drug Deliv. Formul.* **2012,** *6*(2), 149–155.

92. Ozbolat, I. T.; Yu Y. Bioprinting Toward Organ Fabrication: Challenges and Future Trends. *IEEE Trans Biomed Eng.* **2013,** *60*(3), 691–699.

93. How About Them Gams: 3D Printing Custom Legs. http://www.businessweek.com/articles/2012-05-03/how-about-them-gams-3d-printing-custom-legs.

94. The Future Is Here: 3D Printed Prosthetics. http://dealingwithdifferent.com/3d-printing-prosthetics/ (accessed Aug 14, 2016).

95. Li, X. M.; Huang, Y.; Zheng, L. S. et al. Effect of Substrate Stiffness on the Functions of Rat Bone Marrow and Adipose Tissue-Derived Mesenchymal Stem Cells In Vitro. *J. Biomed. Mater. Res. A* **2014,** *102A,* 1092–1101.

96. Li, X.; Fan, Y.; Watari, F. Current Investigations into Carbon Nanotubes for Biomedical Application. *Biomed. Mater.* **2010,** *5*(2), Article ID 022001.

97. Thoma, C. R.; Zimmermann, M.; Agarkova, I.; Kelm, J. M.; Krek, W. 3D Cell Cul ture Systems Modeling Tumor Growth Determinants in Cancer Target Discovery. *Adv. Drug Delivery Rev.* **2014,** *69–70,* 29–41.

98. Puschmann, T. B.; Zandén, C.; De Pablo, Y.; Kirchhoff, F.; Pekna, M.; Liu, J.; Pekny, M: Bioactive 3D Cell Culture System Minimizes Cellular Stress and Maintains the In Vivo-Like Morphological Complexity of Astroglial Cells. *Glia* **2013,** *61,* 432–440.

99. Ma, X.; Qu, X.; Zhu, W.; Li, Y.-S.; Yuan, S.; Zhang, H.; Liu, J.; Wang, P.; Lai, C. S. E.; Zanella, F. et al.: Deterministically Patterned Biomimetic Human iPSC-Derived Hepatic Model via Rapid 3D Bioprinting. *Proc. Natl. Acad. Sci. U.S.A.* **2016,** *113,* 2206–2211.

100. Langer, R.; Vacanti, J. P. Tissue Engineering. *Science* **1993,** *260*(5110), 920–926.

101. Q. L. Loh and C. Choong, "Three-Dimensional Scaffolds for Tissue Engineering Applications: Role of Porosity and Pore Size." *Tissue Eng. Part B Rev.* **2013.**

102. Yang, S.; Leong, K. F.; Du, Z.; Chua, C. K. The Design of Scaffolds for Use in Tissue Engineering. Part I. Traditional Factors. *Tissue Eng.* **2001,** *7*(6), 679–689.

103. Yang, S.; Leong, K. F.; Du, Z.; Chua, C. K. The Design of Scaffolds for Use in Tissue Engineering. Part II. Rapid Prototyping Techniques. *Tissue Eng.* **2002,** *8*(1), 1–11.

104. Leong, K. F.; Cheah, C. M.; Chua, C. K. Solid Freeform Fabrication of Three Dimensional Scaffolds for Engineering Replacement Tissues and Organs. *Biomaterials* **2003,** *24*(13), 2363–2378.

105. Yeong, W. Y.; Chua, C. K.; Leong, K. F.; Chandrasekaran, M. Rapid Prototyping in Tissue Engineering: Challenges and Potential. *Trends Biotechnol.* **2004,** *22*(12), 643–652.

106. Boland, T. et al. Rapid, Prototyping of Artificial Tissues and Medical Devices. *Adv. Mater. Process.,* **2007,** *165*(4), 51–53.

107. Bártolo, P. J.; Chua, C. K.; Almeida, H. A.; Chou, S. M.; Lim, A. S. C. Biomanufacturing for Tissue Engineering: Present and Future Trends. Virtual and Physical Prototyping, **2009,** *4*(4), 203–216.

108. Naumann, A.; Aigner, J.; Staudenmaier, R.; Seemann, M.; Bruening, R.; Englmeier, K. H., et al. Clinical Aspects and Strategy for Biomaterial Engineering of an Auricle Based on Three-Dimensional Stereolithography. *Eur. Arch. Otorhinolaryngol.* **2003,** *260*(10), 568e75.

109. Sodian, R.; Fu, P.; Lueders, C.; Szymanski, D.; Fritsche, C.; Gutberlet, M. et al. Tissue Engineering of Vascular Conduits: Fabrication of Custom-Made Scaffolds Using Rapid Prototyping Techniques. *Thorac. Cardiovasc. Surg.* **2005,** *53*(3), 144e9.

110. Matsuda, T.; Mizutani, M. Liquid Acrylate-Endcapped Biodegradable Poly(e-caprolactone-*co*-trimethylene carbonate). II. Computer-Aided Stereolithographic Microarchitectural Surface Photo Constructs. *J. Biomed. Mater. Res.* **2002,** *62*(3), 395e403.

111. Northen, T. R.; Brune, D. C.; Woodbury, N. W. Synthesis and Characterization of Peptide Grafted Porous Polymer Microstructures. *Biomacromolecules* **2006,** *7*(3), 750e4.

112. Schuster, M.; Turecek, C.; Weigel, G.; Saf, R.; Stampfl, J.; Varga, F. et al. Gelatin-Based Photopolymers for Bone Replacement Materials. *J. Polym. Sci. Pol. Chem.* **2009,** *47*(24), 7078e89.

113. Seitz, H.; Deisinger, U.; Leukers, B.; Detsch, R.; Ziegler, G. Different Calcium Phosphate Granules for 3-D Printing of Bone Tissue Engineering Scaffolds. *Adv. Eng. Mater.* **2009**, *11*, B41–B6.

114. Detsch, R.; Schaefer, S.; Deisinger, U.; Ziegler, G.; Seitz, H.; Leukers, B. In Vitro-Osteoclastic Activity Studies on Surfaces of 3D Printed Calcium Phosphate Scaffolds. *J. Biomater. Appl.* **2010**.

115. Warnke, P. H.; Seitz, H.; Warnke, F.; Becker, S. T.; Sivananthan, S.; Sherry, E., et al. Ceramic Scaffolds Produced by Computer-Assisted 3d Printing and Sintering: Characterization and Biocompatibility Investigations. *J. Biomed. Mater. Res., B: Appl. Biomater.* **2010**, *93*, 212–217.

116. Abarrategi, A.; Moreno-Vicente, C.; Martínez-Vázquez, F. J.; Civantos, A.; Ramos, V.; Sanz-Casado, J. V., et al. Biological Properties of Solid Free Form Designed Ceramic Scaffolds with BMP-2: In Vitro and In Vivo Evaluation. *PLoS One* **2012**, *7*, e34117.

117. Becker, S. T.; Bolte, H.; Krapf, O.; Seitz, H.; Douglas, T.; Sivananthan, S., et al. Endocultivation: 3D Printed Customized Porous Scaffolds for Heterotopic Bone Induction. *Oral Oncol.* **2009**, *45*, e181–e188.

118. Tamimi, F.; Torres, J.; Gbureck, U.; Lopez-Cabarcos, E.; Bassett, D. C.; Alkhraisat, M. H., et al. Craniofacial Vertical Bone Augmentation: A Comparison Between 3D Printed Monolithic Monetite Blocks and Autologous Onlay Grafts in the Rabbit. *Biomaterials* **2009**, *30*, 6318–6326.

119. Butscher, A.; Bohner, M.; Roth, C.; Ernstberger, A.; Heuberger, R.; Doebelin, N., et al. Printability of Calcium Phosphate Powders for Three-Dimensional Printing of Tissue Engineering Scaffolds. *Acta Biomater.* **2012**, *8*, 373–385.

120. Tarafder, S.; Balla, V. K.; Davies, N. M.; Bandyopadhyay, A.; Bose, S. Microwave-Sintered 3D Printed Tricalcium Phosphate Scaffolds for Bone Tissue Engineering. *J. Tissue Eng. Regen. Med.* **2013**, *7*, 631–641.

121. Kalita, S. J.; Bose, S.; Hosick, H. L.; Bandyopadhyay, A. Development of Controlled Porosity Polymer–Ceramic Composite Scaffolds via Fused Deposition Modeling. *Mater. Sci. Eng. C* **2003**, *23*, 611–620.

122. Rai, B.; Teoh, S. H.; Ho, K. H.; Hutmacher, D. W.; Cao, T.; Chen, F., et al. The Effect of rhBMP-2 on Canine Osteoblasts Seeded onto 3D Bioactive Polycaprolactone Scaffolds. *Biomaterials* **2004**, *25*, 5499–5506.

123. Korpela, J.; Kokkari, A.; Korhonen, H.; Malin, M.; Närhi, T.; Seppälä, J. Biodegradable and Bioactive Porous Scaffold Structures Prepared Using Fused Deposition Modeling. *J. Biomed. Mater. Res. B: Appl. Biomater.* **2013**, *101*, 610–619.

124. H. J. Yen, C. S. Tseng, S. H. Hsu and C. L. Tsai, Evaluation of Chondrocyte Growth in the Highly Porous Scaffolds Made by Fused Deposition Manufacturing (FDM) Filled with Type II Collagen. *Biomed. Microdevices* **2009**, *11*, 615–624.

125. Teo, E. Y.; Ong, S.-Y.; Khoon Chong, M. S.; Zhang, Z.; Lu, J.; Moochhala, S., et al. Polycaprolactone-Based Fused Deposition Modeled Mesh for Delivery of Antibacterial Agents to Infected Wounds. *Biomaterials* **2011**, *32*, 279–287.

126. Probst, F.; Hutmacher, D.; Müller, D.; Machens, H.; Schantz, J. Calvarial Reconstruction by Customized Bioactive Implant. *Handchir. Mikrochir. Plast. Chir.* **2010**, *42*, 369–373.

127. Shim, J.-H.; Moon, T.-S.; Yun, M.-J.; Jeon, Y.-C.; Jeong, C.-M.; Cho, D.-W., et al. Stimulation of Healing within a Rabbit Calvarial Defect by a PCL/PLGA Scaffold Blended with TCP Using Solid Freeform Fabrication Technology. *J. Mater. Sci. Mater. Med.* **2012,** *23,* 2993–3002.

128. Kim, J. Y.; Cho, D-W. Blended PCL/PLGA Scaffold Fabrication Using Multi-Head Deposition System. *Microelectron. Eng.* **2009,** *86,* 1447–1450.

129. Van Bael, S.; Desmet, T.; Chai, Y. C.; Pyka, G.; Dubruel, P.; Kruth, J.-P., et al. In Vitro Cell-Biological Performance and Structural Characterization of Selective Laser Sintered and Plasma Surface Functionalized Polycaprolactone Scaffolds for Bone Regeneration. *Mater. Sci. Eng. C* **2013,** *33,* 3404–3412.

130. Kang, S.-W.; Bae, J.-H.; Park, S.-A.; Kim, W.-D.; Park, M.-S.; Ko, Y.-J., et al. Combination Therapy with BMP-2 and BMSCs Enhances Bone Healing Efficacy of PCL Scaffold Fabricated Using the 3D Plotting System in a Large Segmental Defect Model. *Biotechnol. Lett.* **2012,** *34,* 1375–1384.

CHAPTER 5

PROGRESS AND PROSPECTS OF POLYMERIC NANOGEL CARRIER DESIGNS IN TARGETED CANCER THERAPY

PRASHANT SAHU[1], SUSHIL K. KASHAW[1*], SAMARESH SAU[2], and ARUN K. IYER[2]

[1]*Department of Pharmaceutical Sciences, Dr. Harisingh Gour University (A Central University), Sagar (MP), India*

[2]*Use-inspired Biomaterials & Integrated Nano Delivery (U-BiND) Systems Laboratory, Department of Pharmaceutical Sciences, Wayne State University, Detroit, Michigan, USA*

*Corresponding author. E-mail: sushilkashaw@gmail.com

ABSTRACT

Nanogel has emerged as one of the most flexible drug delivery systems exclusively for site-specific or time-controlled delivery of bioactive agents owing to their combining features of hydrogel and nanoparticles (NPs). Nanogel can swell by absorption of a large amount of solvent, but they do not dissolve owing to the arrangement of the chemically or physically cross-linked polymeric network. Structurally, nanogel is a class of cross-linked colloidal particles, which exhibits a behavior that extends from polymeric solutions (swollen form) to solid particles (collapsed form). They can respond to chemical stimuli (temperature, ionic strength, magnetic or electric fields, etc.), physical stimuli (temperature, electric or magnetic fields, ionic fields, etc.), and biochemical stimuli (ions, pH, specific molecules, etc.). Effective nanogel–nanogel and nanogel–drug interactions can be reversed by changing the degree of swelling of the particles, giving rise to a broad diversity of possibilities. Nanogel serves as an ideal drug delivery carrier, exhibiting high stability, drug loading capacity, biologic consistency and reacts to a broad range of environmental stimuli. Nowadays, targeted delivery of drug has become an issue in research with the major focus of conjugation with proteins and anti-tumor agents. Comprehensively, we attempt to provide synthetic methods, characterization parameters an in-depth description of elementary hypothesis, method of synthesis, characterization parameters, and mechanism of action and clinical implementation of polymeric nanogels as targeted anti-cancer drug delivery systems.

5.1 INTRODUCTION

Cancer is derived from the Greek word "carcinoma," chronicled as an old ailment that has transformed into solutions most feared determination with such entering figurative, medicinal, logical, and political intensity that the infection is frequently portrayed as the sovereign of all illnesses. Pathologically, cancer is a mind-boggling malady including various tempospatial changes in cell physiology, which eventually prompt to malignant tumors.

Abnormal cell development (neoplasia) is the natural endpoint of the infection. Tumor cell intrusion of encompassing tissues and far-off organs is the essential driver of morbidity and mortality for most patients. The natural procedure by which ordinary cells are changed into harmful growth cells has been the subject of a huge research exertion in the biomedical sciences for a long time.[1] Excessive diving by the examination clique has built up the way that unpredictability in hereditary and phenotypic levels has prompted to the malady showing clinical assorted qualities and restorative resistance. Thus, an assortment of methodologies is being polished for the treatment of growth each with some noteworthy confinements and reactions.[2] Disease treatment incorporates surgical expulsion, chemotherapy, radiation, and hormone treatment. Chemotherapy, an extremely regular treatment, conveys anticancer medications systemically to patients for quenching the uncontrolled expansion of carcinogenic cells.[3]

Unfortunately, because of nonspecific focusing by anticancer agents, many symptoms happen, and poor medication conveyance of these agents cannot draw out the fancied result in the greater part of the cases. Malignancy tranquilize improvement includes an extremely complex method which is connected with cutting-edge polymer science and electronic building. The primary test of malignancy therapeutics is to separate the dangerous cells and the ordinary body cells. That is the reason the primary target gets to be designing the medication in such a route as it can distinguish the malignancy cells to decrease their development and multiplication. Ordinary chemotherapy neglects to focus on the destructive cells specifically without interfacing with the typical body cells. It works by devastating quickly partitioning cells, which is the primary property of neoplastic cells. This is the reason chemotherapy additionally harms ordinary sound cells that gap quickly, for example, cells in the bone marrow, macrophages, stomach-related tract, and hair follicles. Hence, this causes genuine reactions incorporating organ harm bringing about debilitated treatment with lower measurements and at last low survival rates.[4]

The limitations associated with conventional medication conveyance frameworks have set off the exploration on to the field of nanotechnology which since its approach has made an incredible insurgency in particular diseases focusing on. Nanotechnology is the science that makes arrangements with the size range from a couple of nanometers (nm) to a few hundred nm, depending on their intended use.[5] NPs can be outlined through different adjustments, for example, changing their size, shape, substance and physical properties, and so forth, to program them for focusing on the fancied cells. It is exceptionally encouraging both in disease finding and treatment since it can enter the tissues at atomic level. Cancer nanotechnology is in this manner being energetically assessed and executed in malignancy treatment demonstrating a noteworthy progress in identification, determination, and treatment of the disease.[6]

Among the rising class of nanodevice for medication conveyance, the limited system of polymeric nanogel holds vast promise. The term "nanogel" (nano gel) was initially acquainted by us with characterized, cross-connected bifunctional systems of a polyion and a nonionic polymer for conveyance of polynucleotides [cross-connected polyethyleneimine (PEI) and poly(ethylene glycol) (PEG) or PEG-cl-PEI]. They have been implicated as managed and controlled discharge gadgets for the conveyance of anticancer medications. Henceforth, lately, nanogel transporter frameworks are by and large broadly, looked into, assessed and upgraded for interpretation of lab disease treatment into effective remedial practice.[7]

5.2 PREEMINENCE OF NANOGEL

Nanogels are hydrogels made up of cross-linked polymer systems having size in nanorange. Easy drug loading, stimuli-responsive nature, physical stability, stability of entrapped drug, versatility in design, and controlled release of the anti-inflammatory, antimicrobial, protein, peptide, and anticancer drugs are the major advantages of nanogels. As the diseases cancer and inflammation are associated with acidic pH, heat generation, and change in ionic content, so stimuli-responsive nature of nanogel is of particular significance in anti-inflammatory and anticancer drug (Fig. 5.1).[8]

Prolonged residence time and increased ocular availability of loaded drugs is achieved by muco-adhesive polymers in nanogels. Nanogels can

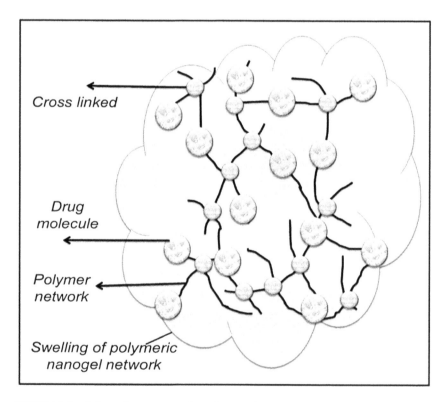

FIGURE 5.1   Schematic representation of nanogel system.

increase the stability and activity of protein/peptide drugs by forming suitably sized complex with proteins or by acting as artificial chaperones, thus help to keep the enzymes and proteins in proper confirmation necessary for exerting biological activity. Enhanced drug penetrations are achieved by extended contact with skin, which contribute much in transdermal drug delivery. The presence of multiple interactive functional groups in nanogels is of supreme importance in cancer drug delivery, as it aids the conjugation of different targeting agents for selective delivery of the untargeted drugs.[9]

Rapid clearance of nanogels by phagocytic cells can be manipulated by changing the particle size and surface properties and allowing both passive and active drug targeting. Achieving relatively high drug loading without chemical reactors can be preserved by encapsulating drug in the polymeric matrix of the nanogel.[10]

The action of the encapsulated medication in the polymeric grid of the nanogel can be saved by accomplishing moderately high-medication stacking without synthetic reactors. Nanogels possess the ability to reach the smallest capillary vessels, due to their tiny volume, and to penetrate the tissues either the paracellular or the transcellular pathways apart from being highly biocompatible and biodegradable.[11]

5.3   METHODS OF NANOGEL SYNTHESIS

For the preparation of nanogels, two methods are most commonly used, that is, chemical cross-linking of performed polymers and physical self-assembly of polymers. Controlled aggregations of hydrophilic polymers capable of bonding with each other are involved in making of physical self-assembly.[12] Non-covalent attractive forces, such as hydrophobic–hydrophobic, hydrophilic–hydrophilic, hydrogen bonding, and ionic interactions, are involved in nanogel formation by physical cross-linking. After appropriate organization of amphiphilic block copolymers and complexation of opposite charged polymeric chains, nanogel formation occurs within few minutes.[13] The various methods of nanogel synthesis are described in Fig. 5.2.

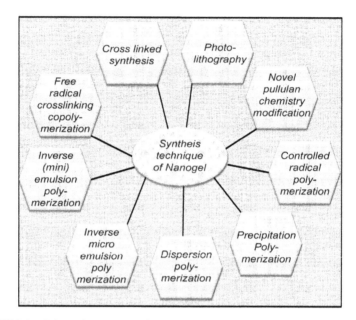

FIGURE 5.2   Schematic representation of various techniques involved in synthesis of nanogel.

5.3.1   CROSS-LINKED NANOGELS

For preparation of nano-scopic gels with micellar network or porous struc-tures, cross-linking reactions are suitable preparation methods. Diverse cross-linking methods, such as crystallization, cross-linking polymeriza-tion, ionic cross-linking, radiation cross-linking, and functional cross-linking, can be used to prepare nanogels[14] (Fig. 5.3). The cross-linked nanogel can be further subcategorized as follows.

5.3.1.1   PHYSICALLY CROSS-LINKED NANOGELS

Physical cross-linked nanogel formation occurs via crystallization, cross-linking polymerization, ionic cross-linking, radiation cross-linking, and functional cross-linking, can be used to prepare nanogels. These systems are sensitive, and this sensitivity depends on temperature, polymer composition, concentration of the polymer, ionic strength of the medium, and of the cross-linking agent. Micro- and nanogels are formed in few

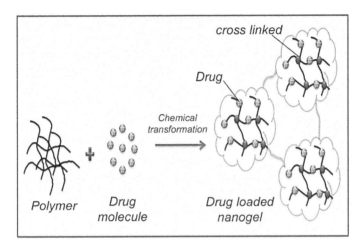

FIGURE 5.3   Schematic representation of cross-linked synthesis technique of nanogel.

minutes by association of amphiphilic block copolymers and complexation of oppositely charged polymeric chains.[15] Modified polysaccharides with hydrophobic groups (e.g., cholesterol group-modified pullulans) instantaneously form highly monodispersed nanogel structures, 20–30 nm in size, by intermolecular self-aggregation in water. Various hydrophilic polymers such as cholesterol, deoxy-cholic acid, bile acid, and so on are also modified by cholesterol bearing pullulan (CHP). Supra-molecular self-assembly formed by CHP-modified nanogels can entrap or capture enzymes, protein, drug molecules, and DNA. Curcumin in β-cyclodextrin or poly(β-cyclodextrin) has been encapsulated to show increase curcumin uptake, resulting in greater therapeutic effects in cancer cells through various molecular mechanism.[16]

5.3.1.2   CHEMICALLY CROSS-LINKED NANOGELS

Chemically cross-linked nanogels are constructed with numerous cross-linking points throughout the backbone of polymeric chains. The cross-linkers have a crucial role in tailoring the swelling, morphology, and pore size of the gel macromolecules to make ideal matrices that, in turn, are responsible for obtaining predetermined release kinetics of the entrapped drug molecules. Hydrophilic functional nanogels have been produced by chain growth polymerization. During the past two decades, numerous cross-linked nanogels have been produced by a variety of monomer, co-monomers, and cross-linkers. Commonly used cross-linkers are N, N'-methylene-bis-acrylate, di-allyl phthalate, di-vinyl-benzene, and PEG di-acrylate. Different strategies have been followed to control the formation of gel NPs from 5 to 400 nm.[17] Various strategies for preparation of chemically cross-linked includes top-down nanotechnologies, such as template-assisted nanofabrication (imprint photolithographic), or bottom-up approaches such as micromolding and microfluidic methods.[18]

5.3.2 PHOTOLITHOGRAPHY

Fabrication of 3D hydrogel particles and microgel or nanogels rings for drug delivery has been explored by photolithography. This method requires the development of techniques for surface treatment of the stamps or new materials for replica molds to allow the release of molded gels from stamps or replica molds.[19] Photolithography consists of five steps: in the first step, the UV cross-linkable polymer, which possesses low surface energy, is released as a substrate on the prebaked resist-coated water. In the second step, molding the polymer into patterns on the silicon water by pressing the quartz template onto the polymer has exposed it to the intense UV light. In the next step, the particles with a thin remaining interconnected film layer are exposed by removing the quartz template. Consequently, this remaining thin layer is removed by a plasma-containing oxygen that oxidizes it. In the last step, the fabricated particles are directly collected by dissolution of the substrate in water of buffer (Fig. 5.4).[20]

5.3.3 SYNTHESIS OF NANOGELS/MICROGELS BY FREE RADICAL CROSS-LINKING COPOLYMERIZATION

Synthetic pathways that necessitate minimum of precise equipment are cross-linking reactions carried out on preformed colloidal self-assemblies of synthetic or natural polymers and finally "free radical cross-linking copolymerization" (RCC) of monovinylic monomers with di- or multifunctional comonomers (cross-linker). Typically, di-vinyl benzene, di(methyl) acrylates, such as ethylene glycol di(meth)acrylate and 1,4-butanediol

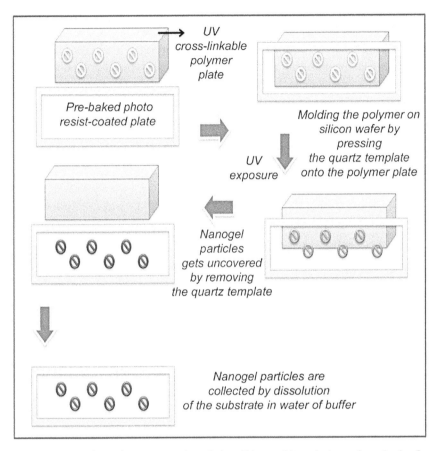

FIGURE 5.4   Schematic representation of photolithographic technique of synthesis of nanogel.

di-acrylate, di-vinyl benzene, or degradable cross-linker are used as cross-linking agents. The main challenge relies on identifying strategies that permit avoiding the formation of long-range networks as colloidal networks are limited in size.[21] To solve this problem, several strategies have been developed in order to attain nanometric or micrometric gels as an alternative of macroscopic networks, that is, macroscopic gelation. They usually rely on the control of the distance between growing polymer chains.

The approach is based on RCC performed in greatly diluted solution. Decreasing the monomer concentration increases the distance between propagating chains, which limits the intermolecular cross-linking and increases the probability of intramolecular cross-linking; consequently, macroscopic gelation can be prevented. Soluble branched polymers preparation can be favored by the utilization of low amounts of cross-linker or stopping the polymerization at low monomer conversion.[22]

5.3.4   INVERSE (MINI) EMULSION POLYMERIZATION

Inverse (mini) emulsion polymerization is a W/O polymerization method that contains aqueous droplets (including water-soluble monomers) firmly

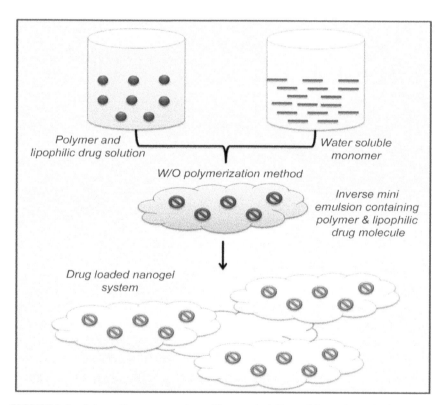

FIGURE 5.5   Schematic representation of inverse (mini) emulsion polymerization synthesis technique of nanogel.

dispersed with the aid of oil-soluble surfactants in a continuous organic medium. Mechanical stirring technique for inverse emulsion process results in formation of stable dispersion and by sonication for inverse miniemulsion polymerization (Fig. 5.5). Then after addition of radical initiators, polymerization occurs within the aqueous droplets producing colloidal particle.[23]

5.3.5   INVERSE MICROEMULSION POLYMERIZATION

Kinetically stable macroemulsions at around or below the critical micellar concentration are prepared by inverse (mini) emulsion polymerization, thermodynamically stable microemulsion upon addition of emulsifier above the critical threshold, prepared by inverse microemulsion

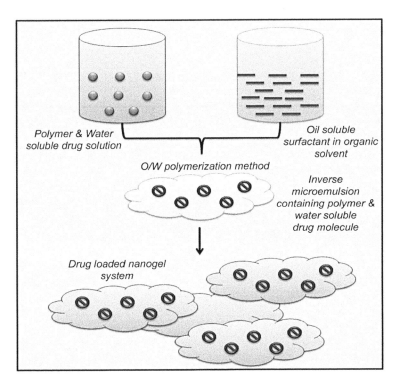

FIGURE 5.6   Schematic representation of inverse microemulsion polymerization synthesis technique of nanogel.

polymerization. In this process, aqueous droplets are stably dispersed with the assistance of a huge amount of oil-soluble surfactants in a continuous organic medium; polymerization occurs within the aqueous droplets (Fig. 5.6), producing firm hydrophilic and water-soluble colloidal NPs having a diameter of less than 50–100 nm.[24] Synthesis of well-defined nanogels was explored by inverse microemulsion polymerization. Water-soluble macromolecular carbohydrate drug nanogels were prepared by incorporation of Poly(vinylpyrrolidone) with Dex.[25]

5.3.6 DISPERSION POLYMERIZATION

In this process, continuous phase is used as organic solvent in which most ingredients are soluble including initiator, stabilizers, and monomers. Homogenous reaction mixture is formed at the onset of polymerization; however, the formed polymers become insoluble in the continuous medium, which leads to the formation of stable dispersion of polymeric

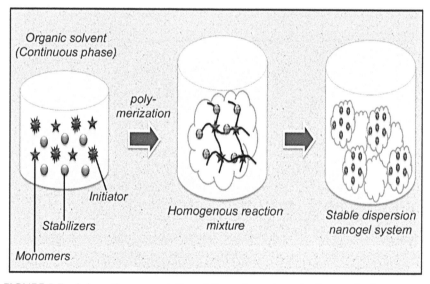

FIGURE 5.7   Schematic representation of dispersion polymerization synthesis technique of nanogel.

particle with the aid of colloidal stabilizers (Fig. 5.7). Drugs and magnetic NPs are either chemically attached or physically incorporated to micro-gels. The resulting microgels were effective as drug delivery carriers and for DNA applications.[26]

5.3.7   PRECIPITATION POLYMERIZATION

Initially, this type of polymerization takes place in the homogenous mixture. If the polymers are not swellable in the medium, then the use of cross-linker is needed to cross-link the polymer chains for separa-tion of particles. High polydispersity of cross-linked polymer often has an irregular shape. By using poly(methacrylic acid-g-ethylene glycol) P-(MAA-g-EG) nanogels can be synthesized for the delivery of proteins. By increasing the concentration of cross-linker during polymerization, decrease in the swelling equilibrium of nanospheres is observed.[27]

5.3.8   SYNTHESIS OF NANOGELS/MICROGELS USING CONTROLLED RADICAL POLYMERIZATION TECHNIQUES

In the mid-1990, use of controlled radical polymerization (CRP) tech-niques has been considered as a breakthrough toward the easy synthesis of complex macromolecular structures with a high degree of functionality and compositional variety. They are tolerant to a wide variety of functional groups and solvents and can be performed using simple polymerization conditions. The CRP techniques that have been implemented so far are mainly catalytic atom (group) transfer radical polymerization, nitroxide-mediated polymerization, degenerative chain transfer polymerization represented by iodine-mediated polymerization, reversible addition-fragmentation chain transfer, and polymerization/macromolecular design via the interchange of xanthates.[28] The fundamental mechanism of CRP comprises four elementary reactions: initiation, propagation, transfer, and termination. In contrast to conventional RP, initiation process is slow and continuous, and propagation and termination reactions are fast, leading to "dead" chains of broad molecular weight distributions with essentially no control over composition and macromolecular structure, CRP is essen-tially on a fast initiation step and a dynamic equilibrium between a low concentration of propagating radicals and a large amount of dormant reactive species. This results ideally in a nearly constant number of chains throughout the polymerization, which are initiated and grown at the same time with the same rate allowing control over molar mass distribution and architecture.[29]

5.3.9   NOVEL PULLULAN CHEMISTRY MODIFICATION

Cholesterol iso-cynate in dimethyl sulfoxide and pyridine was reacted to synthesize cholesterol-based pullulan nanogel (CHP) (Fig. 5.8). Pullulan was substituted with 1.4 cholesterol moieties per 100 anhydrous glucoside units. The preparation was freeze dried, and in aqueous phase, it formed nanogel which was complexed with W-9 peptide for delivery in osteological

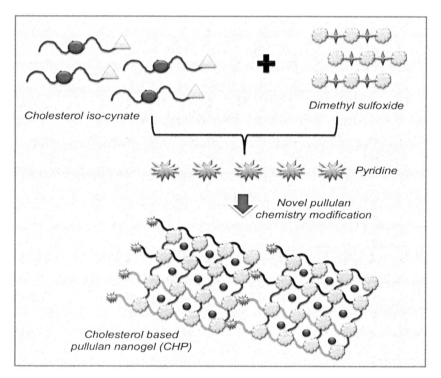

FIGURE 5.8   Schematic representation of novel pullulan chemistry modification synthesis technique of nanogel.

disorders. For drug delivery from nanogel formulation, pullulan has been known to act as good protein.[30] By adopting Michael addition reaction, CHP has been further modified with PEG, which allowed reduction in mesh size to 40 nm encapsulating 96% interleukin-12. Recently, foliate receptor targeted system has been used with pullulan in which foliate was substituted to pullulan by 1.6 glucose units. Further coupling of pullulan and photosensitizer (pheo-A) was done with carbodimide to produce the conjugate which was transformed to nanogel by dialysis in DMSO against deionized water, investigated for photodynamic therapy (PDT), and were successfully localized at cancerous cells to cause cell death by photo destruction.[31]

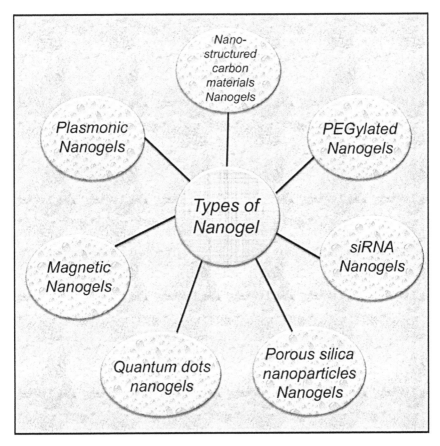

FIGURE 5.9 Schematic representation of various types of nanogel system.

5.4   TYPES OF NANOGEL

There are various types of nanogel system which totally depend on their potential and on-demand drug delivery. The important types of nanogel system are discussed in Figure 5.9.

5.4.1   PLASMONIC NANOGELS

Photothermal therapy (PTT) and PDT are currently the most promising techniques for treating cancer.[32] This is because of the possibility of devices in the nanometric scale that present photothermal transductors (PTs) with absorption in the biological window. Near infrared (NIR) irradiation is known as a biological window since it is hardly absorbed by water and blood cells and can penetrate deeply into the body. PDT is based on PTs that can use NIR light to produce reactive oxygen species which induce tissue destruction. On the other hand, PTT makes use of specific PTs that can effectively transform NIR light into local heat, surpassing the traditional hyperthermia methods

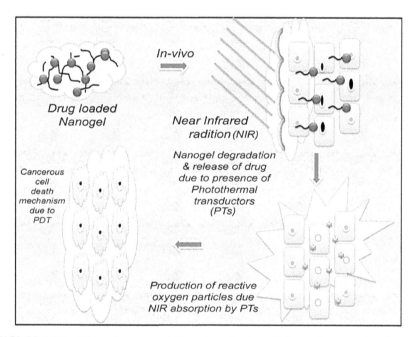

FIGURE 5.10   Schematic representation of mechanism of plasmonic nanogel system.

such as hot-water bath or heated blood perfusion (Fig. 5.10).[33] Nanogels composed of PTs and different polymers are ideal nanodevices for PTT because of their controlled size and architecture biocompatibility degradation ability, physical properties, accumulation in tumors, loading capacity, and postsynthetic modification. Moreover, state-of-the-art drug delivery systems have emerged that make use of thermoresponsive polymers that can swell or shrink with the temperature in combination with PTs.

The local heat produced by PTs upon exposure to specific radiation causes the transition of the thermoresponsive polymer and expels the drug retained in it.[34]

Gold nanoclusters have proven to be useful agents for PTT after they were shown to have an absorption in the NIR region four times higher than conventional photoabsorbing dyes. More specifically, gold nanogel showed better results in the ablation of T47D cancer cells than the bare nanogels. Gold nanogels formed by thermoresponsive polymers controlled by PTs were used as smart drug delivery systems in combination with hyperthermia. Thus, these nanogels are theranostic systems with drug delivery, hyperthermia, and bioimaging capabilities. Poly(N-iso-propylacrylamide) (PNIPAm) is a thermoresponsive polymer with a lower critical solution temperature of around 32°C which is suitable for medical applications. All these properties favor the use of plasmonic hybrid nanogels as theranostic agents in chemo- and thermotherapy as well as in in vitro in vivo imaging.[35]

5.4.2 MAGNETIC NANOGELS

Within nanomaterial–nanogel composites, magnetic NPs have attracted interest because of their application as PTs and in imaging. In the past, the most common methodology for preparing super paramagnetic iron NPs consisted of reduction and coprecipitation of ferrous and ferric salts in aqueous media in the presence of stabilizers.[36] This procedure has led to several commercial products like Feridex (Ferumoxides), Resovist (Ferucarbotran), and Ferumoxtran-10 (Combidex).[37] As their biodistribution depended on their size and surface modification, much effort has been made in developing magnetic nanomaterial with increased colloidal stability under physiological conditions. Moreover, the research focus was on improving contrast properties while lowering cytotoxicity, maintaining longer blood half-life improving biocompatibility, and enhancing tissue/organ targeting ability as well as pharmaceutical efficacy due to surface functionalization.[38]

5.4.3   QUANTUM DOTS NANOGELS

To fulfill the need for tracking certain biomolecules or cells by in vitro/in vivo imaging, development of specific biocompatible labels is an inevitable task for the study and understanding of the role of novel drug delivery systems. In this regard, quantum dots (QDs) containing nanogel composites have been reckoned as desirable inorganic–organic hybrid materials in optical sensing and imaging due to their intriguing luminescence characteristics and electronic properties of QDs such as bright fluorescence, broad absorption with narrow symmetric emission spectra, remarkable photostability, and biocompatibility.[39] QDs are nanometer-order semiconductor crystals that exhibit discrete energy levels, and their electronic and optical properties are determined only by their size. Depending on their diameter, QDs therefore shine in different colors when exposed to ultraviolet light (e.g., blue = 2 nm QDs; red = 6 nm QDs).[40]

In the last few years, different strategies have been used to obtain stimuli-sensitive QDs-based fluorescent probe nanogel composites as drug delivery systems with simultaneous imaging and biosensing. A simple approach is to incorporate surface-modified monodisperse hybrid NPs of 38 nm by mixing protein-coated QDs with amino-modified cholesterol-bearing pullulan nanogels ($CHPNH_2$-QD) in phosphate-buffered saline solution for 30 min at room temperature. Although the incorporation of QDs into unmodified, neutral cholesterol-bearing pullulan (CHP) nanogels was suppressed, electrostatic interaction between negatively charged QDs and positively charged amino moieties containing nanogels induced the successful formation of the hybrid nanogel complex (Fig. 5.11). Furthermore, the QDs@$CHPNH_2$ nanogel composite did not form aggregation after its internalization into various human cells. In a similar way, scientific community reported the preparation of multifunctionalized biodegradable nanogels by simply mixing mercapto-propionic-acid-capped CdTe QDs and chitin nanogels (CNGs) in an aqueous solution. The unreacted –OH and NH_2 groups of the repeating N-acetylglucosamine units, that composed the chitin chains, acted as anchor points to sequester Cd^{2+} ions and further immobilized CdTe QDs in the nanogel matrix.

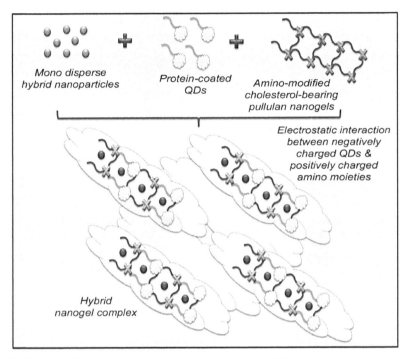

FIGURE 5.11 Schematic representation of mechanism of QDs nanogels system.

The obtained CdTe QDs CNG composite as a fluorescent probe and drug delivery system could be easily loaded with bovine serum albumin (BSA) showing promising applications for simultaneous bioimaging during drug release under local environmental conditions.[41]

5.4.4 POROUS SILICA NANOPARTICLES NANOGELS

Porous silica nanoparticles (PSNPs)-loaded nanogel have received significant attention in the last decade because of their superior and tunable physiochemical properties making them ideal in many fields of application such as catalysis, adsorption, sensors, and biomedical technology (Fig. 5.12). PSNPs are particularly attractive as drug delivery systems owing to their large surface area, tunable pore size, nontoxic nature, facile surface functionalization, high loading and controlled drug release, and excellent biocompatibility.[42]

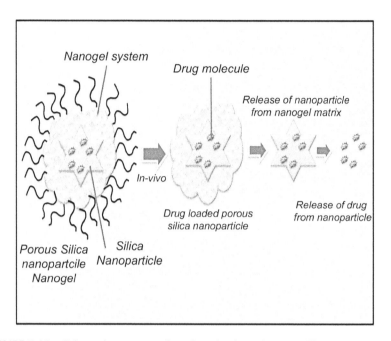

FIGURE 5.12   Schematic representation of mechanism of porous silica nanoparticles nanogels system.

Although the stimuli-responsive silica-based nanogels can be a fascinating subject of study, a few studies in this topic have become known in the recent years. Chai et al. reported a surfactant-free emulsion polymerization and biomimetic template approach for the synthesis of thermoresponsive hybrid nanogel–silica core–shell particles. The N-isopropylacrylamide (NIPAM)-co-2-(di-methyl-amino)ethyl methacrylate, methyl-chloride quaternized nanogel particles were in situ covered with silica, and the obtained nanogel–silica system particles were further loaded with aspirin. The drug delivery nanogel–silica system showed a controlled slow drug release of more than 24 h, while the non-template silica nanogel practically released its cargo within 5 h. This work demonstrated that nanogel particles covered with a silica shell can effectively retard or control the drug release.[43]

Motivated by the potential application of spherical hollow silica matrices, Liu et al. constructed a thermoresponsive drug release system based on hollow silica nanospheres coated with a PNIPAm nanogel shell. The prepared hollow silica nanogels showed good biocompatibility under in vitro cytotoxicity evaluation as well as excellent thermoresponsive con-trolled-release behavior of Rhodamine B during release studies.[44]

5.4.5 NANOSTRUCTURED CARBON MATERIALS NANOGELS

Over the past several decades, nanostructured carbon materials in various allotropic forms (e.g., fullerenes, nanotubes, graphene, and diamonds) have received overwhelming attention because of their unique physical and chemical properties tunable in a wide range, such as high specific surface area, narrow pore width, large pore volume, low density, excel-lent electronic conductivity, and high thermal and mechanical stability.[45] Considering such features, carbon-based nanostructured materials have been extensively studied in many different fields, including energy conver-sion and storage, catalysis support, adsorption, gas storage, water treat-ment, and more recently biomedicine. The use of nanostructured carbon materials with tunable functionalities promises great advances in drug delivery and therapeutic treatment.[46]

5.4.6 PEGYLATED NANOGELS

PEGylation refers to the modification of a particle surface by covalently grafting, entrapping, or adsorbing PEG. PEGylated NPs are long circulating making the drug available for a prolonged period of time. PEGylation of nanogels not only improves their circulation time but also delivers their drug load into tumors following intravenous injection. PEGylated nanogels are also prepared by chemically cross-linking poly(2-N,N-(di-ethyl-amino) ethyl methacrylate) (PEAMA) gel cores surrounded by PEG palisade layers. The PEGylated nanogels showed high stability under extremely dilute and high-salt conditions, in contrast to self-assembled nanocarrier. Moreover, the nanogels showed pH-dependent swelling/deswelling transitions across the pKa of the PEAMA gel core around pH 7.0. Such nanogels swell under acidic conditions and deswell under alkaline conditions.[47]

Recent research suggested chemically cross-linked nanogels with PEG derivatives to overcome the instability of physically cross-linked nano-gel in vivo. Polysaccharide-PEG hybrid nanogels were cross-linked by both covalent ester bonds and physical interactions by the reaction of a thiol-modified PEG (PEGSH) with acryl-oyl-modified cholesterol-bearing pullulan (CHPOA). The formulations were injected intravenously in mice to study their blood clearance.[48] CHP nanogels were eliminated from the blood within 6 h, whereas the CHPOA–PEGSH nanogels had a significantly longer circulation time: approximately, 40–50% of the NPs remained in circulation for 6 h in flowing injections, and after 24 h, 20–30% of the NPs remained in the blood. The half-life of CHPOA-PEGSH NPs was about 15-fold greater than that of CHP nanogels indicating long circulation behavior of PEGylated nanogels.[49]

5.4.7   NANOGELS FOR LOADING siRNA

A small interfering RNA (siRNA) is a class of double-stranded RNA molecules consisting of 21–23 nucleotides, involved in inhibition of protein synthesis encoded by the messenger RNAs (mRNA).[50] The siRNA mediates posttranscriptional gene silencing of a specific target protein by disrupting mRNA when introduced into cells. They show promise to be used for any disease-causing gene as well as for targeting any tissue. siRNA as a gene-regulating tool has a tremendous therapeutic potential in the areas of cancer treatment.[51]

However, the clinical application of siRNA is hindered by its poor stability, degradation by endogenous enzymes, low cellular uptake efficiency, low endosomal escape efficiency, and short half-life in blood. Also, the naked siRNA is unable to penetrate cellular membranes due to its large size and high negative charge. Such obstacles restrict the delivery of siRNA in vivo and require a suitable delivery carrier. Among different carriers, nanogels show promise as a novel transport medium for siRNA.[52]

Many researchers suggest targeted delivery of siRNAs by nanogels may be a promising strategy to increase the efficacy of chemotherapy drugs for the treatment of cancer.[56] It is difficult to load siRNA into nanogel carrier with high encapsulation efficiency as it easily leaks from the carrier due to its hydrophilic character (Fig. 5.13). To increase siRNA-loading efficiency, it is complexed with cationic excipients to enhance the affinity between the siRNA and the particle matrix. Research community has used PEI nanogels as an effective siRNA carrier.

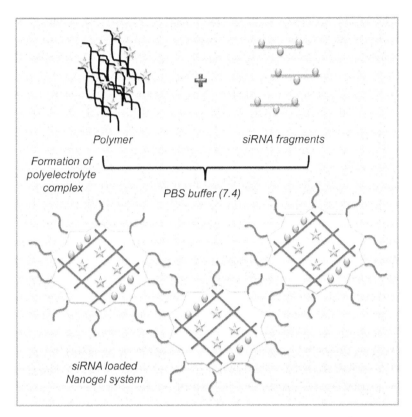

FIGURE 5.13   Schematic representation of mechanism of siRNA nanogel system.

The negative charge of siRNA allows it to form a strong electrostatic interaction with the positively charged PEI. The consequential poly-ionic complexes also protect siRNA against enzymatic degradation. Other negatively charged complexing agents used are di-oleyltrim ethyl ammonium propane and polyamines.[53]

Chitosan (CS) has been shown to be useful as a carrier for improving the cellular uptake of naked siRNA both in vitro and in vivo via different administration routes. CS is also useful for preventing the rapid degradation of siRNA in vivo. Recent studies explored the potential possibility of hyaluronic acid (HA) as a biocompatible and biodegradable nanogel for delivery of siRNA. These nanogels cross-linked with disulfide linkages showed target-specific intracellular delivery if siRNA-to-HA-specific CD44 receptor overexpresses cancer cells.[59]

Cationic dextran hydroxyl-ethyl methacrylate (dex-HEMA)-based nanogels are promising carriers for siRNA delivery since they can be loaded efficiently with siRNA, taken up by cells in vitro and were able to deliver intact siRNA into the cytosol of cells. The photo-polymerization method is another novel technique to load siRNA into dextran nanogels (dex-HEMA-*co*-TMAEMA) using UV-induced emulsion. These nanogels were used as a siRNA depot from which siRNA released at the desired time to prolong the gene-silencing effect. It was reported that in this way the siRNA dose was more efficiently deployed. Photochemical internalization was used as a trigger to induce endosomal escape of siRNA through the use of amphiphilic photosensitizers.[54]

PEGylate dex-HEMA nanogels produced by covalent attachment of NHS-PEG to the reactive amine groups of the nanogels with an aim to deliver siRNA in vivo. It was shown that dex-(HE)-MA-*co*-AEMA-*co*-TMAEMA nanogels retained their high loading efficiency of siRNA after PEGylation. The diffusion of the negatively charged siRNA molecules inside the gels occurred very slowly and that the siRNA was trapped by the cationic charges in the nanogels. Also, siRNA-loaded PEGylated dex-(HE) MA-*co*-AEMA-*co*-TMAEMA nanogels were able to successfully down regulate enhanced green fluorescent protein (EGFP) without causing severe toxicity in a HuH-7 EGFP cell line.[55]

5.5   NANOGELS IN ANTICANCER DRUG DELIVERY

Nanogels, cross-linked and swellable hydrophilic polymer NPs, have recently gained much interest as promising NPs carriers due to their nanoscale size (50–200 nm) and high stability favorable for intravenous and intracellular drug delivery. Nanogels have opened new approaches for cancer chemotherapy offering advantageous delivery of anticancer drugs with nonspecific toxicity and a narrow therapeutic window. The responsiveness of smart nanogels to temperature and pH and also the provision for attaching more functional groups for targeting the cancer cells made this nano-scopic carrier a better platform for delivering anticancer agents.[56]

5.4.1 NANOTECHNOLOGY VERSUS NANOGEL TECHNOLOGY IN CANCER THERAPEUTICS

Generally, oral or intravenous chemotherapeutic agents, drugs, enzymes, proteins and biomolecules in aqueous solution do not offer ideal pharmacokinetics in the physiological environment. Anticancer drugs have toxic effects on normal tissue and have a very narrow therapeutic window. Prominent nano-therapeutic approaches treat different types of cancer and have opened up a new era of cancer chemotherapy with beneficial contributions to health care. Various types of nano-system are widely used in drug delivery applications.[57] The delivery of NP-mediated anticancer drugs, siRNA, DNA, proteins, and other biomolecules to tumors can be achieved through passive targeting and active targeting pathways. However, stability and short systematic circulation times have often limited their use in vivo, and the continuous development of novel nanocarriers has resulted in a new material, so-called nanogels. The significance of these nanogel systems in cancer chemotherapy has been increasing year on year.[58]

5.4.2 STIMULI-SENSITIVE ANTICANCER NANOGELS

Recently, responsive nanogels have been utilized widely as smart drug delivery system for controlled drug release and cancer therapeutic applications enabling prolonged and better control of drug administration. Thermoresponsive nanogels having network structures with water can be used for biomedical applications because of their temperature response. The triblock copolymer composed of PEG and poly (ethyl ethylene phosphate) was used by Juan Wu et al. to prepare thermosensitive nanogels loaded with Doxorubicin (Dox). The triblock polymer chains self-assembled into NPs under the effect of heat were further cross-linked and swelled to form the nanogels. The uptake and release of Dox were studied in nonsmall cell lung carcinoma (A459) cells.[59] Combined chemo and PTT was the goal of the research conducted by various scientific fraternities where they used the idea of core–shell nanogels. Here, the core was bimetallic consisting of silver and gold, the immediate coating of this core was made of polystyrene and this was then coated with a thin layer of PEG. The core helps in cell imaging as well as help in PTT, while the polystyrene coating was meant to provide strong hydrophobic interactions with curcumin to increase loading efficiency of the drug, and the thermoresponsive external PEG layer is to control drug release. Potent cytotoxicity was seen when the drug-loaded nanogels were administered to mouse melanoma cells (B16F10) cells, and thus, it was concluded that this drug delivery system was capable of increasing the therapeutic efficiency of the drug.[60]

pH-sensitive nanogels that create switching carriers in release kinetics from slow release while circulating to rapid release at the targets have been considered as promising anticancer drug carriers. Employing CS and ovalbumin to prepare stable nanogels using a green method for its synthesis exhibits a very fine biodegradable nanogel system. Their nanogels consisted of ovalbumin nanospheres containing some CS chains interspersed in the nanogel structure and the rest forming its shell. The 100-nm-sized nanogels were pH sensitive, and the hydrodynamic radius was constant in the pH range of 4.3–5, increases in the pH range of 5.3–5.8. They also reported a change in the hydrophobicity/hydrophilicity with respect to pH. They are hydrophilic at neutral and acidic pH and hydrophobic at alkaline pH. The nanogels they prepared thus have the potential to be used in drug delivery where their pH-dependent properties could be utilized. CS was modified with glycidyltrimetylammonium chloride and used to synthesize the nanogels which showed pH-dependent volume phase transition from 180 to 450 nm. Targeting was made possible by conjugation with transferrin. Methotrexate was the anticancer drug chosen for this study, and release from the nanogels was carried out under acidic conditions. In another study, glycol-CS was modified with cholanic acid and self-assembled with hydrophobic anticancer drug docetaxel using a dialysis method. The resulting hydrophobically modified glycol-CS nanogels had a diameter of 350 nm and were stable in the presence of BSA and also showed a sustained drug release profile in vitro and also enhanced antitumor activity in vivo in lung cancer model, suggesting a promising drug delivery for cancer therapy.[61] pH-responsive Dox-loaded CNG was developed and its in vitro anticancer efficacy was evaluated in different cancer cell lines. The study reported that 130–160-nm-sized nanogels showed pH-sensitive controlled release of Dox. They also showed the cellular uptake of these nanogels and toxicity toward variety of cancer cell lines. The fluorescent Dox enabled the imaging of the cancer cells. Nanogel sensitive to endosomal pH with Dox prodrug were studied and investigated for triggered intracellular drug release in cancer cells.[62]

A virus-mimetic nanogel (VM-nanogel) was developed with a hydrophobic polymer core (poly(L-histidine-*co*-phenylalanine)(poly(His32-*co*-Phe6)) and two layers of hydrophilic shell loaded with Dox. PEG forms the inner shell whose one end is linked to the polymeric core and another end to BSA. The VM-nanogel is reported to be pH sensitive, whereby reversible swelling and shrinking of the nanogel was observed when the pH was cycled between the typical cytosolic pH range (7.4 and 6.8).[63] The release rate of Dox is influenced by the pH-induced reversible swelling/deswelling of the core as a result of which enhanced amount of Dox was released at endosomal pH 6.4 compared to that of cytosolic pH 6.8–7.4. These VM-nanogels were found to kill human ovarian cancer cells (A2780).[64]

pH-responsive poly(1-amino acid) nanogels were prepared for the intercellular drug delivery. The nanogels were generated by cross-linking of poly [1-aspartic acid-*g*-(3-diethylaminopropyl)]-*b*-PEG-maleimide [poly(1-Asp-*g*-DEAP)-*b*-PEGal] and poly(1-aspartic acid-*g*-ethylthiol)-*b*-PEG [poly(1-Asp-SH)-*b*-PEG] in an oil/water emulsion. The proton absorption of DEAP occurs in lysosomal pH due to which these nanogels efficiently uptake cell efficiencies in MCF-7 cells.[65] One recent report suggested that reduction-sensitive functional nanogels using inverse miniemulsion atom transfer radical polymerization and the disulfide-thiol exchange reaction. HeLa cells treated with Dox-loaded disulfide cross-linked nanogels; upon addition of 20 wt% GSH, the growth was significantly inhibited.[66] Scientific community also developed HA nanogels (200–500 nm) physically encapsulating siRNA by an inverse water-in-oil emulsion method where in thiol-conjugated HA in aqueous emulsion droplets was ultrasonically cross-linked via the formation of disulfide linkages. CD44+ HCT-116 cells readily took up HA/siRNA nanogels. Release rates of siRNA from the HA nanogels were regulated by the GSH concentration, indicating intracellular reductive agent triggered cleavage of disulfide cross-linked HA nanogels. Similarly, HA nanogels with the disulfide linkages could be used for the specific intracellular delivery of anticancer agents and could be exploited for cancer therapy.[67]

5.4.3 NANOGELS FOR CANCER CELL IMAGING

QDs with bright fluorescence, narrow emission, broad excitation, and high photostability have been developed as potential fluorescent probes for long-term cellular and molecular imaging. One recent report investigated the ability of CHP and amino-modified CHP (CHPNH$_2$) to deliver QDs into cells. They reported that positively charged amino group interacts well with negatively charged QD, and they were to image human cervical cancer cell lines. Here, they have shown the QDs capped with protein A so as to generate stable negatively charged moiety for the interaction with positively charged amine group on the nanogels, and also, neutral nanogels could self-complex with the protein-capped QDs.[68] Thus, these multifunctional nanogels form the basis for sensing, imaging, and controlled drug delivery and offer opportunities for combined diagnosis and therapy.

Nanogels with bimetallic could be used for cell imaging, the temperature dependent volume transition could be utilized for efficient drug loading and controlled release, the targeting ligands were obviously for active targeting of the nanogel to the effect site. The nanogels bimetallic core consisted of gold and silver along with a coating of thermoresponsive non-linear polymer PEG-based hydrogel and semi-interpenetrating targeting ligands.[69] They found that temperature sensing was a possibility, which was because the thermoresponsive volume phase transition of the PEG shell modifies the physiochemical environment around the bimetallic core and changes the fluorescence intensity.

They fabricated photo-thermoresponsive nanogels for use in simultaneous optical temperature-sensing, cell imaging, and combined chemo-photothermal treatment.[70] The nanogels were composed of an interpenetrating network of NIPAM hydrogel with a gold NP (GNP) (50 nm) at its core. The model drug chosen here for studying the temperature controlled release was 5-FU, and cytotoxicity studies were done on HeLa cells. HeLa cells significantly took up these nanogels and got localized to the lysosomes. The scattering properties of the GNPs have been used for cell imaging using dark-field microscopy. They thus proved to be a novel contrast agent for cell imaging.[71]Core–shell-structured hybrid nanogels composed of an Ag NP as core and smart gel of PNIPAm-*co*-acrylic acid as shell was developed, and it facilitated the simultaneous cancer cell imaging and adequate local delivery to tumor site. The study reported that smart hybrid nanogels developed by the incorporation of functional building block with optical and therapeutic functionality, overcome the cellular barriers of the mouse melanoma B16F10 cells, and light up intracellular region, thus making it suitable for biomedical applications. A current report suggests a nanoprobe based on a PEGylated-nanogel-containing GNP (fluorescence quenchers) in the cross-linked polyamine gel core and fluorescein iso-thiocyanate (FITC)-labeled DEVD peptides at the tethered PEG chain ends.[72] The nanogel–nanoprobe system is reported to be biocompatible, caspase-3-responsive, and fluorescence-quenching suitable for monitoring the cancer response to therapy. Apoptotic cells were detected in human hepatocyte (HuH-7) multicellular tumor spheroids (MCTSs) as early as 1 day posttreatment with stauro-sporine (an apoptosis inducing agent), while growth inhibition of the HuH-7 MCTSs was observed after a delay of 3 days (i.e., on day 4). This study demonstrated the effectiveness of the FITC-DEVD-nanogel–GNP probe as a smart nanoprobe for real-time monitoring of response to cancer therapy.[73]

The recent developed biodegradable and biocompatible CNGs which have proved to be advantages for simultaneous drug delivery and imaging of cancer cells. They have studied the efficiency of CNGs to load dye rhodamine-123 and also MPA-capped CdTe QDs for studying the in vitro localization within the cells. The fluorescence emitted from the samples indicates its application for cellular imaging. They proved the cellular uptake of these CNGs with the imaging agents in an array of normal as well as cancer cell lines.[74] CD44⁺-receptor-targeted HA and pH-sensitive poly(-amino ester) (PBAE) were formulated with NIR fluorescent indo-cyanine green (ICG) to form cancer-cell-specific pH-activated polymer nanogels for detecting cancer cells. The probe was internalized by CD44⁺-receptor-mediated endocytosis and accumulated in the late endosomes or lys-osomes followed by the solubilization and disassembly of the polymer nanogel. During endosomal maturation, the encapsulated ICG was released by solubilizing the PBAE due to the acidic pH, thereby generating a highly tumor-specific NIR signal with a reduced background signal.[75]

5.6 OTHER THERAPEUTIC APPLICATIONS OF NANOGEL FORMULATIONS

Nanogel-based drug delivery formulations improve the effectiveness and safety of anticancer drugs, due to their chemical composition, and have been extensively utilized in delivery of wide varieties of therapeutic agents through various routes. The current section describes some of the applications other than cancer therapeutic applications of the nanogel carrier systems.[76]

5.6.1 AUTOIMMUNE DISEASE

Nanogels were fabricated by remotely loading liposomes with mycophenolic acid (MPA) solubilized within cyclodextrin, oligomers of lactic acid-
PEG that were terminated with an acrylate end group and Irgacure 2959
photoinitiator. Particles were then exposed to ultraviolet light to induce
photopolymerization of the PEG oligomers. The nanogels are attractive
because of their intrinsic abilities to enable greater systematic accumulations of their cargo and to bind more immune cells in vivo than free fluorescent tracer, which, we reason, permits high, localized concentrations of
MPA.[77] This new drug delivery system increases the longevity of the patient
and delays the onset of kidney damage, a common complication of lupus.[78]

5.6.2 OPHTHALMIC

pH-sensitive polyvinyl pyrrolidone-poly(acrylic acid) (PVP/PAAc) nanogels prepared by gamma radiation-induced polymerization of acrylic acid
in an aqueous solution of PVP as a template polymer were used to encapsulate pilocarpine in order to maintain an adequate concentration of the
pilocarpine at the site of action for prolonged period of time.[79]

5.6.3 DIABETICS

An injectable nanonetwork that responds to glucose and releases insulin
has been developed. It contains a mixture of oppositely charged NPs that
attract each other. This keeps the gel together and stops the NPs drifting
away once in the body. To make the nanogel respond to increased acidity
dextran, a modified polysaccharide was used. Each NP in the gel holds
spheres of dextran loaded with insulin and an enzyme that converts glucose
into gluconic acid. Glucose molecules can easily enter and diffuse through
the gel.[80] Thus when levels are high, lots of glucose pass through the gel
and trigger release of the enzyme that converts it to gluconic acid. This
increases acidity, which triggers the release of the insulin. There is still
some work to do before the gel is ready for human trials.[81]

5.6.4 NEURODEGENERATIVE

Nanogel is a promising system for delivery of ODN to the brain. A novel system for oligonucleotides delivery to the brain based on nanoscale network of cross-linked PEG and poly-ethylenimine ("nanogel") is used for the treatment of neuro-degenerative diseases. Nanogels bound or encapsulated with spontaneously negatively charged ODN results in formation of stable aqueous dispersion of polyelectrolyte complex with particle size less than 100 nm which can be effectively transported across the blood–brain barriers. The transport efficacy is further increased when the surface of the nanogel is modified with transferrin or insulin.[82]

5.6.5 ANTICOAGULANT

A nanogel composed of protein molecules in solution has been used to stop bleeding, even in severe gashes. The proteins self-assemble on the nanoscale into a biodegradable gel.[83]

5.6.6 ANTI-INFLAMMATORY ACTION

Poly(lactide-*co*-glycolic acid) and CS were used to prepare bilayered NPs, and the surface was modified with oleic acid. Hydroxy-propyl methyl cellulose and carbopol with the desired viscosity were utilized to prepare the nanogels. Two anti-inflammatory drugs, spantide II and ketoprofen, which are effective against allergic contact dermatitis and psoriatic plaque, were applied topically along with nanogel. The result shows that nanogel increases potential for the percutaneous delivery of spentide II and keto-profen to the deeper skin layers for treatment of various skin inflammatory disorders.[84]

5.7 PATENT STATUS OF NANOGELS

Patents on nanogels have been mainly filed for their synthetic procedures. "Ringing gels" (system vibrates with a little tap for returning to original configuration) describes a surfactant-free preparation of oil in water meta-stable nanoemulsion which has mainly silicone as a component prepared by high shear rate devices converting it into nanogel. The PAAc nanogel prepared by free radical polymerization method uses gamma radiations for cross-linking of polymers for dispensing biologically active molecules.[85] Nanogel prepared from metal salts containing solid particulate matter or solid dispersions is utilized for either barrier system or loading system for chemicals, gases, drugs, etc.[86] Invention of water-soluble degradable bioimagining and drug particle conjugating to free carboxyl groups of polymers hydrogel has been made. Targeted delivery of antineoplastic agents noncovalently bound to polymer core in nanogel has been filed.[87]

5.8 CONCLUSION

Nanogel-based drug delivery system discloses as a futuristic drugs carrier administration approach which plays a very important role in delivery of numerous bioactives, medicine, and therapeutics in wide varieties of diseases and disorders. They exhibit substantial physiognomies and circumvent complications related with formulation like toxicity, stability, absorption, and compatibility. Nanogels emerge as an important contender in several targeting areas like skin, brain, liver, lungs, colon, GIT, and heart with vast benefits of several routes of administration. In future, the limitations like tissue selectivity, site specificity, side effects, and therapeutic efficacy can be curtailed by adjusting the precise elements linked with formulation and administration of nanogels which outcomes in intensifying prospect of nanogel delivery system in wide varieties of infections. Extended levels of preclinical and clinical research must be organized for the improved economic production of nanogel at commercial stages.

KEYWORDS

- nanogel
- cross-link
- stimuli-responsive
- drug delivery
- drug interaction
- nanoparticles

REFERENCES

1. Seyfried, T. N.; Shelton, L. M. Cancer as a Metabolic Disease. *Nutr. Metab.* **2010,** *7*(1), 1.
2. Zhao, G.; Rodriguez, B. L. Molecular Targeting of Liposomal Nanoparticles to Tumor Microenvironment. *Int. J. Nanomed.* **2013,** *8*, 61.
3. Jabir, N. R.; Tabrez, S.; Ashraf, G. M.; Shakil, S.; Damanhouri, G. A.; Kamal, M. A. Nanotechnology-Based Approaches in Anticancer Research. *Int. J. Nanomed.* **2012,** *7*, 4391.
4. Mousa, S. A.; Bharali, D. J. Nanotechnology-Based Detection and Targeted Therapy in Cancer: Nano-Bio Paradigms and Applications. *Cancers* **2011,** *3*(3), 2888–2903.
5. Sahu, P.; Das, D.; Mishra, V. K.; Kashaw, V.; Kashaw, S. K. Nanoemulsion: A Novel Eon in Cancer Chemotherapy. *Mini Rev. Med. Chem.* **2017,** *7*, 1–22.
6. Nagahara, L. A.; Lee, J. S.; Molnar, L. K.; Panaro, N. J.; Farrell, D.; Ptak, K.; Grodzinski, P. Strategic Workshops on Cancer Nanotechnology. *Cancer Res.* **2010,** *70*(11), 4265–4268.
7. Sivaram, A. J.; Rajitha, P.; Maya, S.; Jayakumar, R.; Sabitha, M. Nanogels for Delivery, Imaging and Therapy. *Wiley Interdiscip. Rev. Nanomed. Nanobiotechnol.* **2015,** *7*(4), 509–533.
8. Adhikari, B.; Sowmya, C.; Reddy, C. S.; Haranath, C.; Bhatta, H. P.; Inturi, R. N. Recent Advances in Nanogels Drug Delivery Systems. *World J. Pharm. Pharm. Sci.* **2016,** *5*(9), 505–530.
9. Sharma, A.; Garg, T.; Aman, A.; Panchal, K.; Sharma, R.; Kumar, S.; Markandeywar, T. Nanogel—An Advanced Drug Delivery Tool: Current and Future. *Artif. Cells Nanomed. Biotechnol.* **2016,** *44*(1), 165–177.
10. Das, D.; Sahu, P.; Mishra, V. K.; Kashaw, S. K. Nanoemulsion—The Emerging Contrivance in the Field of Nanotechnology. *Nanosci. Nanotech. Asia.* **2017,** *2*, 1–22.
11. Oh, J. K.; Siegwart, D. J.; Lee, H. I.; Sherwood, G.; Peteanu, L.; Hollinger, J. O.; Matyjaszewski, K. Biodegradable Nanogels Prepared by Atom Transfer Radical Polymerization as Potential Drug Delivery Carriers: Synthesis, Biodegradation, In Vitro Release, and Bioconjugation. *J. Am. Chem. Soc.* **2007,** *129*(18), 5939–5945.
12. Yallapu, M. M.; Jaggi, M.; Chauhan, S. C. Design and Engineering of Nanogels for Cancer Treatment. *Drug Discov. Today* **2011,** *16*(9), 457–463.
13. Soni, G.; Yadav, K. S. Nanogels as Potential Nanomedicine Carrier for Treatment of Cancer: A Mini Review of the State of the Art. *Saudi Pharm. J.* **2014,** *24*(2), 133–139.
14. Yallapu, M. M.; Jaggi, M.; Chauhan, S. C. Poly(β-cyclodextrin)/Curcumin Self-Assembly: A Novel Approach to Improve Curcumin Delivery and Its Therapeutic Efficacy in Prostate Cancer Cells. *Macromol. Biosci.* **2010,** *10*(10), 1141–1151.
15. Nomura, Y.; Sasaki, Y.; Takagi, M.; Narita, T.; Aoyama, Y.; Akiyoshi, K. Thermoresponsive Controlled Association of Protein with a Dynamic Nanogel of Hydrophobized Polysaccharide and Cyclodextrin: Heat Shock Protein-Like Activity of Artificial Molecular Chaperone. *Biomacromolecules* **2005,** *6*(1), 447–452.
16. Chaurasia, A.; Das, D.; Sahu, P. Evaluation, Assessment & Screening of Antidiabetic Activity of Stevia Rebaudiana leaves extract in Alloxan-Induced Diabetic Rat Model. *J. Int. Res. Medical Pharma. Sci.* **2016,** *9*, 107–112.

17. Kim, S.; Chung, E. H.; Gilbert, M.; Healy, K. E. Synthetic MMP-13 Degradable ECMs Based on Poly(*N*-isopropylacrylamide-*co*-acrylic acid) Semi-Interpenetrating Polymer Networks. I. Degradation and Cell Migration. *J. Biomed. Mater. Res. A* **2005,** *75*(1), 73–88.

18. Das, D.; Chaurasia, A.; Sahu, P.; Mishra, V. K.; Kashaw, S. K. Antihypercholes-trolemic Potential of Omega-3-Fatty Acid Concentrate in Alloxan Induced Diabetic Rodent. *Int. J. Pharma. Sci. Res.* **2015,** *6*, 3634–3640.

19. Oh, J. K.; Drumright, R.; Siegwart, D. J.; Matyjaszewski, K. The Development of Microgels/Nanogels for Drug Delivery Applications. *Progr. Polymer Sci.* **2008,** *33*(4), 448–477.

20. Sahu, P.; Bhatt, A.; Chaurasia, A.; Gajbhiye, V. Enhanced Hepatoprotective Activity of Piperine Loaded Chitosan Microspheres. *Int. J. Drug Dev. Res.* **2012,** *4*, 259–262.

21. Gao, H.; Matyjaszewski, K. Synthesis of Functional Polymers with Controlled Architecture by CRP of Monomers in the Presence of Cross-linkers: From Stars to Gels. *Prog. Polym. Sci.* **2009,** *34*(4), 317–350.

22. Xu, S.; Nie, Z.; Seo, M.; Lewis, P.; Kumacheva, E.; Stone, H. A.; Whitesides, G. M. Generation of Monodisperse Particles by Using Microfluidics: Control over Size, Shape, and Composition. *Angew. Chem. Int. Ed.* **2005,** *117*(5), 734–738.

23. Oh, J. K.; Bencherif, S. A.; Matyjaszewski, K. Atom Transfer Radical Polymerization in Inverse Miniemulsion: A Versatile Route Toward Preparation and Function-alization of Microgels/Nanogels for Targeted Drug Delivery Applications. *Polymer* **2009,** *50*(19), 4407–4423.

24. Bannister, I.; Billingham, N. C.; Armes, S. P.; Rannard, S. P.; Findlay, P. Development of Branching in Living Radical Copolymerization of Vinyl and Divinyl Monomers. *Macromolecules* **2006,** *39*(22), 7483–7492.

25. Shah, R. K.; Kim, J. W.; Agresti, J. J.; Weitz, D. A.; Chu, L. Y. Fabrication of Monodisperse Thermosensitive Microgels and Gel Capsules in Microfluidic Devices. *Soft Matter* **2008,** *4*(12), 2303–2309.

26. Dorwal, D. Nanogels as Novel and Versatile Pharmaceuticals. *Int. J. Pharm. Pharm. Sci.* **2012,** *4*(3), 67–74.

27. Gonçalves, C.; Pereira, P.; Gama, M. Self-Assembled Hydrogel Nanoparticles for Drug Delivery Applications. *Materials* **2010,** *3*(2), 1420–1460.

28. Ouchi, M.; Terashima, T.; Sawamoto, M. Transition Metal-Catalyzed Living Radical Polymerization: Toward Perfection in Catalysis and Precision Polymer Synthesis. *Chem. Rev.* **2009,** *109*(11), 4963–5050.

29. Sanson, N.; Rieger, J. Synthesis of Nanogels/Microgels by Conventional and Controlled Radical Crosslinking Copolymerization. *Polym. Chem.* **2010,** *1*(7), 965–977.

30. Robinson, D. N.; Peppas, N. A. Preparation and Characterization of pH-Responsive Poly(methacrylic acid-g-ethylene glycol) Nanospheres. *Macromolecules* **2002,** *35*(9), 3668–3674.

31. Bae, B. C.; Na, K. Self-Quenching Polysaccharide-Based Nanogels of Pullulan/Folate-Photosensitizer Conjugates for Photodynamic Therapy. *Biomaterials* **2010,** *31*(24), 6325–6335.

32. Alles, N.; Soysa, N. S.; Hussain, M. A.; Tomomatsu, N.; Saito, H.; Baron, R.; Ohya, K. Polysaccharide Nanogel Delivery of a TNF-α and RANKL Antagonist Peptide Allows Systemic Prevention of Bone Loss. *Eur. J. Pharm. Sci.* **2009,** *37*(2), 83–88.

33. Jaque, D.; Maestro, L. M.; Del Rosal, B.; Haro-Gonzalez, P.; Benayas, A.; Plaza, J. L.; et al. Nanoparticles for Photothermal Therapies. *Nanoscale* **2014**, *6*(16), 9494–9530.

34. Sahu, P.; Kashaw, S. K.; Jain, S.; Sau, S.; Iyer, A. K. Assessment of Penetration Potential of pH Responsive Double Walled Biodegradable Nanogels Coated with Eucalyptus Oil for the Controlled Delivery of 5-Fluorouracil: In Vitro and Ex Vivo Studies. *J Control. Rel.* **2017**, *253*, 122–136.

35. Fomina, N.; Sankaranarayanan, J.; Almutairi, A. Photochemical Mechanisms of Light-Triggered Release from Nanocarriers. *Adv. Drug Delivery Rev.* **2012**, *64*(11), 1005–1020.

36. Liang, R.; Wei, M.; Evans, D. G.; Duan, X. Inorganic Nanomaterials for Bioimaging, Targeted Drug Delivery and Therapeutics. *Chem. Commun.* **2014**, *50*(91), 14071–14081.

37. Reddy, L. H.; Arias, J. L.; Nicolas, J.; Couvreur, P. Magnetic Nanoparticles: Design and Characterization, Toxicity and Biocompatibility, Pharmaceutical and Biomedical Applications. *Chem. Rev.* **2012**, *112*(11), 5818–5878.

38. Wu, W.; Mitra, N.; Yan, E. C.; Zhou, S. Multifunctional Hybrid Nanogel for Integration of Optical Glucose Sensing and Self-Regulated Insulin Release at Physiological pH. *ACS Nano* **2010**, *4*(8), 4831–4839.

39. Molina, M.; Asadian-Birjand, M.; Balach, J.; Bergueiro, J.; Miceli, E.; Calderón, M. Stimuli-Responsive Nanogel Composites and Their Application in Nanomedicine. *Chem. Soc. Rev.* **2015**, *44*(17), 6161–6186.

40. Menéndez, G. O.; Pichel, M. E.; Spagnuolo, C. C.; Jares-Erijman, E. A. NIR Fluorescent Biotinylated Cyanine Dye: Optical Properties and Combination with Quantum Dots as a Potential Sensing Device. *Photochem. Photobiol. Sci.* **2013**, *12*(2), 236–240.

41. Li, Z.; Xu, W.; Wang, Y.; Shah, B. R.; Zhang, C.; Chen, Y.; Li, B. Quantum Dots Loaded Nanogels for Low Cytotoxicity, pH-Sensitive Fluorescence, Cell Imaging and Drug Delivery. *Carbohydr. Polym.* **2015**, *121*, 477–485.

42. Hasegawa, U.; Shin-ichiro, M. N.; Kaul, S. C.; Hirano, T.; Akiyoshi, K. Nanogel–Quantum Dot Hybrid Nanoparticles for Live Cell Imaging. *Biochem. Biophys. Res. Commun.* **2005**, *331*(4), 917–921.

43. Rejinold N, S.; Chennazhi, K. P.; Tamura, H.; Nair, S. V.; Rangasamy, J. Multifunctional Chitin Nanogels for Simultaneous Drug Delivery, Bioimaging, and Biosensing. *ACS Appl. Mater. Interfaces* **2011**, *3*(9), 3654–3665.

44. Yang, P.; Gai, S.; Lin, J. Functionalized Mesoporous Silica Materials for Controlled Drug Delivery. *Chem. Soc. Rev.* **2012**, *41*(9), 3679–3698.

45. Hu, X.; Hao, X.; Wu, Y.; Zhang, J.; Zhang, X.; Wang, P. C.; Liang, X. Multifunctional Hybrid Silica Nanoparticles for Controlled Doxorubicin Loading and Release with Thermal and pH Dual Response. *J. Mater. Chem. B* **2013**, *1*(8), 1109–1118.

46. Chai, S.; Zhang, J.; Yang, T.; Yuan, J.; Cheng, S. Thermoresponsive Microgel Decorated with Silica Nanoparticles in Shell: Biomimetic Synthesis and Drug Release Application. *Colloids Surf. A: Physicochem. Eng. Aspects* **2010**, *356*(1), 32–39.

47. Liu, G.; Zhu, C.; Xu, J.; Xin, Y.; Yang, T.; Li, J.; Liu, W. Thermo-Responsive Hollow Silica Microgels with Controlled Drug Release Properties. *Colloids Surf. B: Biointerfaces* **2013**, *111*, 7–14.

48. Balach, J.; Wu, H.; Polzer, F.; Kirmse, H.; Zhao, Q.; Wei, Z.; Yuan, J. Poly(ionic liquid)-Derived Nitrogen-Doped Hollow Carbon Spheres: Synthesis and Loading with Fe_2O_3 for High-Performance Lithium Ion Batteries. *RSC Adv.* **2013**, *3*(21), 7979–7986.

49. Mehdipoor, E.; Adeli, M.; Bavadi, M.; Sasanpour, P.; Rashidian, B. A Possible Anticancer Drug Delivery System Based on Carbon Nanotube–Dendrimer Hybrid Nanomaterials. *J. Mater. Chem.* **2011**, *21*(39), 15456–15463.

50. Yadav, K. S.; Jacob, S.; Sachdeva, G.; Chuttani, K.; Mishra, A. K.; Sawant, K. K. Long Circulating PEGylated PLGA Nanoparticles of Cytarabine for Targeting Leukemia. *J. Microencapsulation* **2011**, *28*(8), 729–742.

51. Shimoda, A.; Sawada, S. I.; Kano, A.; Maruyama, A.; Moquin, A.; Winnik, F. M.; Akiyoshi, K. Dual Crosslinked Hydrogel Nanoparticles by Nanogel Bottom-Up Method for Sustained-Release Delivery. *Colloids Surf. B: Biointerfaces* **2012**, *99*, 38–44.

52. Dykxhoorn, D. M.; Palliser, D.; Lieberman, J. The Silent Treatment: siRNAs as Small Molecule Drugs. *Gene Ther.* **2006**, *13*(6), 541–552.

53. Reischl, D.; Zimmer, A. Drug Delivery of siRNA Therapeutics: Potentials and Limits of Nanosystems. *Nanomed. Nanotechnol. Biol. Med.* **2009**, *5*(1), 8–20.

54. Cun, D.; Foged, C.; Yang, M.; Frøkjær, S.; Nielsen, H. M. Preparation and Characterization of Poly(DL-lactide-*co*-glycolide) Nanoparticles for siRNA Delivery. *Int. J. Pharm.* **2010**, *390*(1), 70–75.

55. Yadav, S.; Sahu, P.; Chaurasia, A. Role of Cyamopsis Tetragonoloba Against Cisplatin Induced Genotoxicity: Analysis of Micronucleus and Chromosome Aberrations In vivo. *Int. J. Bio Innov.* **2013**, *2*, 184–193; Morrissey, D. V.; Blanchard, K.; Shaw, L.; Jensen, K.; Lockridge, J. A.; Dickinson, B.; Polisky, B. A. Activity of Stabilized Short Interfering RNA in a Mouse Model of Hepatitis B Virus Replication. *Hepatology* **2005**, *41*(6), 1349–1356.

56. Dickerson, E. B.; Blackburn, W. H.; Smith, M. H.; Kapa, L. B.; Lyon, L. A.; McDonald, J. F. Chemosensitization of Cancer Cells by SiRNA using Targeted Nanogel Delivery. *BMC Cancer* **2010**, *10*(1), 1.

57. Mimi, H.; Ho, K. M.; Siu, Y. S.; Wu, A.; Li, P. Polyethyleneimine-Based Core–Shell Nanogels: A Promising SiRNA Carrier for Argininosuccinate Synthetase mRNA Knockdown in HeLa Cells. *J. Control. Release* **2012**, *158*(1), 123–130.

58. Mao, S.; Sun, W.; Kissel, T. Chitosan-Based Formulations for Delivery of DNA and SiRNA. *Adv. Drug Delivery Rev.* **2010**, *62*(1), 12–27.

59. Lee, H.; Mok, H.; Lee, S.; Oh, Y. K.; Park, T. G. Target-Specific Intracellular Delivery of SiRNA Using Degradable Hyaluronic Acid Nanogels. *J. Control. Release* **2007**, *119*(2), 245–252.

60. Raemdonck, K.; Naeye, B.; Høgset, A.; Demeester, J.; De Smedt, S. C. Prolonged Gene Silencing by Combining SiRNA Nanogels and Photochemical Internalization. *J. Control. Release* **2010**, *145*(3), 281–288.

61. Naeye, B.; Raemdonck, K.; Remaut, K.; Sproat, B.; Demeester, J.; De Smedt, S. C. PEGylation of Biodegradable Dextran Nanogels for SiRNA Delivery. *Eur. J. Pharm. Sci.* **2010**, *40*(4), 342–351.

62. Wani, T. U.; Rashid, M.; Kumar, M.; Chaudhary, S.; Kumar, P.; Mishra, N. Targeting Aspects of Nanogels: An Overview. *Int. J. Pharm. Sci. Technol.* 7(4), 2612–2630.

63. Look, M.; Stern, E.; Wang, Q. A.; DiPlacido, L. D.; Kashgarian, M.; Craft, J.; Fahmy, T. M. Nanogel-Based Delivery of Mycophenolic Acid Ameliorates Systemic Lupus Erythematosus in Mice. *J. Clin. Invest.* **2013,** *123*(4), 1741–1749.

64. Abd El-Rehim, H. A.; Swilem, A. E.; Klingner, A.; Hegazy, E. S. A.; Hamed, A. A. Developing the Potential Ophthalmic Applications of Pilocarpine Entrapped into Polyvinylpyrrolidone–Poly(acrylic acid) Nanogel Dispersions Prepared by γ Radiation. *Biomacromolecules* **2013,** *14*(3), 688–698.

65. Vinogradov, S. V.; Batrakova, E. V.; Kabanov, A. V. Nanogels for Oligonucleotide Delivery to the Brain. *Bioconjugate Chem.* **2004,** *15*(1), 50–60.

66. Shah, P. P.; Desai, P. R.; Patel, A. R.; Singh, M. S. Skin Permeating Nanogel for the Cutaneous Co-Delivery of Two Anti-Inflammatory Drugs. *Biomaterials* **2012,** *33*(5), 1607–1617.

67. Haley, B.; Frenkel, E. Nanoparticles for Drug Delivery in Cancer Treatment. *Urol. Oncol.: Semin. Original Invest.* **2008,** *26*(1), 57–64.

68. Wu, W.; Shen, J.; Banerjee, P.; Zhou, S. Water-Dispersible Multifunctional Hybrid Nanogels for Combined Curcumin and Photothermal Therapy. *Biomaterials* **2011,** *32*(2), 598–609.

69. Sahu, P.; Kashaw, S. K.; Kushwah, V.; Sau, S.; Jain, S.; Iyer, A. K. pH Responsive Biodegradable Nanogels for Sustained Release of Bleomycin. *Bioorg. Med. Chem.* **2017,** *25*, 4595–4613.

70. Hwang, H. Y.; Kim, I. S.; Kwon, I. C.; Kim, Y. H. Tumor Targetability and Antitumor Effect of Docetaxel-Loaded Hydrophobically Modified Glycol Chitosan Nanoparticles. *J. Control. Release* **2008,** *128*(1), 23–31.

71. Jayakumar, R.; Nair, A.; Rejinold, N. S.; Maya, S.; Nair, S. V. Doxorubicin-Loaded pH-Responsive Chitin Nanogels for Drug Delivery to Cancer Cells. *Carbohydr. Polym.* **2012,** *87*(3), 2352–2356.

72. Chaurasia, A.; Sahu, P.; Gajbhiye, G. Improved Anticancerous Activity of Indian Aloe Loaded Chitosan Microspheres. *Int. J. Pharm. Arch.* **2013,** *3*, 71–76.

73. Ryu, J. H.; Chacko, R. T.; Jiwpanich, S.; Bickerton, S.; Babu, R. P.; Thayumanavan, S. Self-Cross-Linked Polymer Nanogels: A Versatile Nanoscopic Drug Delivery Platform. *J. Am. Chem. Soc.* **2010,** *132*(48), 17227–17235.

74. Oh, J. K.; Bencherif, S. A.; Matyjaszewski, K. Atom Transfer Radical Polymerization in Inverse Miniemulsion: A Versatile Route toward Preparation and Functionalization of Microgels/Nanogels for Targeted Drug Delivery Applications. *Polymer* **2009,** *50*(19), 4407–4423.

75. Lee, H.; Mok, H.; Lee, S.; Oh, Y. K.; Park, T. G. Target-Specific Intracellular Delivery of SiRNA Using Degradable Hyaluronic Acid Nanogels. *J. Control. Release* **2007,** *119*(2), 245–252.

76. Das, D.; Sahu, P.; Kashaw, V.; Kashaw, S. K. Formulation and Assessment of In Vivo Anti-Inflammatory Potential of Omega-3-Fatty Acid Loaded Self Emulsifying Nanoemulsion. *Curr. Nanomedicine.* **2017,** *7*, 47–58.

77. Hasegawa, U.; Shin-ichiro, M. N.; Kaul, S. C.; Hirano, T.; Akiyoshi, K. Nanogel–Quantum Dot Hybrid Nanoparticles for Live Cell Imaging. *Biochem. Biophys. Res. Commun.* **2005,** *331*(4), 917–921.

78. Maya, S.; Sarmento, B.; Nair, A.; Rejinold, N. S.; Nair, S. V.; Jayakumar, R. Smart Stimuli Sensitive Nanogels in Cancer Drug Delivery and Imaging: A Review. *Curr. Pharm. Design* **2013,** *19*(41), 7203–7218.

79. Xue-qin Z Tian-xiao W, Wen L, et al. Multifunctional Au@IPNpNIPAAm Nanogels for Cancer Cell Imaging and Combined Chemophotothermal Treatment. *J. Mater. Chem.* **2011,** *21*, 7240–7247.

80. Weitai, W.; Ting, Z.; Alexandra, B.; Probal, B.; Shuiqin, Z. Smart Coreshell Hybrid Nanogels with Ag Nanoparticle Core for Cancer Cell Imaging and Gel Shell for pH-Regulated Drug Delivery. *Chem. Mater.* **2010,** *22*, 1966–1976.

81. Oishi, M.; Tamura, A.; Nakamura, T.; Nagasaki, Y. A Smart Nanoprobe Based on Fluorescence-Quenching PEGylated Nanogels Containing Gold Nanoparticles for Monitoring the Response to Cancer Therapy. *Adv. Funct. Mater.* **2009,** *19*(6), 827–834.

82. Sahu, P.; Kashaw, S. K.; Sau, S.; Iyer, A. K. Stimuli-Responsive Bio-Hybrid Nanogels: An Emerging Platform in Medicinal Arena. *Global J. Nanomed.* **2017,** *1*, 1–3.

83. Mok, H.; Jeong, H.; Kim, S. J.; Chung, B. H. Indocyanine Green Encapsulated Nanogels for Hyaluronidase Activatable and Selective Near Infrared Imaging of Tumors and Lymph Nodes. *Chem. Commun.* **2012,** *48*(69), 8628–8630.

84. Bowen-leaver, H. A.; Tadlock, C. C.; George, L. Ringing Nanogel Compositions. U.S. Patent 7153516B2, Dec 26, 2006.

85. Somasundaran, P.; Chakraborty, S. Polymeric Nanoparticles and Nanogels for Extraction and Release of Compounds. U.S. Patent WO/2006/052285, 2006.

86. Helling, G. Metal Salt Nanogel-Containing Polymers. U.S. Patent 20100178270, 2010.

87. Radosz, M.; Shen, Y.; VanKirk, E. A.; Murdoch, W. J. Degradable Nanogel for Drug Delivery. U.S. Patent Application 20070224164, 2005.

CHAPTER 6

PIEZOELECTRICAL MATERIALS FOR BIOMEDICAL APPLICATIONS

NEELAKANDAN M. S.[1], YADU NATH V. K.[1], BILAHARI ARYAT[1], VISHNU K. A.[2], JIYA JOSE[1], NANDAKUMAR KALARIKKAL[1], and SABU THOMAS[1,2*]

[1]*International and Inter University Centre for Nanoscience and Nanotechnology, Mahatma Gandhi University, Kottayam 686560, Kerala, India*

[2]*School of Chemical Sciences, Mahatma Gandhi University, Kottayam 686560, Kerala, India*

Corresponding author. E-mail: sabuthomas@mgu.ac.in

ABSTRACT

The importance of electric field in cell functioning was a major question that emerged from the discovery of transmembrane potential and endogenous electric field in biological tissues. It has thrived the research in developing new technologies based on the biological electric effects. Cell growth and differentiation are greatly affected by electrical stimulations. Stimulations are also used in the treatment of various diseases like Parkinson's disease, urinary incontinence, chronic pain, and cardiac arrhythmia. Piezoelectric materials can be utilized to achieve indirect stimulation of biological tissues. Various piezoelectric materials have been employed for different tissue repair applications, particularly in bone repair, where charges induced by mechanical stress can enhance bone formation; and in neural tissue engineering, in which electric pulses can stimulate directional neurite outgrowth to fill gaps in nervous tissue injuries.

6.1   INTRODUCTION

The discovery of piezoelectric effect has revolutionized material science. Piezoelectric materials can produce electricity in response to even minute mechanical deformations. When these materials are subjected to external mechanical stress, the positive and negative charges in the material undergo an asymmetric reorganization; this results in electrical polarization inside the material. This phenomenon is first observed and explained by Pierre and Jacques Curie in 1880.[1] They extensively studied piezoelectricity in materials like crystals of tourmaline, quartz, topaz, cane sugar, and Rochelle salt. Initial studies showed that quartz and Rochelle salt have most piezoelectric ability at that time (Fig. 6.1).

One of the unique and useful characteristics of the piezoelectric material is that the effect is reversible, that is, it can exhibit both direct and converse piezoelectric effects. Converse piezoelectricity is a generation of mechanical deformation when a piezoelectric material is subjected to the external electric field.

The renaissance in science actually happened during the period of First World War. The field of science acquired a new dimension since the time. As part of that concept like piezoelectricity, this was earlier confined to walls of laboratories which now had found its way to practical application. In the sonar devices, this invention attracted the attention of the scientific community to the development of piezoelectric devices for various kind of application. Nowadays, piezoelectric materials find lots of applications

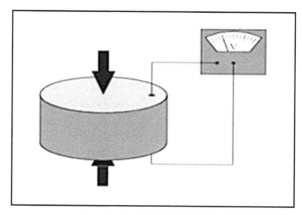

FIGURE 6.1   Schematic representation of piezoelectricity on mechanical deformation.

from cigarettes lighters to highly sophisticated atomic-scale characterization instruments like scanning tunnelling microscopy (STM) and atomic force microscopy (AFM). Piezoelectric materials can deliver electrical stimulus without an external power source; this makes it a suitable candidate for biomedical applications because electrical stimulation is an important therapeutic technique. Electrical stimulations are used for the treatment of a diverse range of conditions like Parkinson's disease, urinary incontinence, chronic pain, and cardiac arrhythmia. Electrical stimulation can also promote tissue regeneration.

6.2 PIEZOELECTRIC EFFECT AND PIEZOELECTRIC MATERIALS

In general case, some of the materials such as ceramics, crystals, and biological materials like bone, DNA, and proteins are able to accumulate electric charge due to some applied mechanical stress; this effect is known as the piezoelectric effect. The word piezoelectricity denotes electricity resulting from pressure. It is derived from the Greek word "piezō" (πιέζω) or "piezein" (πιέζειν), which betokens to constrict or press, and "ēlektron" (ήλεκτρον), which denotes amber, an archaic source of electric charge. Piezoelectricity was discovered in 1880 by French physicists Jacques and Pierre Curie. The piezoelectric effect is understood as the linear electromechanical interaction between the mechanical and the electrical state in crystalline materials with no inversion symmetry. The piezoelectric effect is a reversible process; in that, materials exhibiting the direct piezoelectric effect (the internal generation of electrical charge resulting from an applied mechanical force) additionally exhibit the inversion piezoelectric effect (the internal generation of a mechanical strain resulting from an applied electrical field). For example, lead zirconate titanate crystals will engender quantifiable piezoelectricity when their static structure is deformed by about 0.1% of the pristine dimension. Conversely, those same crystals will transmute about 0.1% of their static dimension when an external electric field is applied to the material. The inverse piezoelectric effect is utilized in the engendering of ultrasonic sound waves.

Piezoelectricity is found in utilizable applications, such as the engendering and detection of sound, generation of high voltages, electronic frequency generation, microbalances, to drive an ultrasonic nozzle, and ultrafine focusing of optical assemblies. It is withal the substratum of a number of scientific instrumental techniques with atomic resolution, the scanning probe microscopies, such as STM, AFM, scanning Near-field Optical microscopy SNOM, etc., and everyday uses, such as acting as the ignition source for cigarette lighters, and push-start propane barbecues, as well as the time reference source in quartz watches.

There are many materials, both natural and man-made, that exhibit a range of piezoelectric effects. Some naturally piezoelectric occurring materials include berlinite (structurally identical to quartz), cane sugar, quartz, Rochelle salt, topaz, tourmaline, and bone (dry bone exhibits some piezoelectric properties due to the apatite crystals, and the piezoelectric effect is generally thought to act as a biological force sensor). An example of man-made piezoelectric materials includes barium titanate (BT) and lead zirconate titanate.

In recent years, due to the growing environmental concern regarding toxicity in lead-containing devices and the RoHS directive followed within the European Union, there has been a push to develop lead-free piezoelectric materials. To date, this initiative to develop new lead-free piezoelectric materials has resulted in a variety of new piezoelectric materials which are more environmentally safe.

A recent application of piezoelectric ultrasound sources is piezoelectric surgery, withal kenned as piezosurgery. Piezosurgery is a minimally invasive technique that aims to cut a target tissue with little damage to neighbouring tissues. For example, Hoigne et al. reported its use in hand surgery for the cutting of bone, utilizing frequencies in the range 25–29 kHz, causing micro vibrations of 60–210 μm. It has the faculty to cut mineralized tissue without cutting neurovascular tissue and other soft tissue, thereby maintaining a blood-free operating area, better overtones, and more preponderant precision.

6.3   PIEZOELECTRICITY IN BIOLOGICAL TISSUES

The invention of piezoelectricity in materials, endogenous electric fields, and transmembrane potentials in biological tissues raised the question, what is the importance of electric fields in cell function? It has kindled research and the development of technologies in emulating biological electricity for tissue regeneration. Promising effects of electrical stimulation on cell growth and differentiation and tissue growth has led to an interest in using piezoelectric scaffolds for tissue repair. Piezoelectric materials can generate electrical activity when deformed. Hence, an external source to apply electrical stimulation or implantation of electrodes is not needed. Various piezoelectric materials have been employed for different tissue repair applications, particularly in bone repair, where charges induced by mechanical stress can enhance bone formation; and in neural tissue engineering, in which electric pulses can stimulate neuritis directional outgrowth to fill gaps in nervous tissue injuries.

The biological piezoelectricity is first observed by Martin in 1940.[2] He could detect electric potential from a bundle of wool enclosed in shellac while compressed by two brass plates. α-Keratin is a protein which is the major constituent in wool, mammalian hair, and horn.[3,4] It has a spiral α-helix structure.[5] The piezoelectric effect exhibited by such tissues can be accounted by closely packed alignment of these highly ordered α-helices and their characteristic polarization. The α-helix structure is stabilized by hydrogen bonding in the molecule, and it has a right-handed coiled structure shown in Figure 6.2.

Yasuda described that piezoelectricity is not formed by a living cell after observing piezoelectricity in boiled bone.[6] This leads to the conclusion that bone is a piezoelectric material and attributed in the application of shear on collagen. Piezoforce microscopy (PEM) is used for understanding the piezoelectricity of a material. PFM is a modification of AFM), which has been recently used to study the piezoelectricity of nanomaterials.[7] An

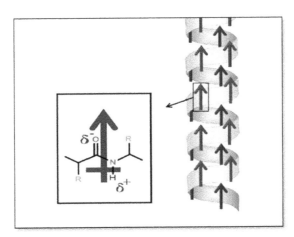

FIGURE 6.2   The α-helix structure is stabilized by hydrogen bonding in the molecule.

AC bias between the conductive AFM tip and the substrate beneath the sample applies an electric field to the sample, causing a deformation of the piezoelectric material. The contacting AFM tip detects the deformation, which is subsequently translated to the amplitude of the piezoresponse.[8]

Piezoelectric effects have been produced in a number of soft tissues, as well as hard, and appear to be associated with the presence of oriented fibrous proteins such as collagen. Thus, piezoelectricity may be a universal property of living tissue and may play a significant part in several physiological phenomena. Biological structures, at different organization levels (macromolecules, tissues, etc.), present typical spiral shape. Spirals do not have a center of symmetry, and hence, almost all biological matter possesses piezoelectricity properties. Thus, the possibility to convert mechanical signals into electric ones, and vice versa, is not only ascribed to minerals or ceramics. The piezoelectric effect even in hard tissues can be attributed to the main organic constituent of the tissue. For example, in the case of bone, it is collagen and cellulose in wood. The piezoelectric effect is highly directional, shear stress is maximum, and tensile stress is minimum. The requirement for piezoelectricity in living tissue is the presence of well-ordered asymmetric fibrous molecules. These can be polarized by shearing stress if it is in a cross-linked uniaxial system.[9] When we studied the piezoelectricity of the single collagen by piezoelectric microscope, PFM images of collagen fibrils will show lateral piezoresponse along the fibril axis and negligible vertical and radial piezoresponse, revealing the unidirectional polarization along the collagen fibril axis. In a biological manner, piezoelectricity has very potential implication such as regenerative tissues, bone, etc. In this chapter, we are discussing the streaming potential due to the piezoelectric effect, application of piezoelectric effect and materials.[10,11]

6.4 PIEZOELECTRICITY AND STREAMING POTENTIAL

In 1892, Wolff[12] suggested that bone remodels its architecture in response to stress. The statement is termed as Wolff's law. After the discovery of piezo-response in dry bone, the mechanism used to describe bone growth, and resorption in response to stress was piezoelectricity. Then, Bassett[13] observed that unlike the unreformed samples, periodically deformed cultivated chick embryonic tibiae developed large periosteal chondroid masses after 7 days and explained Wolff's law as a negative feedback loop: the applied load on the bone will cause strain in less dense regions, while denser and consequently stiffer regions will remain unstrained. The explanation is that the strain is transformed into an electric field that will aggregate and align the macromolecules and ions in the extracellular matrix, which will stimulate the cells to remodel the bone architecture until the signal (strain) is switched off. As piezoelectric measurements expanded to hydrated bone[14] and hydrated collagen[15] to simulate physiological conditions, there was some dissimilarity observed in the amplitude and behavior of the stress-generated potentials between wet and dry samples. It was discovered that the induced electric potential was dependent on the strain rate, and the relaxation time of the induced potential was much higher in hydrated samples. To justify this inconsistency, different hypotheses were proposed. One such hypothesis was the p–n junction characteristic of apatite-collagen,[16] which suggested that bone was piezoelectric in one direction and piezo-resistive in another. However, the hypothesis that drew more recognition proposed that while piezoelectricity was accountable for stress-induced potentials in dry bone, streaming potential was the mechanism responsible for wet bone.[17,18] The potential streaming theory suggested that the stress-generated potential in wet bone was due to the flow of ion-containing interstitial fluid through bone as a result of pressure. The streaming potential theory was popularized as Pollack et al.[19] reported that the conductivity of the saturating fluid was significantly influential on the amplitude and polarity of the generated potentials, and the longer relaxation time was due to the viscosity of the fluid in the bone. A new theory on mechanosensation in bone suggests that the applied stress on the bone is translated into biochemical signals by the interstitial fluid in the canaliculi lacunae space and supplying bone cells with nutrients as well as conveying the shear stress to cells.[20] In spite of these new theories, the debate exists whether one can entirely exclude piezoelectricity from mechanosensation. As illustrated in Figure 6.3, Ahn and Grodzinsky[21] suggested that piezoelectricity will increase the surface charge density of collagen fibres and will consequently increase the zeta potential, which intensifies the streaming potential as the following equation:

$$V = \frac{\zeta P k}{4\pi\sigma\eta}$$

where ζ, P, κ, σ, and η are the zeta potential, the pressure on the bone, the dielectric permittivity, conductivity, and viscosity of the interstitial fluid, respectively.

6.5 CELLULAR RESPONSE TO ELECTRIC STIMULATION

In addition to the piezoelectricity and streaming potentials in bone and other fibrous tissues, electric fields up to 500 mV/mm have been reported to be generated in living tissues.[22] The transport of ionic species and macromolecules induced by these endogenous electric fields play crucial roles in processes such as embryonic development,[23] wound healing,[24] and neuronal regeneration.[25] The difference in intracellular and extracellular ionic concentrations results in transmembrane potential of −10 to −90 mV in different types of cells. Shifts in the transmembrane potential will alter cellular proliferation and differentiation,[26] and an increase in transmembrane potential in the neurons could trigger self-propagation of an action potential along the axon.[27] There have been numerous investigations to mimic biological piezoelectricity and endogenous electric fields and to

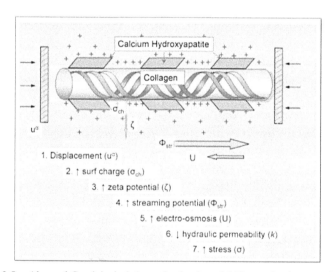

FIGURE 6.3 Ahn and Grodzinsky's hypothesized model illustrating how applied stress on bone results in greater surface charges, which consequently increases zeta potential, streaming potential, electroosmosis, and dynamic stiffness as well as decreases hydraulic permeability in the open-circuit condition. (Reproduced with permission from Reference [21]. © 2009 Elsevier.)

manipulate transmembrane potentials by external electrical stimulation which will enhance cellular growth and differentiation. These studies are due to the realization of the vital role of electricity in tissues and cells. The studies in the area of neural regeneration have resulted in repairing peripheral nerve injuries through improved neuronal differentiation and directional outgrowth of neuritis mm in embryonic chick dorsal root ganglions toward the cathode,[28] facilitated by direct electric fields as low as 70 mV/mm. Applying a direct electric field of 250 mV/mm or higher on Xenopus neurons resulted in more neurite-bearing cells generated with longer neurites directed toward the cathode and contracted neurites on the anode side (Fig. 6.4).[29]

Promising results are not just limited to neurons, as a study on the effect of electrical stimulation on bone formation showed that implanting insulated batteries in the medullary canal of canine femora caused the formation of endosteum near the cathode within a 14–21-day period.[30] Implanting poled sintered hydroxyapatite disks in canine cortical bone resulted in the filling of a 0.2-mm gap between the negatively charged hydroxyapatite surface and the cortical bone in 14 days even in the absence of electrical stimulation.[31] Electrical stimulation can be applied to the substrate or the medium. While DC electrical stimulation through the medium will align the Schwann cells perpendicular to the direction of the electric field, AC

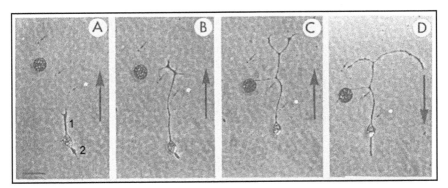

FIGURE 6.4 Direction and outgrowth of the neurites of a bipolar neuron of stimulation by a 500-mV/mm electric field. The black arrow shows the direction of the electric field (A); after 2 h of exposure to the electric field, neurite 1 has noticeably grown toward the cathode (B); 4 h of stimulation in the same direction resulted in further extension of neurite 1 as well as its branching, while neurite 2 has almost diminished (C); and 2 h after changing the direction of the electric field, the tips of neurite 1 curved and neurite 2 grew toward the new cathode (D).

electric field through the media will change the cellular morphology from bipolar spindle shape to flat and spread with more processes.[32] There is also an alternative method of applying currents through conductive polymers such as polypyrrole (PPy) and polyaniline (PANi) for electrical stimulation. Even without electric current, cells have shown proliferation and extension on substrates consisting of these conductive polymers in comparison with control samples.[33–35] Figure 6.5 shows how applying a direct electric potential of 100 mV across PPy film doubles the length of neurites of PC12 cells in comparison with the nonstimulated films.[36]

Applying a direct current of 10 μA through PPy films also triggered higher adsorption of fibronectin (FN) onto the surface, specifically in highly concentrated FN solutions and at early stages of exposure. The neurites in PC12 cells seeded on these films elongated 50% than on unstimulated films.[37] For incorporating three-dimensional (3D) features of fibrous scaffolds and electrical stimulation, Schmidt et al.[57] applied a direct electric field of 100 mV/mm through an electrospun 3D scaffold coated with PPy, to observe a significant improvement in the number of neurite-bearing PC12 cells and their neurite lengths. Another method to construct 3D conductive scaffolds is electrospinning the blends of nonconductive and conductive polymers. The direct electrical stimulation of electrospun composites of 30 wt% PANi and poly(L-lactide-*co*-ε-caprolactone) copolymer with 20 mA current increased the proliferation of NIH-3T3 fibroblasts. Neural stem cells seeded on nanofibrous scaffolds which are comprised of poly(ε-caprolactone) (PCL), gelatin, and 15 wt% PANi, also showed

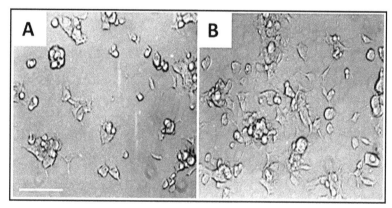

FIGURE 6.5   Differentiation of PC12 cells on PPy films without (A) and with (B) the application of 100 mV across the film. The scale bar is 100 μm.

significantly higher proliferation as well as growing elongated neurites in response to 1 h of direct electrical stimulation with 100 mV/mm.[38] Electrospun scaffolds composed of PANi and poly-L-lactic acid also increased the neurite lengths of C17.2 rat stem neural stem cells when stimulated with the same voltage. In most of these direct electrical stimulation studies, there is an optimum electric field in the range of the endogenous electric fields or transmembrane potential, and over these values, there is either no significant improvement or the field is harmful to the cells.

Alternating electric fields were found to cause morphological changes and the significantly increased number of processes in Schwann cells but did not lead to any directional outgrowth. Since endogenous electric fields and transmembrane potentials are direct, many neuronal studies use direct electrical stimulation. For bones, the periodic nature of the stress applied to bone has inspired researchers to focus on alternating electric fields to enhance osteoblast proliferation and activity. Also, to avoid electrode implantation and consequently the electrolytic byproducts, noninvasive stimulation for bone fracture healing drew attention. These noninvasive bone growth stimulators are approved by the U.S. Food and Drug Administration (FDA)[39,40] and are currently marketed for healing fractures and nonunions. For research, techniques implemented for electrical stimulation in bones vary from capacitively coupled stimulation[41–43] to applying electromagnetic waves using Helmholtz[44–46] and solenoid coil.[47,48] Capacitively coupled electrical stimulation is known to increase the proliferation and matrix mineralization of osteoblast-like cells. The mechanism in which the electrical stimulation induces cellular migration alters proliferation and differentiation is far from being understood, though it is speculated that the electric field effect is either directly with intracellular components such as ions, growth factors, and receptors or indirectly with agglomeration or conformational change of extracellular ions and proteins.[48] Free calcium cations (Ca^{2+}) are considered to be a major factor in both direct and indirect mechanisms of electrical stimulation. It is thought that electric fields will redistribute Ca^{2+} in the extracellular matrix or on the substrate,[51] inducing regeneration. Also, the intracellular Ca^{2+} concentrations are reported to increase due to electrical stimulation.[49] Figure 6.6 illustrates an adaptation of the galvanotaxis of cells as reviewed by Mycielska et al.

Direct electric field depolarizes the cathode side of the cell and hyperpolarizes the anode side. This leads to the diffusion of extracellular Ca^{2+} through the anodal side into the cell. Increase in the Ca^{2+} causes actin

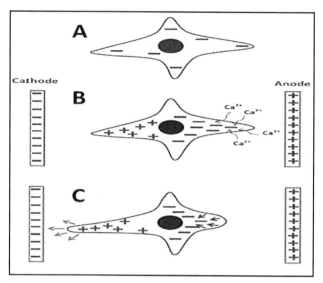

FIGURE 6.6 A cell with insignificant voltage-gated Ca^{2+} channels at resting transmembrane potential (A); application of a direct electric field redistributes the intracellular charges resulting in the depolarization and hyperpolarization of the cathodal and anodal sides of the cell, respectively. Extracellular Ca^{2+} is consequently diffused through the anodal side (B). The increase in the intracellular Ca^{2+} on the anodal side depolymerizes actin. The result is the contraction of the anodal side and protrusion of the cathodal side (C). (Adapted from Reference [22].)

depolymerization and will consequently cause contraction on the anode side, which will push the cell forward to make the cathode side of the cell protrude. This could explain the phenomenon observed by Patel and Poo. The outgrowth of the neurites on the cathodal side of the cell and the diminishing of the neurites on the anodal side are shown in Figure 6.4. Patel and Poo found that neither blocking Na^+ channels nor nullifying intercellular Ca^{2+} gradient stopped directional neurite outgrowth in cells exposed to the direct electric field.[29] However, there was a larger distribution of *concanavalin. A* receptors on the cathode side of the cell than on the anode side showing that the effect of electrical stimulation on the directional growth of neurons could be direct by preferential migration of membrane receptors. Schmidt et al. reported that electrical stimulation could result in favorable conformational changes in FN, facilitating the adsorption of more proteins onto the biomaterial. Since some extracellular matrix proteins play critical roles in cellular attachment, adhered proteins on the surface play an important role in cellular adhesion and outgrowth.

The effect of electrical stimulation on enhanced bone formation was initially based on the hypothesis that the piezoelectricity of bone would generate electric fields that could aggregate charged ions and macromolecules in the interstitial bone fluid, which would result in enhanced osteoblast activity. Direct electric fields will mobilize Ca^{2+} and Mg^{2+} toward the cathode side or negatively charged surface and will cause apatite formation, which becomes a scaffold for bone formation by osteoblasts.[50] Increased levels of gene expression for bone morphogenic proteins (BMP-2 and -4) as a result of electromagnetic stimulation were noted by Bodamyali et al.[46] Zhuang et al. also found that capacitively coupled electrical stimulation will increase the TGF-β1 gene expression as well as the proliferation of osteoblasts. TGF-β1 expression was modulated by the calcium-calmodulin pathway.

6.6 PIEZOELECTRIC MATERIALS IN BIOMEDICAL APPLICATIONS

6.6.1 ZNO NANOWIRES ARRAYS AS SUBSTRATE FOR CELL PROLIFERATION STUDIES

Zinc oxide (ZnO) nanostructures are widely used in biomedical applications, but rather few results found in the area of cell proliferation studies. The exact reason behind the focused work on nanostructured zinc oxide is because they exhibit unique structure and size-dependent electrical, optical, and mechanical properties.[51,52] The researchers mainly concentrated on the wurtzite structure of ZnO nanostructure where Zn cations and O anions are arranged with tetrahedral coordination; due to this, specialty of the surface interaction of polar charges gives rise to a wide variety of nanostructures (e.g., nanobelts, nanosprings, nanorings, and nanohelices).[53] Spearheading the advancement and portrayal of ZnO nanowires nanogenerators, Dr Wang and collaborators initially exhibited that these nanostructured wires can be driven by ultrasonic waves to deliver direct-current yield. Nanogenerators were created with vertically adjusted ZnO nanowires arrays, set underneath proper metal terminals/electrodes. Ultrasonic waves drive the electrode all over to bend as well as vibrate the nanowires and the last change over mechanical energy into electricity because of the piezoelectric semiconducting coupling.[54–57] The piezoelectricity responses of nanomaterials are considerably higher than the conventional materials due to the huge surface to volume proportion, and atoms in nanoscale ZnO structures are able to assume different positions, due to the free boundary, which may further enhance the piezoelectric effect. Different shapes, moreover, will have a different piezoelectric response when mechanically deformed, and in particular, elongated nanostructures may be more easily deformed to produce a piezoelectric effect.[58,59]

The study of proliferation and differentiation of PC12 neuronal-like cells and H9c2 myoblasts over ZnO nanowire arrays was very interesting. The ZnO nanowires are incubated with collagen solution to imitate the typical chemical composition of the extracellular environment. The PC12 is derived from a transplantable rat pheochromocytoma and H9c2 cells, a subclone of the original colonel cell line derived from embryonic BD1X rat heart tissue. In the proliferation study performed on collagen-coated ZnO nanowire arrays, the viability was investigated. The incubation with collagen does not affect any features of the nanoarrays as shown in Figure 6.7. SEM images of proliferation studies of ZnO nanowires are shown in Figure 6.8.

Figure 6.9 shows live/dead staining and nuclei are counterstaining of PC12 cells over ZnO arrays. The nanostructured substrates supported high cell viability: most of the cells (>95%) were stained in green and displayed the typical round morphology at 72 h from seeding over standard substrates.

As from the results obtained, it is clear that both neuronal-like PC12 cells and myoblastic H9c2 cells showed high viability and normal proliferation capability on ZnO nanowire arrays. ZnO nanowires are about an

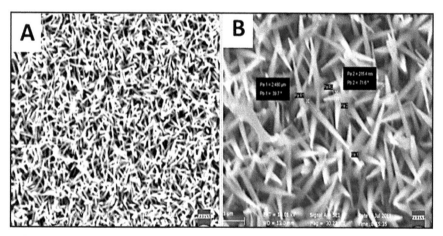

FIGURE 6.7 ZnO nanowire arrays used as cell substrates: low (A) and high (B) magnification SEM images.

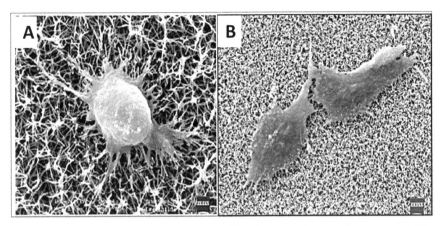

FIGURE 6.8   SEM imaging of (A) PC12 proliferation on ZnO nanowire arrays and (B) H9c2 proliferation on ZnO nanowire arrays.

order of magnitude smaller in diameter than mammalian cells, and their size is comparable to that one of various intracellular components. Along with the size, ZnO nanowire array geometry could readily be accessible to the interiors of living cells, thus facilitating operations of stimulation and sensing.

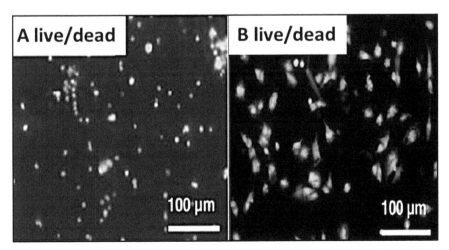

FIGURE 6.9   Live/dead and nucleus staining: (A) PC12 proliferation on ZnO nanowire arrays and (B) H9c2 proliferation on ZnO nanowire arrays.

6.6.2   PIEZOELECTRIC SURGERY

Piezoelectric surgery was defined as using piezoelectric vibration in the application of cutting bone tissue. The basic principle is the separation of positive and negative electrical charges in molecules that cause dipole moment; when mechanical stress is applied, the dipole moments get reoriented and cause a change in the surface density resulting in a voltage.[60] This is demonstrated in Figure 6.10; when the electric field is applied across a piezoelectric medium, there will be a change in dipole moment as well as the dimension of the medium. Piezoelectric surgery was introduced in 1975, in which the use of ultrasonic vibrations is used for a cutting-bone tissue. This technique is minimally invasive that only cuts the targeted tissues. For example, Hoigne et al.[61] used 25–29 kHz range frequency to cutting bones in the neurovascular hand tissue and other soft tissues, thereby maintaining a blood-free operating area, better visibility, and greater precision.[62] Figure 6.11 displays some naturally occurring piezoelectric materials, which produce electric effect according to the mechanical deformations.

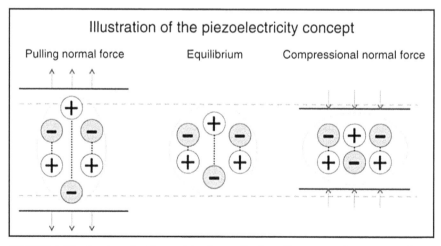

FIGURE 6.10   In piezoelectric crystals, the application of a compressive or expansive force results in a change corresponding to the distribution of dipole moments. This causes a change in the surface charge, which is proportional to the applied force. This phenomenon is called piezoelectricity. Similarly, when a voltage pulse is applied to the surface, a displacement occurs causing an acoustic wave to be generated. This is the inverse piezoelectric effect.

6.6.3 PERIODONTOLOGY AND ENDODONTIC SURGERY

In dentistry, ultrasonic surgery became established in periodontology [63,64,] and endodontics[65] after initial reports by Catuna in 1953 on the use of high-frequency sound waves to cut hard dental tissue.[66] Ultrasonic oscillations can also be used to scale subgingival plaque and to remove root canal fillings and fractured instruments from root canals.[67]

6.6.4 NEUROSURGERY

Operations on the neurocranium may injure the dura and cause the development of a fistula of cerebrospinal fluid. Schaller et al. reported the successful use of piezoelectric surgery in the cranial base and spine in children. He showed that the technique spared soft nerve tissue, was avoided coagulative necrosis, improved the visualization of the surgical

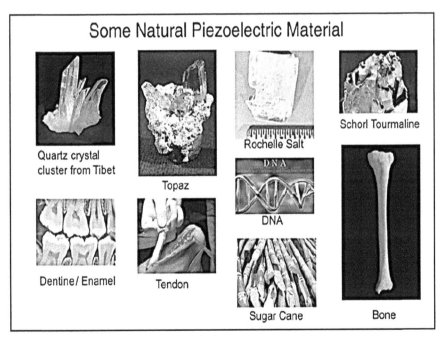

FIGURE 6.11 Examples of materials that produce electricity under pressure. The impacts of mechanical instruments on the structure of bone and the practicality of cells are vital in regenerative surgery. Generally, high temperatures, connected notwithstanding for a brief timeframe, are unsafe to cells and cause necrosis of tissues.

field, and resulted in a more precise cut. Traditional oscillating saws that use macro vibrations do not give the same control of cutting depth at the sides as they do in the center. Craniotomy, orbitotomy, and posterior spinal laminotomy in selected cases were all successful with uneventful postoperative recovery.[68]

6.6.5   CELLULOSE NANOFIBRIL FILM AS A PIEZOELECTRIC SENSOR MATERIAL

Nanocellulose is a renewable biobased nanomaterial with potential application in different fields. The nano dimension and porous structure make them more suitable for fabrication of membranes and films. The piezoelectricity of wood, that is, the change of electrical polarization in a material in response to mechanical stress has been known for decades.[69,70] The piezoelectric effect is exponentially increased when the building block is nano cellulose crystals.[71,72] Nanocellulose is extracted from wood-cellulose fibers obtained from birch wood by sulfate digestion, and subsequent bleaching was suspended in water and then processed through a Masuko grinder using three consecutive passes and further homogenized using six passes at 2000 bar pressure by using a microfluidizer equipped with two chambers, the composite extracted from this process contains ordered crystalline (cellulose I), less ordered cellulose, residual heteropolysaccharides, and lignin. Even in a few, cellulose nanofibers are enough to form a hydrogel.[73] The cellulose nanofibrils films were prepared by pressure filtering and followed by pressing and dried in a hot press (Figure 6.12). Cross-sectional image of CNF is shown in the Figure 6.13.

Electrode substrates for the CNF sensor assembly were fabricated with polyethene terephthalate, the fabrication of CNF sensor is shown in Figure 6.12b, and the sensor is prepared and characterized. Image-based analysis, scanning electron microscopy, and ferroelectric and electromechanical characterizations were carried out for the CNF films. The obtained results suggest that nanocellulose is a promising new piezoelectric material for sensors, actuators, and energy generators, which has obvious applications in fields such as sensors, electronics, and biomedical diagnostics.

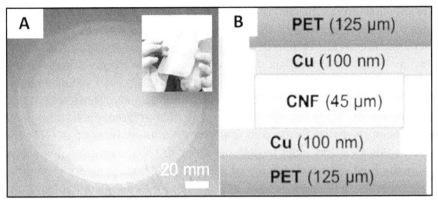

FIGURE 6.12 (A) Photographs of a fabricated self-standing CNF film and (B) a schematic side view of the assembled sensor.

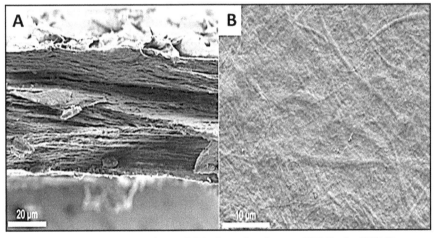

FIGURE 6.13 (A) Cross-section and (B) plane view images (SEM) of a CNF film.

6.6.6 BORON NITRIDE NANOTUBES: AN INNOVATIVE TOOL FOR NANOMEDICINE

Boron nitride nanotube (BNNT) is a critical intrigue to established researchers in view of their imperative properties perfect for basic and electronic applications. They are structurally similar to the carbon nanotube. Alternating B and N atoms entirely substitute for C atoms in a graphite-like sheet with almost no change in atomic spacing. However,

despite this similarity, carbon and BNNTs exhibit many different properties. Our investigations of the cytocompatibility of BNNTs with human
neuroblastoma cells exhibit that their cell uptake does not affect suitability,
metabolism, and replication of this cell line. While having a high Young
modulus like CNTs, BNNTs have thermal stability and have magnificent piezoelectric properties, better than those of piezoelectric polymers.
Whereas many applications of CNTs in the field of biomedical technology
have been proposed in the past few years, the translation of BNNTs for
use in this field has been largely unexplored. One reason for this stems
from the high chemical stability of BNNTs, which accounts for their poor
dispersibility in the aqueous solvents required for biological applications.
This problem has been recently resolved using a technique of noncovalent
polymeric wrapping that allows aqueous dispersion of BNNTs and hence
biocompatibility, enabling studies on their interaction with and effects
on living cells. Moreover, fluorescent labelling of BNNTs with quantum
dots enables us to track the cellular uptake and internalization of polymer-
wrapped BNNTs via the endocytosis pathway.[74]

6.6.7   PIEZO CERAMICS MATERIALS FOR TISSUE ENGINEERING

Piezoelectric materials used as tissue engineering scaffolds enable electrical stimulation without the need for electrodes, an external source of electricity or implanting batteries, and it vanishes the chance of accumulation
of electrolysis products. An electric pulse can be produced by piezoelectric materials as a result of transient deformations, which can be imposed
by attachment and migration of cells or body movements. A biocompatible and piezoelectric resonant coefficient material should be used in
these applications. PZT or lead zirconate titanate is the most used material
(PZT)[75] owing to its notable piezoelectric and electromechanical coupling
coefficients. PZT has been used to build a wirelessly powered nerve-cuff,
in which the implanted piezoceramic was deformed by ultrasound radi-
ated through the skin and consequently generated electric pulses that could
cause muscle-twitch in a rat's hind limb.[76,77] Nevertheless, 60 wt% of PZT
is lead, which even in low doses causes serious health problems such as
neurotoxicity, pregnancy complications attention deficit hyperactivity,
and slow growth rate in children. Several another material used to replace
PZT with lead-free piezoceramics to avoid toxicity. Zinc oxide (ZnO), BT,

potassium–sodium–niobate (KNN), lithium–sodium–potassium–niobate (LNKN), and BNNTs have substantial piezoelectric coefficients which were used as a lead-free piezoceramic material for biological applications. Major concerns about using piezoceramics in tissue engineering applications are the cytotoxicity of these materials. Most piezoelectric ceramics, such as PZT,[78] ZnO,[79] BT,[80] KNN, and LNKN,[81] exhibit ion dissolution in biological fluids. While some of the released ions, for instance, Pb^{2+} could be toxic, others could be relatively safe or even favorable at low doses[82] and cytotoxic at higher concentrations.[83] Therefore, in some studies involving piezoceramics, the piezoelectric particles are embedded in polymer[84,85] or ceramic matrix composites to control ion dissolution. Electrospun fibrous composite scaffolds made of ZnO particles and polyurethane (PU) resulted in improved attachment and proliferation of mouse fibroblasts on the composite scaffolds in comparison to pure PU scaffolds.[85]

6.7 CONCLUSION

In living tissues, electricity exists in several forms such as stress-generated potentials, endogenous electric fields, and transmembrane potentials. Several studies have been carried out to get an idea of these electric fields in the cell cycle. Some of these searches successfully completed, which resulted in medical treatments and clinical needs. Piezoelectric materials have encouraged the repairing of nerve injuries, bone formation, and wound healing, which can be attributed to charge generation as a consequence of body movement and physiological stress on the piezoelectric material. These piezoelectric scaffolds such as nanocellulose and chitins can be used by in vitro method, which led to neurite extension, enhanced adhesion, differentiation, and faster cellular migration. Also, these piezoelectric scaffolds have exhibited favorable protein adsorption, cellular attachment, and proliferation; these can be possible due to stable polarization effect and surface charges of piezoelectric materials or transient deformation caused by the contraction and protrusion of the attached cells. Piezoelectrical materials are also used in neurosurgery, orthodentistry, tissue engineering, and another wide area of biomedical applications. So, it is concluded that piezoelectric materials have potential applications of next-generation biomedical and electrical fields.

KEYWORDS

- piezoelectricity
- streaming potential
- cellular response
- cell proliferation
- piezoelectric surgery

REFERENCES

1. Curie, P.; Curie, J. Développement, par pression, de l'électricité polaire dans les cristaux hémièdres à faces inclinées. Comptes rendus de l'Académie des sciences **1880**, *91,* 294–295.
2. Martin A. J. P. Tribo-Electricity in Wool and Hair. *Proc. Phys. Soc.* **1941**, *53,* 186.
3. Marshall, C. R.; Gillespie, J. The Keratin Proteins of Wool, Horn and Hoof from Sheep. *Aust. J. Biol. Sci.* **1977**, *30,* 389–400.
4. Wu, D. D.; Irwin, D. M.; Zhang, Y. P. Molecular Evolution of the Keratin-Associated Protein Gene Family in Mammals, Role in the Evolution of Mammalian Hair. *BMC Evol. Biol.* **2008**, *8,* 1471–2148.
5. Menefee, E. Thermocurrent from Alpha-Helix Disordering in Keratin. In *Electrets, Charge Storage, and Transport in Dielectrics: Dielectrics and Insulation Division*; Perlman, M. M., Ed.; Electrochemical Society, Princeton 1973; pp 661.
6. Yasuda, I. On the Piezoelectric Activity of Bone. *J. Jpn. Orthop. Surg. Soc.* **1954**, *28,* 267–271.
7. Ramachandran, G. N.; Kartha, G. Structure of Collagen. *Nature* **1955**, *176,* 593–595.
8. Momeni, K.; Asthana, A.; Prasad, A.; Yap, Y.; Shahbazian-Yassar, R. Structural Inhomogeneity and Piezoelectric Enhancement in ZnO Nanobelts. *Appl. Phys. A* **2012**, *109,* 95–100.
9. Shamos, M. H.; Lavine L. S. Piezoelectricity as a Fundamental Property of Biological Tissues. *Nature* **1967**, *213,* 267–269.
10. Minary-Jolandan, M.; Yu, M. F. Nanoscale Characterization of Isolated Individual Type I Collagen Fibrils: Polarization and Piezoelectricity. *Nanotechnology* **2009**, *20,* 0957–4484.
11. Minary-Jolandan, M.; Yu, M.-F. Uncovering Nanoscale Electromechanical Heterogeneity in the Subfibrillar Structure of Collagen Fibrils Responsible for the Piezoelectricity of Bone. *ACS Nano* **2009**, *3,* 1859–1863.
12. Wolff, J. Das gesetz der transformation der knochen. DMW-Deutsch Medizinische Wochenschrift **1892**, *19,* 1222–1224.
13. Bassett, C. A. L. Biologic Significance of Piezoelectricity. *Calcified Tissue Int.* **1967**, *1,* 252–272.
14. Cochran, G. V. B. *Electromechanical Characteristics of Moist Bone*; M. Sc. Thesis. Columbia College of Physicians and Surgeons, New York, N.Y. 1966.

15. Anderson, J. C.; Eriksson, C. Electrical Properties of Wet Collagen. *Nature* **1968,** *218*, 166–168.
16. Becker, R. O.; Brown, F. M. Photoelectric Effects in Human Bone. *Nature* **1965,** *206*, 1325–1328.
17. Johnson, M. W.; Chakkalakal, D. A.; Harper, R. A.; Katz, J. L. Comparison of the Electromechanical Effects in Wet and Dry Bone. *J. Biomech.* **1980,** *13*, 437–442.
18. Gross, D.; Williams, W. S. Streaming Potential and the Electromechanical Response of Physiologically-Moist Bone. *J. Biomech.* **1982,** *15*, 277–295.
19. Pienkowski, D.; Pollack, S. R. The Origin of Stress-Generated Potentials in Fluid-Saturated Bone. *J. Orthop. Res.* **1983,** *1*, 30–41.
20. Riddle, R. C.; Donahue, H. J. From Streaming-Potentials to Shear Stress: 25 Years of Bone Cell Mechanotransduction. *J. Orthop. Res.* **2009,** *27*, 143–149.
21. Ahn, A. C.; Grodzinsky, A. J. The relevance of Collagen Piezoelectricity to "Wolff's Law": A Critical Review. Med. Eng. Phys. **2009,** *31*, 733–741.
22. Mycielska, M. E.; Djamgoz, M. B. Cellular Mechanisms of Direct-Current Electric Field Effects: Galvanotaxis and Metastatic Disease. *J. Cell Sci.* **2004,** *117*, 1631–1639.
23. Nuccitelli, R. Endogenous Electric Fields in Embryos During Development, Regeneration and Wound Healing. *Radiat. Prot. Dosim.* **2003,** *106*, 375–383.
24. Nuccitelli, R. A Role for Endogenous Electric Fields in Wound Healing. *Curr. Topics Develop. Biol.* **2003,** *58*, 1–26.
25. McCaig, C. D.; Rajnicek, A. M.; Song, B, Zhao, M. Has Electrical Growth Cone Guidance Found Its Potential? *Trends Neurosci.* **2002,** *25*, 354–359.
26. Sundelacruz, S.; Levin, M.; Kaplan, D. L. Role of Membrane Potential in the Regulation of Cell Proliferation and Differentiation. *Stem. Cell Rev.* **2009,** *5*, 231–246.
27. McCormick, D. A. Chapter 12—Membrane Potential and Action Potential. In *From Molecules to Networks, 3rd ed.*; Byrne, J. H.; Heidelberger, R.; Waxham, M. N., Eds.; Academic Press: Boston, 2014; pp 351–376.
28. Jaffe, L. F.; Poo, M.-M. Neurites Grow Faster Towards the Cathode than the Anode in a Steady Field. *J. Exp. Zool.* **1979,** *209*, 115–127.
29. Patel, N.; Poo, M. M. Orientation of Neurite Growth by Extracellular Electric Fields. *J. Neurosci.* **1982,** *2*, 483–496.
30. Bassett, C. A. L.; Pawluk, R. J.; Becker, R. O. Effects of Electric Currents on Bone In Vivo. *Nature* **1964,** *204*, 652–654.
31. Kobayashi, T.; Nakamura, S.; Yamashita, K. Enhanced Osteobonding by Negative Surface Charges of Electrically Polarized Hydroxyapatite. *J. Biomed. Mater. Res.* **2001,** *57*, 477–484.
32. Nguyen, H. T.; Wei, C.; Chow, J. K.; Nguy, L.; Nguyen, H. K.; Schmidt, C. E. Electric Field Stimulation Through a Substrate Influences Schwann Cell and Extracellular Matrix Structure. *J. Neural. Eng.* **2013,** *10*, 1741–2560.
33. Schmidt, C. E.; Shastri, V.; Furnish, E. J.; Langer, R. In *Electrical Stimulation of Neurite Outgrowth and Nerve Regeneration.* Biomedical Engineering Conference, 1998, Proceedings of the 17th Southern: IEEE; 1998. p 117.
34. Prabhakaran, M. P.; Ghasemi-Mobarakeh, L.; Jin, G.; Ramakrishna, S. Electrospun Conducting Polymer Nanofibers and Electrical Stimulation of Nerve Stem Cells. *J. Biosci. Bioeng.* **2011,** *112*, 501.

35. Jeong, S. I.; Jun, I. D.; Choi, M. J.; Nho, Y. C.; Lee, Y. M.; Shin, H. Development of Electroactive and Elastic Nanofibers That Contain Polyaniline and Poly(l-lactide-*co*-epsilon-caprolactone) for the Control of Cell Adhesion. *Macromol. Biosci.* **2008,** *8,* 627–637.

36. Schmidt, C. E.; Shastri, V. R.; Vacanti, J. P.; Langer, R. Stimulation of Neurite Outgrowth Using an Electrically Conducting Polymer. *Proc Natl. Acad. Sci.* **1997,** *94,* 8948–8953.

37. Kotwal, A.; Schmidt, C. E. Electrical Stimulation Alters Protein Adsorption and Nerve Cell Interactions with Electrically Conducting Biomaterials. *Biomaterials* **2001,** *22,* 1055–1064.

38. Ghasemi-Mobarakeh, L.; Prabhakaran, M. P.; Morshed, M.; Nasr-Esfahani, M. H.; Ramakrishna, S. Electrical Stimulation of Nerve Cells Using Conductive Nanofibrous Scaffolds for Nerve Tissue Engineering. *Tissue Eng. A* **2009,** *15,* 3605–3619.

39. FDA. *Premarket Approval*; 1986.

40. FDA. *Premarket Approval*; 1999.

41. Zhuang, H.; Wang, W.; Seldes, R. M.; Tahernia, A. D.; Fan, H. Brighton, C. T. Electrical Stimulation Induces the Level of TGF-1 mRNA in Osteoblastic Cells by a Mechanism Involving Calcium/Calmodulin Pathway. *Biochem. Biophys. Res. Commun.* **1997,** *237,* 225–229.

42. Hartig, M.; Joos, U.; Wiesmann, H.-P. Capacitively Coupled Electric Fields Accelerate Proliferation of Osteoblast-Like Primary Cells and Increase Bone Extracellular Matrix Formation In Vitro. *Eur. Biophys. J.* **2000,** *29,* 499–506.

43. Wiesmann, H. –P.; Hartig, M.; Stratmann, U.; Meyer, U.; Joos, U. Electrical Stimulation Influences Mineral Formation of Osteoblast-Like Cells In Vitro. *Biochim. Biophys. Acta (BBA)—Mol. Cell Res.* **2001,** *1538,* 28–37.

44. Cane, V.; Botti, P.; Soana, S. Pulsed Magnetic Fields Improve Osteoblast Activity During the Repair of an Experimental Osseous Defect. *J. Orthop. Res.* **1993,** *11,* 664–670.

45. Lohmann, C. H.; Schwartz, Z.; Liu, Y.; Guerkov, H.; Dean, D. D.; Simon, B., et al. Pulsed Electromagnetic Field Stimulation of MG63 Osteoblast-Like Cells Affects Differentiation and Local Factor Production. *J. Orthop. Res.* **2000,** *18,* 637–646.

46. Bodamyali, T.; Bhatt, B.; Hughes, F. J.; Winrow, V. R.; Kanczler, J. M.; Simon, B., et al. Pulsed Electromagnetic Fields Simultaneously Induce Osteogenesis and Upregulate Transcription of Bone Morphogenetic Proteins 2 and 4 in Rat Osteoblasts In Vitro. *Biochem. Biophys. Res. Commun.* **1998,** *250,* 458–461.

47. Takano-Yamamoto, T.; Kawakami, M.; Sakuda, M. Effect of a Pulsing Electromagnetic Field on Demineralized Bone-Matrix-Induced Bone Formation in a Bony Defect in the Premaxilla of Rats. *J. Dental Res.* **1992,** *71,* 1920–1925.

48. Chang, W. H.-S.; Chen, L.-T.; Sun, J.-S.; Lin, F.-H. Effect of Pulse-Burst Electromagnetic Field Stimulation on Osteoblast Cell Activities. *Bioelectromagnetics* **2004,** *25,* 457–465.

49. Onuma, E. K.; Hui, S. W. Electric Field-Directed Cell Shape Changes, Displacement, and Cytoskeletal Reorganization Are Calcium-Dependent. *J. Cell Biol.* **1988,** *106,* 2067–2075.

50. Yamashita, K.; Oikawa, N.; Umegaki, T. Acceleration and Deceleration of Bone-Like Crystal Growth on Ceramic Hydroxyapatite by Electric Poling. *Chem. Mater.* **1996,** *8,* 2697–2700.

51. Fan, Z.; Lu, J. G. *J. Nanosci. Nanotechnol.* **2005,** *5,* 1561–1573.
52. Djurišić, A. B.; Ng, A. M. C.; Chen, X. Y. *Prog. Quantum Electron.* **2010,** *34,* 191–259.
53. Wang, Z. L. *J. Phys. Condens. Matter* **2004,** *16,* 829–858.
54. Wang, Z. L. *Adv. Funct. Mater.* **2008,** *18,* 3553–3567.
55. Lu, M. P.; Song, J.; Lu, M. Y.; Chen, M. T.; Gao, Y.; Chen, L. J.; Wang, Z. L. *Nano Lett.* **2009,** *9,* 1223–1227.
56. Moon, R. J.; Martini, A.; Nairn, J.; Simonsen, J.; Youngblood, J. Cellulose Nanomaterials Review: Structure, Properties and Nanocomposites. *Chem. Soc. Rev.* **2011,** *40*(7), 3941.
57. Lee, J. Y.; Bashur, C. A.; Goldstein, A. S.; Schmidt, C. E. Polypyrrole-Coated Electrospun PLGA Nanofibers For Neural Tissue Applications. *Biomaterials* **2009,** *30,* 4325–4335.
58. Ciofani, G.; Danti, S.; D'Alessandro, D.; Ricotti, L.; Moscato, S.; Bertoni, G.; Falqui, A.; Berrettini, S.; Petrini, M.; Mattoli, V.; Menciassi, A. *ACS Nano.* **2010,** *4,* 6267–6277.
59. Xiang, H. J.; Yang, J.; Hou, J. G.; Zhu, Q. *Appl. Phys. Lett.* **2006,** *89,* 223111–223113.
60. Trolier-McKinstry, S. Crystal Chemistry of Piezoelectric Materials, Chapter 3 In *Piezoelectric and Acoustic Materials for Transducer Applications*; Safari, A., Akdogan, E. K., Eds.; Springer: New York, 2008.
61. Hoigne, D. J.; Stubinger, S.; Von Kaenel, O.; Shamdasani, S.; Hasenboehler. P. Piezoelectric Osteotomy in Hand Surgery: First Experiences with a New Technique. *BMC Musculoskelet Disord.* **2006,** *7,* 36.
62. Labanca, M.; Azzola, F.; Vinci, R.; Rodella, L. F. Piezoelectric Surgery: Twenty Years of Use. *Br. J. Oral Maxillofac. Surg.* **2008,** *46,* 265–269.
63. Lea, S. C.; Landini, G.; Walmsley, A. D. Ultrasonic Scaler Tip Performance Under Various Load Conditions. *J. Clin. Periodontol.* **2003,** *30,* 876–881.
64. Flemmig, T. F.; Petersilka, G. J.; Mehl, A.; Hickel, R.; Klaiber, B. The Effect of Working Parameters on Root Substance Removal Using a Piezoelectric Ultrasonic Scaler in Vitro. *J. Clin. Periodontol.* **1998,** *25,* 158–163.
65. Walmsley, A. D.; Laird, W. R.; Lumley, P. J. Ultrasound in Dentistry. Part 2—Periodontology and Endodontics. *J. Dent.* **1992,** *20,* 11–17.
66. Catuna, M. C. Sonic Surgery. *Ann. Dent.* **1953,** *12,* 100.
67. Ward, J. R.; Parashos, P.; Messer, H. H. Evaluation of an Ultrasonic Technique to Remove Fractured Rotary Nickel-Titanium Endodontic Instruments from Root Canals: Clinical Cases. *J. Endod.* **2003,** *29,* 764–767.
68. Schlee, M. Ultraschallgestützte Chirurgie-grundlagen und Möglichkeiten. *Z Zahnärztl. Impl.* **2005,** 21 (1), 48–59.
69. Fukada, E. Piezoelectricity of Wood. *J. Phys. Soc. Jpn.* **1955,** *10*(2), 149–154.
70. Fukada, E. Piezoelectricity as a Fundamental Property of Wood. *Wood Sci. Technol.* **1968,** *2*(4), 299–307.
71. Csoka, L.; Hoeger, I. C.; Rojas, O. J.; Peszlen, I.; Pawlak, J. J.; Peralta, P. N. Piezoelectric Effect of Cellulose Nanocrystals Thin Films. *ACS Macro Lett.* **2012,** *1* (7), 867–870.
72. Cheng, H. Flexo Electric Nanobiopolymers (FEPs) Exhibiting Higher Mechanical Strength (7.5 GPa), Modulus (250 GPa), and Energy Transfer Efficiency (75%). *Worldw. Electroact. Polym. (Artificial Muscles) Newsl.* **2008,** *10* (2), 5–7.

73. Pääkko, M.; Ankerfors, M.; Kosonen, H.; Nykänen, A.; Ahola, S.; Österberg, M.; Ruokolainen, J.; Laine, J.; Larsson, P. T.; Ikkala, O.; Lindström, T. Enzymatic Hydrolysis Combined with Mechanical Shearing and High-Pressure Homogenization for Nanoscale Cellulose Fibrils and Strong Gels. *Biomacromolecules* **2007,** *8* (6), 1934–1941.

74. Baghaee-Ravari, S.; Ghazizadeh, M.; and Kelkar, A. D. Boron Nitride Nanotubes for Biomedical Applications: Challenges Ahead. *Austin J. Nanomed. Nanotechnol.* **2015,** *3*(1), 1040.

75. Shirane, G.; Suzuki, K.; Takeda, A. Phase Transitions in Solid Solutions of PbZrO$_3$ andPbTiO$_3$ (II) X-ray Study. *J. Phys. Soc. Jpn.* **1952,** *7,* 12–18.

76. Larson, P. J.; Towe, B. C. In *Miniature Ultrasonically Powered Wireless Nerve Cuff Stimulator*, Neural Engineering (NER), 2011 5th International IEEE/EMBS Conference on2011. p 265–258.

77. Wen, J.; Liu, M. Piezoelectric Ceramic (PZT) Modulates Axonal Guidance Growth of Rat Cortical Neurons via RhoA, Rac1, and Cdc42 Pathways. *J. Mol. Neurosci.* **2014,** *52,* 323–330.

78. Chen, W. P.; Chan, H. L. W.; Yiu, F. C. H.; Ng, K. M. W.; Liu, P. C. K. Water-Induced Degradation in Lead Zirconate Titanate Piezoelectric Ceramics. *Appl. Phys. Lett.* **2002,** *80,* 3587–3589.

79. Zhou, J.; Xu, N. S.; Wang, Z. L. Dissolving Behavior and Stability of ZnO Wires in Biofluids: A Study on Biodegradability and Biocompatibility of ZnO Nanostructures. *Adv. Mater.* **2006,** *18,* 2432–2435.

80. Nesbitt, H. W.; Bancroft, G. M.; Fyfe, W. S.; Karkhanis, S. N.; Nishijima, A.; Shin S. Thermodynamic Stability and Kinetics of Perovskite Dissolution. *Nature* **1981,** *289,* 358–362.

81. Yu, S.-W.; Kuo, S.-T.; Tuan, W.-H.; Tsai, Y.-Y.; Wang, S.-F. Cytotoxicity and Degradation Behavior of Potassium Sodium Niobate Piezoelectric Ceramics. *Ceram. Int.* **2012,** *38,* 2845–2850.

82. Bagchi, A.; Meka, S. R. K.; Rao, B. N.; Chatterjee, K. Perovskite Ceramic Nanoparticles in Polymer Composites for Augmenting Bone Tissue Regeneration. *Nanotechnology* **2014,** *25,* 485101.

83. Lin, W.; Xu, Y.; Huang, C.-C.; Ma, Y.; Shannon, K.; Chen, D.-R., et al. Toxicity of Nano- and Micro sized ZnO Particles in Human Lung Epithelial Cells. *J. Nanopart. Res.* **2009,** *11,* 25–39.

84. Seil, J. T.; Webster, T. J. Decreased Astroglial Cell Adhesion and Proliferation on Zinc Oxide Nanoparticle Polyurethane Composites. *Int. J. Nanomed.* **2008,** *3,* 523–531.

85. Amna, T.; Hassan, M. S.; Sheikh, F. A.; Lee, H. K.; Seo, K. S.; Yoon, D., et al. Zinc Oxide-Doped Poly(Urethane) Spider Web Nanofibrous Scaffold via One-Step Electrospinning: A Novel Matrix for Tissue Engineering. *Appl. Microbiol. Biotechnol.* **2013,** *97,* 1725–1734.

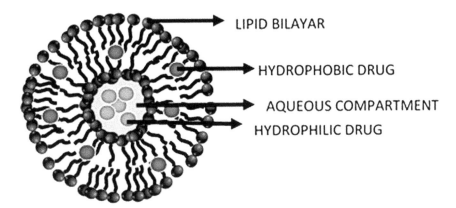

FIGURE 1.4   Structure of liposome.

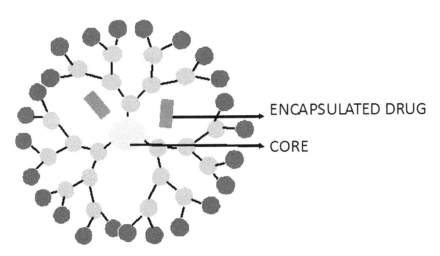

FIGURE 1.6   Structure of dendrimers.

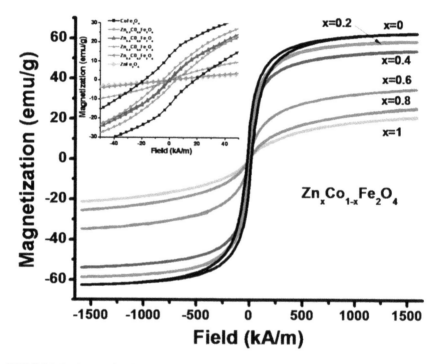

FIGURE 3.2   Magnetization curves of the $Zn_xCo_{1-x}Fe_2O_4$ magnetic nanoparticles. Reprinted from Reference [28] with permission from AIP Publishing.

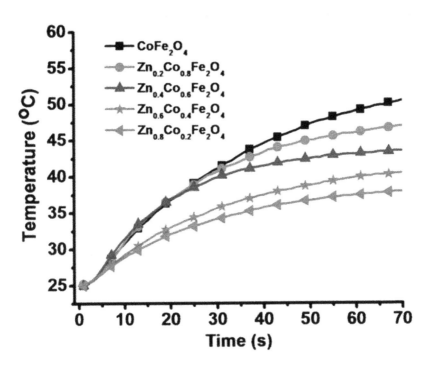

FIGURE 3.3 Heating curves of $Zn_xCo_{1-x}Fe_2O_4$ magnetic nanoparticles at 1.95 MHz frequency and 4.44 kA/m field. Reprinted from Reference [28] with permission from AIP Publishing.

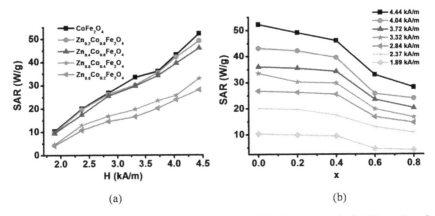

(a)   (b)

FIGURE 3.4 Variation of specific absorption rate with (a) ac magnetic field intensity of aqueous $Zn_xCo_{1-x}Fe_2O_4$ ferrite solution (b) with Zn content at different ac magnetic field. Reprinted from Reference [28] with permission from AIP Publishing.

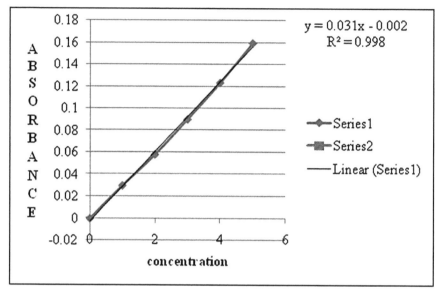

FIGURE 7.1   Calibration curve of aspirin in ethanol (λ_{max} = 275 nm).

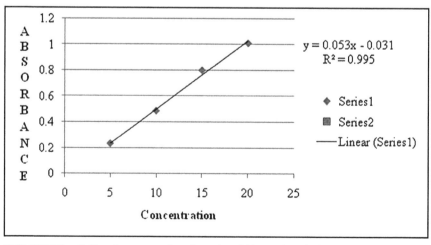

FIGURE 7.2   Calibration curve of mefenamic acid in methanol (λ_{max} = 285 nm).

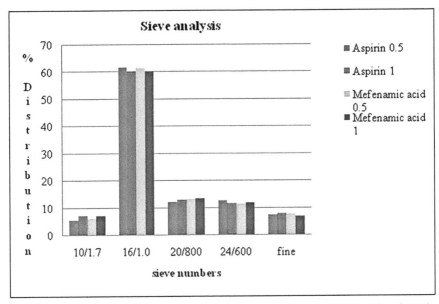

FIGURE 7.7   Histographic representation showing pellet yield of aspirin and mefenamic acid formulations.

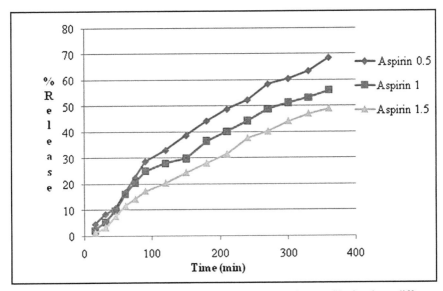

FIGURE 7.8   Graphical representation of aspirin dissolution profile for three different formulations.

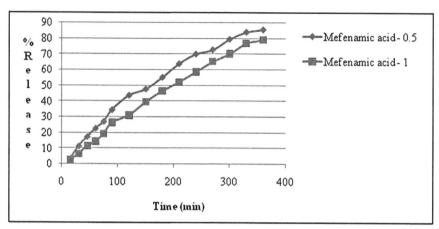

FIGURE 7.9 Graphical representation of mefenamic acid dissolution profile for two different formulations.

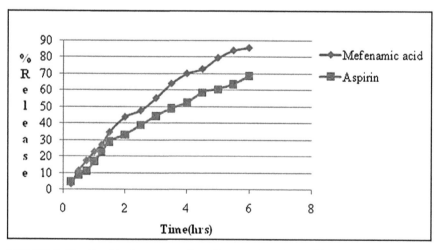

FIGURE 7.10 Graphical representation of comparison for aspirin and mefenamic acid dissolution profile for best formulations [drug:GMS ratio (1:0.5)].

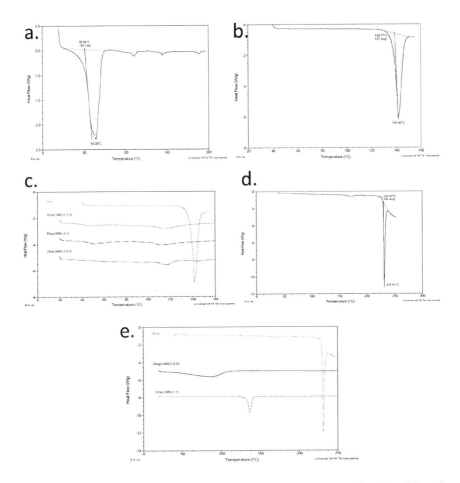

FIGURE 7.12　DSC Thermograms of (a) lipid carrier (GMS), (b) aspirin, (c) aspirin and its SE formulations, (d) mefenamic acid, (e) mefenamic acid and its SE formulations.

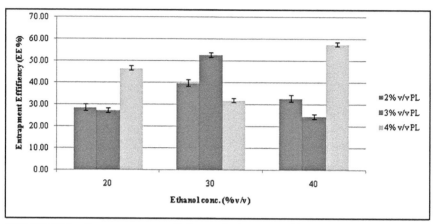

FIGURE 12.5   Drug entrapment efficiency of ethosomes with varying concentrations of ethanol and phospholipids.

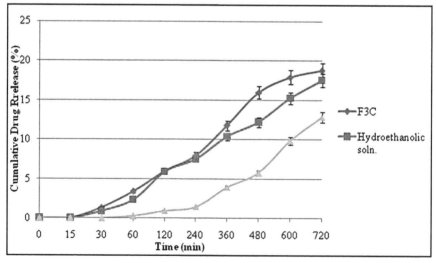

FIGURE 12.7   Comparison of in vitro skin permeation of drug from various formulations in phosphate buffer saline pH 7.4 ($n = 3$).

CHAPTER 7

FORMULATION AND COMPARISON OF LIPOPHILIC DRUGS THROUGH SELF-EMULSIFYING PELLETS USING EXTRUSION–SPHERONIZATION TECHNIQUE

JAWAHAR NATARAJAN[1*] and
VEERA VENKATA SATYANARAYANA REDDY KARRI[2]

[1]*Department of Pharmaceutics, JSS College of Pharmacy, Ootacamund, Rocklands, Udhagamandalam 643001, Tamil Nadu, India*

[2]*JSS Academy of Higher Education and Research, Mysuru, Karnataka, India*

Corresponding author. E-mail: jawahar.n@jssuni.edu.in

ABSTRACT

Multiparticulate dosage forms are receiving an immense attention as alternative drug delivery system for oral route of administration, even when the single-unit dosage forms have been widely used for decades. In the current research, the model drugs chosen for preparation of self-emulsifying pellets are aspirin and mefenamic acid, with which series of batches are prepared and various concentrations of glycerol monostearate (lipid carrier) are checked for better solubility and release. The evaluation studies of the prepared pellets are carried out with the traditional morphological assessment, and the results state that pellets formed are of soft, spherical in shape, and yields pellets with self-emulsification. Pellets with a reasonably good yield, acceptable sphericity, and low friability were obtained.

They also show good flow parameters like bulk density, tapped density, Carr's index. Hausner's ratio and angle of repose show flow behavior as excellent. The in vitro dissolution studies reveal that higher lipophilic drug (mefenamic acid, log P: 4.2) shows better release compared to less lipophilic drug (aspirin, log P: 1.4). Further studies using differential scanning calorimetry (DSC) and X-ray powder diffractometer are carried out to know the nature of drug. DSC studies clearly indicate the decrease in drug crystalline nature for both the drugs; hence, it enhances dissolution rate. These studies further revealed that the conversion from crystalline to amorphous is more in case of higher lipophilic drug (mefenamic acid) when compared to lower lipophilic drug (aspirin).

7.1 INTRODUCTION

Since the time of immemorial centuries, oral drug administration has been one of the most convenient and widely accepted routes of delivery for most therapeutic agents. Traditionally, oral dosage forms are classified as single-unit and multiple-unit dosage forms (MUDFs).[1,2] Multiparticulate dosage forms are receiving an immense attention as alternative drug delivery system for oral route of administration, even when the single-unit dosage forms have been widely used for decades.[3–6] The most commonly used pharmaceutical solid dosage forms today include granules, pellets, tablets, and capsules, out of which tablets being the most popular dosage form, accounting for 70% of all ethical pharmaceutical preparations produced.[7] But soon it was sensed that some of the formulating and clinical problems (free flowing property, dose dumping, dysphagia, etc.) come along with the single-dose formulations. This soon led to the division of monolithic dosage forms into multiples. Thus, there came the existence of idea of multiparticulates. MUDFs or multiparticulates are formulated as granules, pellets, or mini tablets (1 mm in diameter). The concept of this MUDFs solves many formulating problems and with a common strategy to control the release of drug as showing the reproducible release profiles when compared to single-unit dosage forms. These MUDFs can either be filled into hard capsules or compacted into bigger tablets or can be dispensed in a dose pouches or packets. The most increasingly interesting area in the development of MUDFs is incorporation into tablets instead of hard gelatin capsules in order to make it more economical to the consumers and gaining more attention currently.[8] The current research work enlightens on the pelletized form of multiple units that are being prepared using the process of pelletization which is referred to as a size-enlargement process and the final product is referred as pellets. Pelletization involves the process of renovation of fine powder or granules of bulk drugs and the excipients into small, free flowing, spherical units. Thus being a consumer-friendly alternative, over the single-unit dosage forms, many pharmaceutical companies are switching their products franchise to improve the technology.

7.2   EXPERIMENTAL

Aspirin and mefenamic acid are obtained from Balaji Chemicals, Chennai, India. Avicel PH 101 was obtained from Signet Chemicals, Mumbai, India. Glycerol monostearate (GMS) and Tween 80 were procured from Sigma, Mumbai, India. Lactose, mannitol, ethanol, and glacial acetic acid are purchased from SD Fine Chemicals, Chennai, India.

7.2.1   PREFORMULATION STUDIES

7.2.1.1   PHYSICAL PROPERTIES

Physical properties are used to determine the color, purity of the drug, and excipients.

7.2.1.2   MELTING POINT

Melting point test helps for the identification of the drug and its degree of purity. Capillary tube melting point apparatus was used. A small quantity (1 mg) of powder was placed into a capillary fusion tube after it is sealed from one end with the help of the burner. The capillary tube is placed in the melting point determining apparatus-containing liquid paraffin. The temperature of the liquid paraffin gradually increased automatically, and the temperature at which powder started to melt and the temperature when all powders got melted were observed and noted. The temperature at which melting of the drug occurs is known as melting point of the drug.

7.2.1.3   COMPATIBILITY STUDIES

Infrared spectra matching approach was used for detection of any possible chemical interaction between the drug and the excipients. Selection of the samples depends on stability, compatibility, and therapeutic acceptability. FTIR spectra of aspirin and mefenamic acid were obtained using SHIMADZU FTIR 8400 S spectrophotometer using diffuse reflectance technique (KBr disc technique) as a part of qualitative analysis by comparing it with the spectra of aspirin and mefenamic acid USP standards in the region of 4000–500 cm^{-1}. Background spectrum was collected under identical conditions. The spectrum is recorded for the pure drug sample and formulation (physical mixture) separately and for the physical mixture of the pure drug and excipients mixed in suitable quantity. Required amount of this mixture was compressed to form a transparent pellet using a hydraulic press at 15 t pressure. It is scanned from 4000 to 500 cm^{-1} in a SHIMADZU FTIR spectrophotometer. The IR spectrum of the physical mixture is compared with those of pure drug and excipients, and matching was done to detect any appearance or disappearance of peaks.

7.2.1.4   PREPARATION OF CALIBRATION CURVE

Calibration curve of aspirin and mefenamic acid was plotted by spectroscopic method. The test was carried out in clean glassware of five 10-mL volumetric flasks. The stock solution of the drug is prepared in 1 mg/mL concentrated solution of the drug, by dissolving 100 mg of the drug in 100 mL (1000 µg/mL) suitable solvent in a volumetric flask (A). From this stock solution, further dilution was done to get 100 µg/mL, by diluting 10 mL to 100 mL (B) using the suitable solvent. And then, dilutions were continued from stock solution B to get 10 µg/mL following the same dilution pattern of diluting 10 mL of stock solution B to 100 mL (C). From this stock solution of C, 1.0–5.0 µg/mL solutions are prepared by diluting 1.0, 2.0, 3.0, 4.0, and 5.0 mL solutions, respectively, to 10 mL with the suitable solvent and diluting 5.0, 10, 15, and 20 mL, respectively, to 50 mL. These solutions were scanned by UV Spectrophotometer. The λ_{max} of the drug was determined by scanning one of the dilutions between 400 and 200 nm by using a UV–visible spectrophotometer. At this wavelength, the absorbance of all other solutions was measured against a blank. Standard curve between concentration and absorbance was plotted.

7.2.1.5   POWDER PARAMETERS

Bulk density, tapped density, angle of repose, Carr's index, and Hausner's ratio were calculated.

7.2.2   PREPARATION OF PELLETS

In the process of preparation of the pellets, several batches have been prepared to get the desired property to be imparted into the final ideal formula.[9] During this process of preparation, the initial parameters to be set are the processing variables like spheronizing speed and time. Therefore, initially certain dummy batched are prepared to just set the process variables.

These batches are tabulated (Table 7.1) and carried out to optimize spheronizing speed and spheronizing time. These batches are prepared without drug to optimize the process parameters. The spheronizing is done at different speeds starting from 1000 RPM and proceeding with 1500, 2000, 2500, and 3000 RPM. And for each speed assigned, the experiment speed is optimized by checking the shape of the pellets formed at the end of every 5 min. Among the above speeds, 3000 RPM gives the satisfactory results to achieve the good pellets. So, we have chosen 3000 RPM for formulation of self-emulsifying (SE) pellets. These batches are prepared to optimize the lipid and binder quantity to be taken in the formulation. The pellets formed were mostly dumb-bell shaped, and very few successfully attained the spherical shape. These pellets are also hard on touch. Hence, the binder used in this formulation does not show any impressive behavior to be imparted into the multiparticulate forms. In this formulation, usage of binder is not required because lipid acts as a carrier which also imparts binding nature; so in the reaming batches, we did not use binder. These batches mentioned in Table 7.1 showed the use of combination of lipid, surfactant, and solvent for the self-emulsification in pelletization. The excipients used in the above ratio which resulted in both round-shaped and dumbbell-shaped pellets result in desirable self-emulsification property along with the hard nature of pellets. However, these pellets are observed

TABLE 7.1   Batches Prepared for Optimizing Speed and Time for Forming Pellets.

Ingredients	F1	F2	F3	F4	F5	F6	F7	F8	F9	F10	F11	F12	F13	F14	F15	F16	F17	F18	F19	F20	F21	F22	F23
Glycerol monostearate (g)	2	2	2	2	2	1	1.5	2	2	2	2	2	2	2	2	2	2	2	2	2	2	4	5
Tween 80 (Polysorbate 80) (mL)	8	8	8	8	8	8	8	8	8	8	8	7	7	7	7	7	7	7	7	7	7	7	7
Water (mL)	3	3	3	3	3	4	4	4	4	4	4	3	2.5	2.5	2.5	2.5	2.5	2.5	2.5	2.5	2.5	2.5	2.5
Drug (g)	4	4	4	4	4	4	4	4	4	4	4	4	4	4	4	4	4	4	4	4	4	4	4
Avicel PH 101 (g)	12	12	12	12	12	15	15	15	15	15	12	12	12	10	11	12	12	11.5	11.25	12	11.5	13	13
Granulating fluid	4 mL	4 mL	4 mL	4 mL	4 mL	4 mL	4 mL	4 mL	4 mL	4 mL	4 mL	4 mL	4 mL	4 mL	4 mL	3.5 mL	3 mL	3 mL	3 mL	3 mL	3 mL	3 mL	3 mL
Lactose monohydrate	1 g	1 g	1 g	1 g	1 g	500 mg	500 mg	500	500	500	500 mg	500 mg	500 mg	500 mg	500 mg	500 mg	500 mg	500 mg	500 mg	750 mg	750mg	750 mg	750 mg
PEG 2000 (mg)	–	–	–	–	–	50	75	100	50	100	–	–	–	–	–	–	–	–	–	–	–	–	–

Granulating fluid (ethanol:water = 1:1).

to be much smooth and soft than earlier batches due to lack of the binder and as change quantity of solvent. Table 7.1 indicates the usage of various quantities of spheronizing enhancer and granulating fluid for the formation of SE pellets and after decrease in quantity of granulating fluid, which results in the formation of good pellets, but it contains moisture to a little extent, so it results in the formation of little clumps. These batches shown in Table 7.1 are carried out using different quantities of diluent and spheronizing enhancer to formulate SE pellets free from moisture content, after increasing the quantities of spheronizing enhancer and lactose monohydrate results in formation of ideal SE pellets. These batches of SE pellets are prepared with different concentration of GMS to study the dissolution parameter and nature of the drug in formulation by using differential scanning calorimetry (DSC) and X-ray powder diffractometer (XRD).

7.2.3 MORPHOLOGICAL STUDIES

7.2.3.1 COMPARISON OF VISUALIZATION METHODS

This test involves the comparison of different visual snapshots taken under system microscope—motic images plus for different SE pellet formulations prepared by using different concentration polymer.

7.2.3.2 SURFACE PROPERTIES OF THE PELLETS BY SEM

This test states the surface properties of the pellets formed by using particular polymer. It indicates the surface texture that is imparted to the pellet structure due to polymer; however, it also depends on concentration of the polymer used apart from the type and nature of the polymer used.

7.2.3.4 MECHANICAL PROPERTIES

7.2.3.4.1 Pellet Yield

Pellets yield analysis was performed using a set of sieves in increasing order placed vertically. 100 g of pellets were placed on top sieve and agitated for 20 min on a mechanical sieve shaker.

7.2.3.5   SHAPE AND SURFACE ROUGHNESS

In order to obtain good performance of prepared pellets, it is necessary to have spherical and smooth particles usually for achieving fast-release. The commonly used method is the analysis of microscopic or nonmicroscopic snapshots of interest.

7.2.3.6   FRIABILITY (F)

Friability of the pellets is determined using Roche Friabilator. This device subjects the pellets to the combined effect of abrasion and shock in a plastic chamber revolving at 25 rpm and dropping a single dose of pellet at height of 6 in. in each revolution. Preweighted sample of pellets was placed in the Friabilator and were subjected to the 100 revolutions. Pellets are then dusted using a soft muslin cloth and reweighed.

7.2.3.7   BULK DENSITY (D_b)

It is the ratio of total mass of pellets to the bulk volume of pellets occupied. It was measured by pouring the weight quantity of the pellets into a measuring cylinder, and the initial weight was noted. This initial volume is called the bulk volume.

7.2.3.8   TAPPED DENSITY (D_t)

It is the ratio of total mass of the pellets to the tapped volume of the pellets. The volume was measured by tapping the pellets for 100 times, and the tapped volume was noted if the difference between these two volumes is less than 2%. If more, tapping is continued for 400 times, and the tapped volume was noted. Tapping was continued until the difference between successive volumes is less than 2% (in a bulk density apparatus).

7.2.3.9   ANGLE OF REPOSE (Ø)

The friction forces in the pellets can be measured by the angle of repose (Ø). It is an indicative of the flow properties of the powder. It is defined as maximum angle possible between the surface of the pile of pellets and the horizontal plane. The pellet formulation was allowed to flow through the funnel fixed to a stand at definite height (h). The angle of repose was then calculated by measuring the height and radius of the heap of pellet formed. Care was taken to see that the pellets slip and roll over each other through the sides of the funnel.

7.2.3.10   DRUG CONTENT UNIFORMITY

This test involves determination of the content of the active pharmaceutical ingredient (API) by any standard assay method described for the particular API in any of the standard pharmacopoeia. Content uniformity is determined by estimating the API content in individual formulation. This was performed in 0.05 acetate buffer and phosphate buffer pH 6.8. Limit of content uniformity is 85–115%.

7.2.4   DISINTEGRATION PROPERTIES

This test is carried to observe the disintegrating property of the dosage form which helps to evaluate the dissolving property of SE pellets for their rapid disintegration properties, following test was carried out by taking little quantity of SE pellets in capsule and kept in disintegration apparatus and study the time according to its property. The customary procedure of performing disintegration test for delayed dissolving dosage forms has quite a lot of limitations, and they give adequate information for the measurement of disintegration times.

7.2.5   DRUG DISSOLUTION STUDIES

In this study, work was carried out on two different drugs, so that each drug has its own dissolution media as per Indian Pharmacopeia (IP). According to IP, dissolution medium for aspirin is 0.05 acetate buffer, prepared by mixing 2.99 g of sodium acetate trihydrate and 1.66 mL of glacial acetic acid and with water to obtain 1000 mL of solution having a pH of 4.50 ± 0.05. Apparatus USP 1; RPM: 50 maintained at 37 ± 0.5°C. According to IP, dissolution medium for mefenamic acid is phosphate buffer 6.8; place 50 mL of 0.2 M potassium dihydrogen phosphate in a 200-mL volumetric flask, and then add 22.4 mL of 0.2 M NaOH and then add water to the volume. The samples are carried at the intervals of 15, 30, 45, 60, 75, 90, 120, 150, 180, 210, 240, 270, 300, 330, and 360 min. The samples were withdrawn for 6 h to know the release. The samples were withdrawn by a bulb pipette. The missing dissolution volume has to be considered by restoring the same volume into the apparatus. The collected samples are diluted sufficiently prior to the measurement with a UV-spectrophotometer and calculated within the values.

7.2.6 TESTS FOR SELF-EMULSIFICATION

For the preliminary assessment of the SE properties of the formulation, 0.1% Sudan Red was incorporated. Pellets (1 g) were then gently agitated in 50 mL distilled water. Agitation was provided by gentle shaking in an orbitary shaker 50 oscillations per minute at a temperature of 37°C. Samples were taken after 30 min and observed samples under Motic microscope for assessment by using optical zoom and an eye piece of 10×.

7.2.7 DIFFERENTIAL SCANNING CALORIMETRY

The thermograms of the samples were obtained by DSC (Q-200, TA Instruments, USA) using scanning rate of 10°C/min from 40°C to 375°C. About 4–5 mg of samples were taken and sealed in aluminum pans and analyzed under the flow of nitrogen gas at the rate of 50 mL/min.

7.2.8 X-RAY POWDER DIFFRACTOMETER

XRD is one of the most powerful and established techniques for material structural analysis, capable of providing information about the structure of a material at the atomic level. Low and high temperature measurement facilities are available. The temperature attachment enables analysis from −170°C to +450°C under vacuum. Change in unit cell dimensions, structural changes at phase transitions, etc. as a function of temperature can be determined.

Make/model	: Bruker AXS D8 Advance
Configuration	: Vertical, Theta/2 Theta geometry
Measuring circle diameter	: 435, 500, and 600 mm predefined
Angle range	: 360°
Max. Usable angular range	: 3–135°
X-ray source	: Cu, Wavelength 1.5406 Å
Detector	: Si(Li) PSD
Make/Model	: Anton Paar, TTK 450
Temperature range	: −170°C to +450°C

The D8 Advance X-ray Diffractometer can be used for all X-ray powder diffraction applications such as structural analysis of crystalline solids, determination of unit-cell parameters, determination of crystal morphology, etc. Solid samples, ~1 g, are required for analysis. Sensitivity to moisture, air, etc. is to be specific.

7.3 RESULTS AND DISCUSSION

7.3.1 PREFORMULATION STUDIES

7.3.1.1 MELTING POINT

Literature reports give supportive data on melting behavior of the drug substance. US Pharmacopoeia states that aspirin melts at about 135°C. However, the producer of the aspirin used in the present work reported a melting point of 133.5–135.5°C and no measurement conditions were provided. The melting point of aspirin observed was 133.5°C as shown in Table 7.2, relates completely with the specified and reported reading which states that the sample is pure with no impurities.

TABLE 7.2   Results of Melting Point.

Substance	Melting point (°C)	Observed melting point (°C)			Average melting point (°C)
		01	02	03	
Aspirin	135	133.5	133.5	134	133.75
Mefenamic acid	230–231	229	228.5	229	229

US Pharmacopoeia states that mefenamic acid melts at 230–231°C. However, the procedure of mefenamic acid used in the present work reported a melting point of 228.5–230.5°C and no measurement conditions were provided. The melting point of mefenamic acid observed was 229°C as shown in Table 7.2, relates completely with the specified and reported reading which states that the sample is pure with no impurities.

7.3.1.2 COMPATIBILITY STUDIES

The peaks are seen at 2587.56, 1753.35, 1694.52, 1436.05, 755.16, and 667.39 cm^{-1}, respectively, denoting carboxylic acid O–H, ester C=O, carboxylic acid C=O, CH$_3$ (C–H) bending, C–H bending, and aromatic C–H, respectively. These readings are shown in Table 7.3.

TABLE 7.3 FTIR Spectra Range of Drug and Physical Mixture.

Drug	Physical mixture	Formulation	Peaks
2587.59	2588.56	2587.59	O–H (3000–2500)
1753.35	1753.35	1751.42	Ester, C=O (1755–1735)
1694.52	1694.52	1683.91	Carboxylic acid, C=O (1725–1680)
1436.05	1436.05	1436.05	CH$_3$(C–H) (B) (1440)
755.16	755.16	755.16	C–H (B) (770–730)
667.39	668.36	668.36	Aromatic, C–H (700–600)

After interpretation of the above spectra, it was confirmed that there was no major shifting, loss or appearance of functional peaks between the spectra of drug, excipients, and physical mixture of drug and excipients. From the spectra, it was concluded that the drug was blended well with the excipients without any chemical interaction. The peaks are seen at 3011.95, 2642.57, 1731.87, 1278.85, and 662.57 cm^{-1}, respectively, denoting aromatic CH$_3$(C–H), carboxylic acid (O–H) wagging, –COOH (C–O) stretching, aromatic (N–H), and aromatic (C–H), respectively. These readings are shown in Table 7.4.

TABLE 7.4 FTIR Spectra Range of Drug and Physical Mixture.

Drug	Physical mixture	Formulation	Peaks
3011.95	3005.20	3005.09	Aromatic CH$_3$ (C–H) 3000–3030
2642.57	2645.46	2647.39	–COOH ,O–H (W) 2700–2500
1731.87	1732.13	1731.17	–COOH, C–O (S) 1725–1700
1278.85	1279.81	1280.78	Aromatic (N–H) 1350–1250
662.57	668.36	668.36	Aromatic (C–H) 600–750

7.3.2   PREPARATION OF CALIBRATION CURVE FOR ASPIRIN

The scanning of drug solution against blank in UV range showed the maximum absorbance (λ_{max}) at 275 nm; hence, the calibration curve was developed at this wavelength. The calibration curve of the drug thus developed is found to be linear between concentration of drug in the solution and its optical density. It showed the perfect linearity between the concentration and absorbance in the concentration range of 1–5 µg/mL. Figure 7.1 shows the calibration of aspirin using ethanol as solvent. The values of "slope (K)" and "intercept (β)" were found to be 0.031 and 0.002, respectively.

7.3.3   PREPARATION OF CALIBRATION CURVE FOR MEFENAMIC ACID

The scanning of drug solution against blank in UV range showed the maximum absorbance (λ_{max}) at 285 nm; hence, the calibration curve was developed at this wavelength. The calibration curve of the drug thus developed is found to be linear between concentration of drug in the solution and

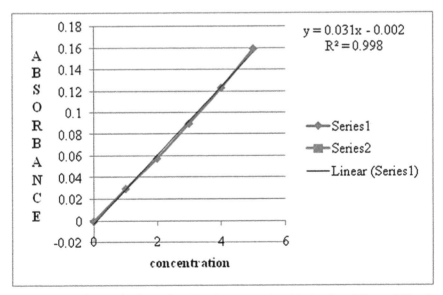

FIGURE 7.1   **(See color insert.)** Calibration curve of aspirin in ethanol (λ_{max} = 275 nm).

its optical density. It showed the perfect linearity between the concentration and absorbance in the concentration range of 5–20 µg/mL. Figure 7.2 shows the calibration of aspirin using ethanol as solvent. The values of "slope (*K*)" and "intercept (*β*)" were found to be 0.053 and 0.031, respectively.

7.3.4 POWDER PARAMETERS

TABLE 7.5   Powder Parameters of Pure Drug (Aspirin).

Powder parameters	Observed values	Flow property
Bulk density	0.6084 g/mL	–
Tapped density	0.8847 g/mL	–
Carr's index	32%	Very poor flow
Hausner's ratio	1.57	Cohesive nature
Angle of repose	52° 8″	Poor flow

The powder parameters of the pure drug indicate that it is having very less flow property and considered as powder with more cohesive nature showing the poor flow (Table 7.5).

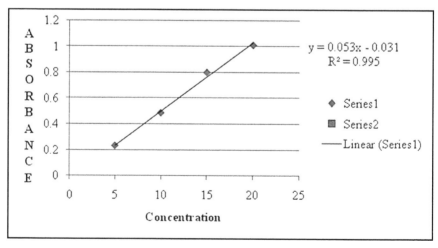

FIGURE 7.2   (See color insert.) Calibration curve of mefenamic acid in methanol (λ_{max} = 285 nm).

7.3.5 MORPHOLOGICAL STUDIES

7.3.5.1 MICROSCOPIC AND IMAGE ANALYSIS

The microscopic Motic images of the SE Pellets show clear picture of the surface texture of the SE Pellets (Fig. 7.3). The above figures show visually how pellets look after formulation. Both of the pellets formulated by aspirin and mefenamic acid have shown smoother surface. The naked eye observation of the pellets shows that all pellet batches show uniform size distribution (Figs. 7.4 and 7.5).

FIGURE 7.3   Motic image of SE pellets.

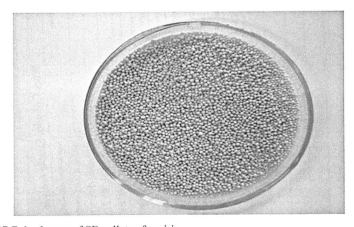

FIGURE 7.4   Image of SE pellets of aspirin.

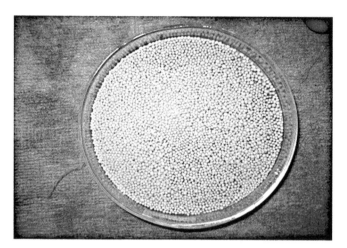

FIGURE 7.5   Image of SE pellets of mefenamic acid.

7.3.5.2   SURFACE PROPERTIES OF SE PELLETS

The SEM images (Fig. 7.6) of the pellets show clear picture of the surface texture and shape of the pellets. With the help of those pictures, SE pellets show smoother surface and spherical shape.

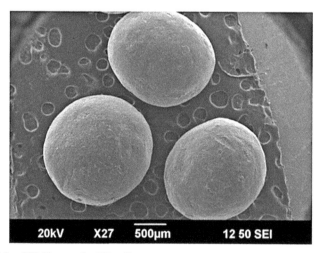

FIGURE 7.6   SEM image for SE pellets structure.

7.3.6   MECHANICAL PROPERTIES

7.3.6.1   PELLET YIELD

The pellet yield was found to be 60.2% and 61.8% in successive unit operations when 100 g of the bulk material is being used which results a standard deviation of 1.3 between the two yield percentages obtained.

7.3.6.2   SIEVE ANALYSIS

Sieve analysis represents (Fig. 7.7) that there is no much variation in pellet yield, but increase in GMS concentration results in increase pellet size and thereby little decrease in pellet yield.

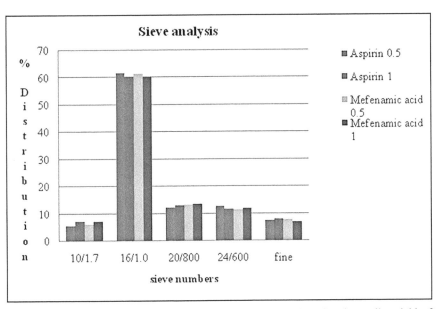

FIGURE 7.7   (See color insert.) Histographic representation showing pellet yield of aspirin and mefenamic acid formulations.

7.3.6.3 FRIABILITY (F)

TABLE 7.6 Friability of SE Pellets of Aspirin and Mefenamic Acid.

Pellets	Initial weight(g)	Final weight (g)	Percentage friability
Aspirin	2.0025	1.980	0.01
	2.0016	1.976	0.01
	2.0020	1.986	0.01
Mefenamic acid	2.0028	1.984	0.001
	2.0012	1.981	0.001
	2.0016	1.975	0.001

The friability test (F) was carried out for both drugs of different batches of pellets bearing different concentrations of GMS. The results in Table 7.6 have shown that the pellets of all batches occur according to the limits. This indicates that the pellets do not show any brittle nature which might not lead to any damage of the pellets during transport or packaging.

7.3.6.4 BULK DENSITY (D_b)

Bulk density of SE pellets of aspirin.

TABLE 7.7 Bulk Density of SE Pellets of Aspirin and Mefenamic Acid.

Pellets	Weight taken (g)	Bulk volume (mL)	Bulk density (D_b)
Aspirin	10.123	16.0	0.632
Mefenamic acid	10.250	16.0	0.640

7.3.6.5 TAPPED DENSITY (D_t)

Tapped density of SE pellets of aspirin.

TABLE 7.8 Tapped Density of SE Pellets of Aspirin and Mefenamic Acid.

Pellets	Weight taken (g)	Bulk volume (mL)	Bulk density (D_b)
Aspirin	10.123	14.5	0.698
Mefenamic acid	10.250	14.65	0.699

The results are obtained for the bulk density and tapped density for both the drugs of the pellets. The details show that both the drugs in SE pellets result in good bulk density and tapped density which result further good flow properties. This can be easily understood by using the data from Tables 7.7 and 7.8.

7.3.6.6 ANGLE OF REPOSE

TABLE 7.9 Angle of Repose of SE Pellets, Formed Using Aspirin and Mefenamic Acid.

SE pellets	Height (cm)	Average diameter (cm)	Average radius (cm)	Angle of repose	Flow property
Aspirin	1.16	7.23	3.61	17°81″	Excellent
Mefenamic acid	1.18	7.19	3.51	18°57″	Excellent

The results obtained for the angle of repose are tabulated in Table 7.9. The interpretation of their numerical data states that the pellets with aspirin and with mefenamic acid show good flow property reflecting their angle of repose to be at 1781′ and 18°57′.

7.3.6.7 CARR'S INDEX OR % COMPRESSIBILITY

TABLE 7.10 Carr's Index of SE Pellets Formed Using Aspirin and Mefenamic Acid.

SE pellets	Carr's index (%)	Flow property
Aspirin	9.45	Excellent
Mefenamic acid	8.45	Excellent

The results of the Carr's index give the relation with the flow properties of the pellets. The above tabulated (Table 7.10) readings have shown that the pellets with aspirin and mefenamic acid showed good results. However, all the batches fall under the limits and showed the excellent flow property.

7.3.6.8 HAUSNER'S RATIO

TABLE 7.11 Hausner's Ratio of Pellets Formed Using Aspirin and Mefenamic Acid.

SE pellets	Hausner's ratio		Flow property
	Specified limit	Observed value	
Aspirin	<1.25	1.04	Free flowing
Mefenamic acid	<1.25	1.09	Free flowing

The results of Hausner's ratio state the same as Carr's index; it also shows high-flow properties (Table 7.11) for SE pellets. Both the formulations observed below the specified limit which results in free flowing of SE pellets.

7.3.7   DRUG CONTENT UNIFORMITY

TABLE 7.12   Drug Content of Pellets Formed Using Aspirin and Mefenamic Acid.

Drug	Medium	Drug content	% Purity
Aspirin	0.05 M acetate buffer	97.1	97.1%
Mefenamic acid	Phosphate buffer pH 6.8	97.07	97.07%

The uniformity in aspirin drug content was found to be 97.1% in the 0.05-M acetate buffer and the drug content uniformity for mefenamic acid was found to be 97.07% in phosphate buffer pH 6.8 medium (Table 7.12). These studies were carried out for the single dose of the pellets. This indicates the drug is uniformly distributed during the preparation.

7.3.8   DISINTEGRATION PARAMETERS

TABLE 7.13   Average Disintegration Time of SE Pellets Formed Using Aspirin.

Drug:lipid ratio	Time 1 (min)	Time 2 (min)	Time 3 (min)	Average time (min)
Drug (1):lipid (0.5)	17.3	18	18	18
Drug (1):lipid (1)	19.3	19.3	19	19.3
Drug (1):lipid (1.5)	21	21.3	21.3	21.3

TABLE 7.14   Average Disintegration Time of SE Pellets Formed Using Mefenamic Acid.

Drug:lipid ratio	Time 1 (min)	Time 2 (min)	Time 3 (min)	Average time (min)
Drug (1):lipid (0.5)	18	17.3	17.3	17.3
Drug (1):lipid (1)	19	18.3	19	19
Drug (1):lipid (1.5)	22	21.3	22	22

The disintegration results (Tables 7.13 and 7.14) give idea regarding the release behavior of the drug through the SE system. Lipid present in the SE system forms a matrix; with increase in lipid concentration, it retards drug release from the system and increases disintegration time.

7.3.9 DISSOLUTION PROFILE OF DRUGS

In both the formulations, less percentage of GMS (0.5) concentration yields good release rate compared to higher GMS concentrations (Figs. 7.8–7.10). SE formulation of aspirin releases 68.68% in 6 h, whereas mefenamic acid released 85.66% at the same time. In higher GMS concentrations, drugs release from SE formulation is slow, which means it retards release behavior of drug, in which, mefenamic acid gives more release than aspirin.

7.3.9.1 RELEASE KINETICS

In order to elucidate mode and mechanism of release pattern for both the drugs, the in vitro release of both drugs data was transformed and interpreted at graphical interface constructed using various kinetic models.

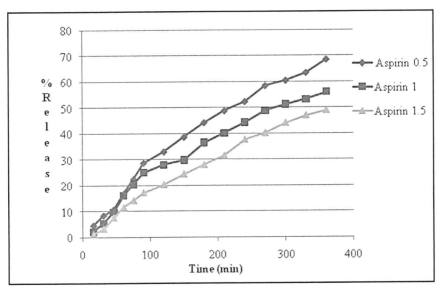

FIGURE 7.8 (See color insert.) Graphical representation of aspirin dissolution profile for three different formulations.

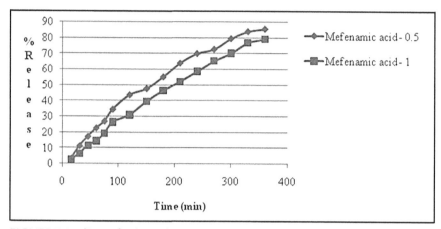

FIGURE 7.9 (See color insert.) Graphical representation of mefenamic acid dissolution profile for two different formulations.

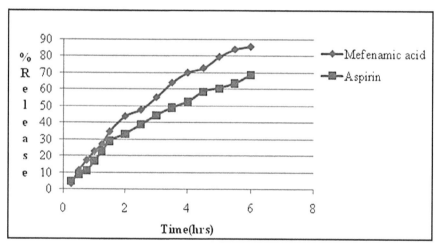

FIGURE 7.10 (See color insert.) Graphical representation of comparison for aspirin and mefenamic acid dissolution profile for best formulations [drug:GMS ratio (1:0.5)].

TABLE 7.15   Regression Value for Various Kinetic Models.

Drug	R^2		
	Zero order	**First order**	**Higuchi's**
Aspirin	0.968	0.788	0.994
Mefenamic acid	0.969	0.722	0.995

The best linearity was obtained in Higuchi's plots for both the SE formulations; it indicates that these formulations belong to diffusion mechanism. The release exponent values n for SE pellets of aspirin and mefenamic acid were found to be 0.850 and 0.930, respectively (Table 7.15), since the release exponent n values were between 0.5 and 1.

7.3.10   SELF-EMULSIFICATION

From Figure 7.11, SE formulation was able to introduce Sudan Red into water within the first few minutes followed by gentle agitation. Whereas coming to reference pellets, formulated without SE system in it, they are not able to deliver the lipophilic dye into the media. Motic absorption of the release media of SE pellets showed lipid droplets, incorporating the

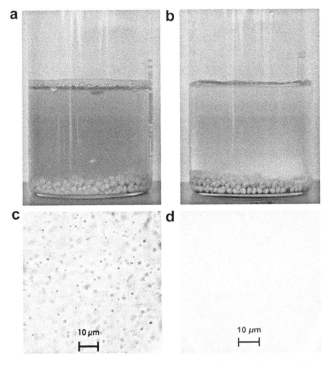

FIGURE 7.11   Photograph showing the release media of self-emulsifying pellets (a). Self-emulsifying pellet formulation (b). Reference pellets containing Sudan red dye after 30 min of dissolution in distilled water at a temperature of 37°C. Motic pictures of (c). Self-emulsifying pellets. (d). Reference pellets.

dye. In which reference pellets, motic examination of the release media of the reference pellets, did not show any signs of droplets.

7.3.11   DSC STUDIES OF ASPIRIN AND MEFENAMIC ACID AND LIPID CARRIER

DSC is a tool to investigate the melting and recrystallization behavior of crystalline material. Figure 7.12 shows DSC curves of aspirin, mefenamic acid, bulk GMS, and both the drugs with its formulations. The DSC curve

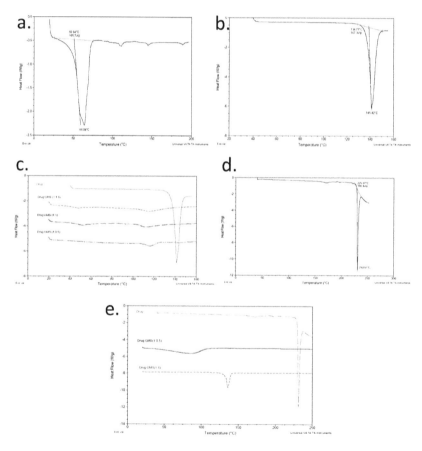

FIGURE 7.12   (See color insert.) DSC thermograms of (a) lipid carrier (GMS), (b) aspirin, (c) aspirin and its SE formulations, (d) mefenamic acid, (e) mefenamic acid and its SE formulations.

of aspirin showed a melting endotherm at 138.77°C. The thermograms of various concentrations of GMS in formulations did not show the melting peak of aspirin. This suggests that aspirin was not in crystalline state but was present in amorphous state. The DSC curve of mefenamic acid showed a melting endotherm at 229°C. The thermograms of various concentrations of GMS in formulations did not show the melting peak of mefenamic acid. This suggests that mefenamic acid was not in crystalline state but present in amorphous state after making into a formulation by using lipid carrier. DSC analysis of aspirin and mefenamic acid prepared by self-emulsification technique showed that both the drugs exist in amorphous state after formulation. As reported earlier, that self-emulsification does not allow the drug molecules dispersed in lipid phase to crystallize. This behavior should improve the solubility of drugs in water. Furthermore, the presence of surfactants could not allow the drug to crystallize. Melting point of lipid GMS were depressed when compared to melting points of corresponding bulk lipids according to DSC endotherm. This melting point depression might be due to change in its nature, occurrence of self-emulsification and the presence of surfactant.

7.3.12   X-RAY DIFFRACTION STUDIES OF ASPIRIN, MEFENAMIC ACID, AND LIPID CARRIER

The results of X-ray powder diffractometry studies for aspirin pure drug and its formulations, which contain various quantities of GMS, show that the pure drug has the intensity of crystallinity about 55.8 at 22.623 in 2-theta angle; in the formulation, the crystallinity intensity decreased is 18.3 due to low GMS concentration that is drug (1):GMS (0.5). X-ray powder diffractometry studies for mefenamic acid pure drug and its formulation which contains GMS concentration of drug (1):GMS (0.5), for the pure drug the intensity of crystallinity, is 100 at 26.08 in 2-theta angle; in the formulation, the crystallinity intensity decreased to 24.6 at 22.742 in formulation which is having GMS concentration that is drug (1):GMS (0.5). Pure aspirin and mefenamic acid show a diffraction peaks corresponding to a separate crystalline drug phase. SE pellets prepared by using both the drugs showed less intensity of diffraction peaks pointing to a transition of aspirin and mefenamic acid from a crystalline to an amorphous

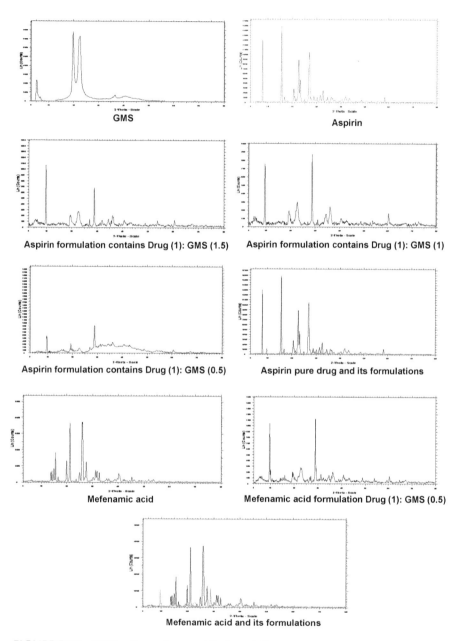

FIGURE 7.13   XRDs of pure drugs and its formulations.

state as a result of SE formulation. The crystallinity intensity of both the pure drugs and its formulations is already shown in Figure 7.13.

7.4 CONCLUSION

Extrusion–spheronization technique was suitable for producing SE pellets. Both the drugs aspirin and mefenamic acid were successfully formulated by using SE excipients. SE pellets are considered more advantageous over the single-unit dosage formulations. The critical parameters for formation of SE pellets were optimization of speed, optimization of residence time, and concentration of GMS. SE pellets with a reasonably good yield, acceptable sphericity, and low friability were obtained. They also showed good flow properties. DSC studies clearly indicate decrease in drug crystalline nature for both the drugs, so that it may enhance dissolution rate. XRD gives the intensity of crystalline nature in active ingredients, and its formulations showed clear change in crystalline form to amorphous in wide range, it may enhance release rate of drug from SE formulation. Studies revealed that the conversion from crystalline to amorphous is more in case of higher lipophilic drug (mefenamic acid) when compared to lower lipophilic drug (aspirin) and also the dissolution studies on these drugs states the same by showing better release for higher lipophilic drug, this may be due to more solubility of the higher lipophilic drug in lipid carrier.

KEYWORDS

- pellets
- aspirin
- mefenamic acid
- XRD
- DSC

REFERENCES

1. Lavanya, K.; Senthil, V.; Rathi, V. Pelletization Technology: A Quick Review. *Int. J. Pharm. Sci. Res.* **2011**, *2*(6), 1337.

2. Shettigar, R.; Damle, A. V. Controlled Release Pellets of Nitrofurantoin. *Ind. J. Pharm. Sci.* **1996,** *5,* 179–185.

3. Cheboyina, S.; Chambliss, W. G.; Wyandt, C. M. A Novel Freeze Pelletization Technique for Preparing Matrix Pellets. *Pharm. Tech.* **2004,** *28,* 98–108.

4. Scott, D. C.; Hollenbeck, R. G. Design and Manufacture of a Zero-Order Sustained-Release Pellet Dosage form through Nonuniform Drug Distribution in a Diffusional Matrix. *Pharm. Res.* **1991,** *8*(2), 156–161.

5. Celik, M. *Multiparticulate Oral Drug Delivery*; Marcel Dekker: New York, 1994, pp 181.

6. Ghebre-sellassie, I. *Pharmaceutical Pelletization Technology*; Marcel Dekker Inc.: New York, USA, 1989; pp 1–13.

7. Gu, L.; Liew, C. V.; Heng, P. W. S. Wet Spheronization by Rotary Processing—A Multistage Single-Pot Process for Producing Spheroids. *Drug Dev. Ind. Pharm.* **2004,** *30,* 111–119.

8. Bechgaard, H.; Nielsen, G. Controlled Release Multiple Units and Single Unit Doses. *Drug. Dev. Ind. Pharm.* **1978,** *4,* 53–57.

9. Harrison, P. J.; Newton, J. M.; Rowe, R. C. The Characterization of Wet Powder Masses Suitable for Extrusion/Spheronization. *J. Pharm. Pharmacol.* **1985,** *37,* 686–691.

CHAPTER 8

NANOCOMPOSITE HYDROGELS FOR BIOMEDICAL APPLICATIONS

KARTHIKA M.[1], ASWATHY VASUDEVAN[1], JIYA JOSE[1], NANDAKUMAR KALARIKKAL[1,2], and SABU THOMAS[1,3*]

[1]*International and Inter University Centre for Nanoscience and Nanotechnology, Mahatma Gandhi University, Kottayam 686560, Kerala, India*

[2]*School of Pure and Applied Physics, Centre for Nanoscience and Nanotechnology, Mahatma Gandhi University, Kottayam 686560, Kerala, India*

[3]*School of Chemical Sciences, Mahatma Gandhi University, Kottayam 686560, Kerala, India*

Corresponding author. E-mail: sabuchathukulam@yahoo.co.uk

ABSTRACT

Nanocomposite hydrogels (NCHs) are superior biomaterials that can efficiently be used for various biomedical and pharmaceutical applications. The application of these hydrogel networks includes sensors, actuators, drug delivery, stem cell engineering, regenerative medicine, other biomedical devices, etc. Compared to conventional polymeric hydrogels, NCHs have exceptional physical, chemical, electrical, and biological properties which make them potential candidates for various biomedical applications. The improved performance of the NCH network is mainly attributed to the enhanced interactions between the polymer chains and nanoparticles. This chapter deals with different types of NCHs and highlights some of the emerging biomedical and biotechnologcal applications of those hydrogels.

8.1 INTRODUCTION

8.1.1 HYDROGELS—AN OVERVIEW

Hydrogels are hydrophilic, three-dimensional (3D) polymeric networks that are able to absorb an outsized quantity of water. Owing to their exceptional potential in a very wide range of applications, hydrogels have received huge interest for the past 50 years. Its ability to imbibe a large quantity of water arises from hydrophilic functional groups connected to the polymeric backbone, whereas their resistance to dissolution arises from cross-links between network chains. These cross-links may be either physical (entanglements or crystallites) or chemical (tie-points and junctions). The classification of hydrogels may be done in many different ways on the basis of supply, polymeric composition, configuration, variety of cross-linking, and then on.[1] The various synthetic ways being adopted for the preparation of hydrogels include physical cross-linking, chemical cross-linking, grafting polymerization, and radiation cross-linking.[2] Their easy acceptance for the medical specialty applications can be closely correlated to the resemblance with living tissue owing to their ability to imbibe water and biological fluids. As compared with alternative artificial biomaterials, they have a soft and rubbery consistency. Hydrogels may also be synthesized as stimuli-sensitive that answer stimuli like temperature, hydrogen ion concentration, and presence of the electrolyte.[3] These stimuli-sensitive hydrogels can display changes in their swelling behavior of the network structure according to the external environments. The water-holding capacity and permeability are the most important characteristic features of a hydrogel. The mechanism of swelling process of hydrogels is as follows. As soon as they come into contact with water, the polar hydrophilic groups are hydrated first, leading to the formation of primary bound water. Thus, the network swells and exposes the hydrophobic groups which are also capable of interacting with the water molecules. This leads to the formation of hydrophobically bound water, called "secondary bound water." The primary and secondary bound water are together called "total bound water." This network will absorb additional water, due to the osmotic driving force of the network chains toward infinite dilution but is opposed by the covalent or physical cross-links resulting in an elastic network retraction force. This leads the hydrogel to an equilibrium swelling level.

Depending on the nature and composition of the hydrogel, the next step is the disintegration or dissolution if the network chain or cross-links are degradable. Biodegradable hydrogels, containing labile bonds, are therefore advantageous in applications such as tissue engineering, wound healing, and drug delivery. Biocompatibility is another important property that should possess hydrogels to be a potential candidate for biomedical applications. Ideally, they should be metabolized into harmless products. Generally, hydrogels possess a good biocompatibility since their hydrophilic surface has a low interfacial free energy when in contact with body fluids, which results in a low tendency for proteins and cells to adhere to these surfaces. Moreover, the soft and rubbery nature of hydrogels minimizes irritation to encompassing tissue. It is additionally possible to alter the chemistry of the hydrogel by controlling their polarity, surface properties, mechanical properties, and swelling behavior.[4] Hydrogels are physically or chemically cross-linked natural or artificial 3D networks, which might be forged into varied shapes and retain high amounts of water (up to 4000% of their dry weight), though they are hardly hydrosoluble. Their water retention properties principally arise from the presence of hydrophilic groups, like amido, amino, carboxyl, and hydroxyl radical, within the polymer chains; the swelling degree depends on the polymer composition and, additionally, the cross-link density and nature. The water content of hydrogels creates a highly porous structure, a soft and elastic consistency, and a low interfacial surface tension in contact with water or biological fluids. These options make hydrogel's properties nearer to those of biological tissues than the other artificial biomaterial. A "smart" or "stimulus responsive" material is ready to alter its properties like shape, with response to a stimuli. The stimuli can be pH, temperature, ionic strength, solvent composition, pressure, electrical potential, radiation, chemical and biological agents, etc. Smart materials respond with sharp, massive property changes in response to small change in physical or chemical conditions. Common stimuli for smart hydrogels in biological applications are hydrogen ion concentration, temperature and ionic strength, etc. Below given are a number of the summarized advantages and characteristic benefits of hydrogels over typical biomaterials that make them a lot of useful for medical specialty applications, particularly in targeted drug delivery, tissue engineering, etc.

8.1.2   GENERAL BENEFITS

1. Biocompatible
2. Can be injected in vivo as a liquid that then gels at body temperature
3. Protect cells
4. Good transport properties
5. Timed delivery of medicines or nutrients
6. Easy to modify
7. Can be perishable or bioabsorbable

8.1.3   GENERAL LIMITATIONS

1. High cost
2. Low mechanical strength
3. Can be arduous to handle
4. Difficult to load with drugs/nutrients
5. May be tough to sterilize
6. Nonadherent

8.1.4   APPLICATIONS

Hydrogels are used naturally by the human body; examples are cartilage, blood clots, mucin (lining the abdomen, cartilaginous tube tubes, and intestines), and vitreous humor of the eye. The following are examples of biomedical applications:

1. Soft contact lenses
2. Disposable diapers
3. Drug delivery (stimulus responsive and unresponsive)
4. Wound dressings
5. Provide absorption, desloughing, and debriding of necrotic and fibrotic tissue
6. Electroencephalography parts
7. Blood compatible surface for medical devices
8. Scaffolds in tissue engineering

8.1.5 NANOCOMPOSITE HYDROGELS

Traditional hydrogels have low mechanical properties and lack multifunctionalities. Nanosized particles/fillers are widely studied as biomaterials or biofunctional materials. Nanocomposite hydrogels (NCHs), that mix the benefits of both nanofillers and gel matrices, could end in improved mechanical and biological properties and find potential applications in medical specialty field. NCHs (NC gels) are nanomaterial incorporated, hydrated, polymeric networks that exhibit high mechanical strength in comparison with the standard hydrogels. They show much higher strength than usual hydrogels. Over the last decade, new applications of hydrogels have emerged, notably in stem cell engineering,[5] immunomodulation,[6] cellular and molecular therapies,[7] and cancer research.[4] Most of the applications require multiple functionalities of the hydrogel network and its interactions between the surrounding cells. By incorporating a nanomaterial, as similar as in the usual polymer nanocomposite, the properties could be improved by the interaction of the nanomaterial with the polymer chains. By controlling the interaction, various properties could be engineered for the hydrogel. A wide range of nanoparticles, such as carbon-based, polymeric, ceramic, and metallic nanomaterials, can be integrated into the hydrogel networks to obtain nanocomposites with superior properties and tailored functionality. The process of synthesizing NCHs is simple and can be easily made in different shapes such as huge blocks, sheets, thin films, rods, hollow tubes, spheres, bellows, and uneven sheets. NCHs can be potentially used for various pharmaceutical and biological applications. The application of these hydrogel networks includes sensors, actuators, drug delivery, stem cell engineering, regenerative medicine, and other biomedical devices. When compared with the conventional polymeric hydrogels, NCHs possess unique physical, chemical, electrical, and biological properties. Electrically conductive scaffolds, mechanically stiff and highly elastomeric network, cell and tissue adhesive matrices, injectable matrices, and controlled drug delivery depots are some examples of NCHs. Research trends are currently focused on the incorporation of many nanoparticulate systems such as carbon-based nanomaterials, polymeric nanoparticles, ceramic nanoparticles, and metal/metal-oxide nanoparticles into the hydrogel network so as to obtain NCHs. The interaction of this system with the polymeric chains of the hydrogel structure results in exceptional properties of the nanocomposite absent in the individual components. Nanoparticle addition may reinforce the starting hydrogels and provide the NCHs with responsiveness to external stimuli such as mechanical, thermal, magnetic, and electric. In particular, the nature of the nanosystems incorporated in the starting hydrogel determines the kind of stimuli to which the NCH is responsive.

8.2 NCHs FOR BIOMEDICAL APPLICATIONS

Over the past time, new applications of hydrogels have developed, particularly in stem cell engineering, cellular and molecular therapies, immunomodulation, and cancer research. A vast range of innovations, in polymer chemistry, micro and nanofabrication technologies, and biomolecular engineering, have wiped away the limits of designing composite hydrogel networks with customized functionality. Recent trends also indicate significant and growing stage in developing NCHs for various biomedical applications. In this chapter, we have focused on the innovations in the design and development of NCHs with tailored physical properties and custom functionality. We will discuss different types of NCHs and highlight some of the emerging biomedical and biotechnological applications of those hydrogels.

As mentioned earlier, we can use different types of nanomaterials such as

- Carbon-based nanomaterials (carbon nanotubes, CNTs, graphene, and nanodiamonds)
- Polymeric nanoparticles (polymer nanoparticles, dendrimers, and hyperbranched polyesters)
- Inorganic/ceramic nanoparticles (hydroxyapatite, silica, silicates, and calcium phosphate)
- Metal/metal-oxide nanoparticles (gold, silver, and iron oxide) are combined with the polymeric network to obtain NCHs

These nanoparticles physically or covalently act with the polymeric chains and end in novel properties of the nanocomposite network. Developing NCHs with modified functionality has opened new potentialities in developing advanced biomaterials for numerous biomedical and biotechnological applications. Smart hydrogels are useful materials that are able to modulate the equilibrium swelling as a function of a change within the external setting, such as pH, ionic strength, temperature, pressure, light, magnetic force, sound fields, etc.[8] Figure 8.1 explains the use of various nanoparticles such as carbon-based nanomaterials, polymeric nanoparticles, inorganic nanoparticles, and metal/metal-oxide nanoparticles combined with the synthetic or natural polymers to obtain NCHs with desired property combinations and their potential applications. The porous network structure of NCHs makes it simple for the nutrients to enter and exit the structure. This porous structure might be effectively made use of for the drug delivery applications wherever the drug will simply escape the gel and absorbed to the body. The elastomeric properties obtained by the incorporation of the nanomaterial would be helpful to use them for creating scaffolds as they will acquire the specified form without having previous molding. Apart from that, researchers also are wanting into incorporating NCHs with silver nanoparticles for antibacterial applications and microorganism elimination in medical and food packing and water treatment.[9] The biomimetic properties of gel attract several scientists within the area of tissue engineering that may build the tissue engineering method more practical and fewer invasive to our body.[10] The various applications developed in the biomedical field with the different types of NCHs as the potential candidate are briefly discussed here.

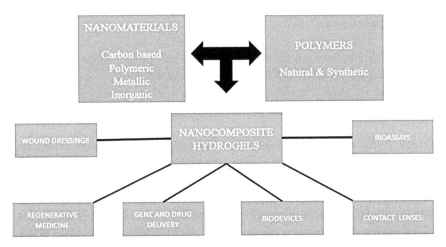

FIGURE 8.1 A variety of nanoparticles combined with the synthetic or natural polymers to obtain nanocomposite hydrogels with desired property for various biomedical applications.

8.2.1 CARBON-BASED NCHs FOR BIOMEDICAL APPLICATIONS

When hydrogels are incorporated with carbon-based materials such as nanotubes,[11] graphene,[12] fullerene (C60),[13] and nanodiamonds,[13] they acquire higher mechanical stability and electrical conduction that might be appropriate for tissue engineering, drug delivery, biosensing, etc. They could be able to mimic the nerve, muscle, and cardiac tissues. Both the single-walled and multiwalled CNTs are investigated for the biomedical applications.[14] However, their low affinity with the hydrophilic polymer might limit the interaction. Major carbon-based nanomaterial suggested for the potential uses in biomedicine are CNTs, graphene, fullerene (C60), and nanodiamonds. Especially, CNTs and graphene are used for incorporating multicharacteristics like high mechanical, electrical conductivity, and optical properties within the synthetic or natural polymeric systems.[15] For example, both CNTs and graphene-based NCHs are used for the applications like actuators, conductive tapes, biosensors, tissue engineering scaffolds, drug delivery systems, and biomedical devices. CNTs are classified as multiwalled or single-walled CNTs. Although both of them are used for medicine applications, there is a limitation related to CNTs is their hydrophobic nature. It ends up in the restricted interaction of the material with hydrophilic polymers. To beat this, many ways are being developed to reinforce the dispersion of the CNTs among the nanocomposite network as follows:

- CNT surfaces are modified using various polar groups like amines (NH$_2$), hydroxyls (OH), carboxyls (COOH), etc.
- Use of single-stranded DNA, proteins, and surfactants to modify the surface properties.

As a result of their high electrical conductivity, nanocomposites strengthened with CNTs are used to engineer a variety of electrically conductive tissues like nerve, muscle, and viscus tissues. As an example, methacrylate gelatin (GelMA) was bolstered with multiwalled COOH-functionalized CNTs to fabricate hybrid NCHs. A nearly threefold increase in the tensile modulus was observed due to the addition of 0.5% CNTs to the GelMA hydrogels. Because of the addition of the multiwalled CNTs, the rise within the mechanical strength of GelMA hydrogels was attributed to the formation of a nanofibrous mesh-like network of GelMA-coated multiwalled CNTs.[13] It additionally will increase the mechanical strength of GelMA hydrogels. It absolutely was found that electrical conductivity of the CNTs was enhanced and a conductive network was formed within the NCHs, due to the cardiomyocytes that are seeded onto CNT-GelMA NCHs (Fig. 8.2).

When the engineered, cardiac tissues are discharged from the substrate, it is actuated spontaneously and moved within the fluid surroundings owing to the cyclical contraction of the cells. The CNT-based hydrogels are able to demonstrate that the lab-grown tissues will mimic the native tissues and their utility within the body as tissue replacements. Another functionalization of the surface of carbon-based nanomaterials is the grafting of polymer chains onto the CNT surfaces.[17] In this approach, the CNT surface is kept secure; surrounding polymeric network mainly perceives the surface-grafted polymer which reinforces the dispersion of CNTs and also the physical interactions between CNTs and therefore the polymer. For example, the surface of the multiwalled CNTs was hybridized with amyloid fibrils to get fibrous hydrogels. It was observed that grafting the amyloid fibers enhances the polymer–nanotube interactions. This

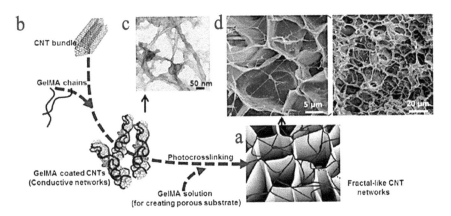

FIGURE 8.2   (a) Schematic diagram illustrating the isolated heart conduction systems showing the Purkinje fibers, which are located in the inner ventricular walls of the heart. Heart muscle with Purkinje fiber networks on the surface of the heart muscle fibers. (b) Preparation procedure of fractal-like CNT networks embedded in GelMA hydrogel. (c) Trsansmission electron microscopy image of GelMA-coated CNT. (Adapted with permission from References [15,16]. ©2013 American Chemical Society.)

approach is applied to varied protein-CNT systems since several proteins can form amyloid fibers. These protein-CNTs NCHs are used to design responsive hydrogels for applications like tissue engineering, drug delivery, biosensing, etc. Another carbon-based nanomaterial widely being employed to provide hydrogels for the medicine purpose is graphene.[17] Graphene is a two-dimensional carbon-based nanomaterial with high mechanical strength and is an excellent conductor of heat and electricity. Graphene sheets are treated with strong oxidizers to obtain graphite oxide (GO), to increase the solubility of graphene in physiological conditions. Graphene is a two-dimensional carbon-based nanomaterial with high mechanical strength and is a superb conductor of heat and electricity. Graphene sheets are treated with strong oxidizers to get GO, to extend the solubility of graphene in physiological conditions. GO will act with hydrophilic polymers to make physically or covalently cross-linked networks. Genetically designed elastin-like polypeptides (ELPs) were noncovalently bolstered with GO to get colloidal gel actuators that are stimuli-responsive. The ensuing hybrid GO-ELP hydrogels are mechanically actuated to bend, stretch, and twist once subjected to different radiation intensities. The covalent cross-linking between compound and nanoparticles permits the transfer of mechanical force among the cross-linked network, leading to increased mechanical strength and toughness. Mechanically stiff and extremely resilient NCHs were fabricated by covalently conjugating GO sheets to polyacrylamide [GO sheets were functionalized via radiation-induced peroxidation to get graphene peroxide (GPO)].[18] It increases the interfacial interactions. The functionalized GPO was covalently cross-linked with polyacrylamide to obtain NCHs. This method could be used to fabricate elastomeric scaffolds for tissues.

8.2.2 POLYMERIC MATERIAL-BASED NCHs FOR BIOMEDICAL APPLICATIONS

Hydrogels incorporated with polymeric nanoparticles are potential candidates for drug delivery and tissue engineering.[19] The addition of polymeric nanoparticles provides these hydrogels a bolstered polymeric network that's stiffer and has the flexibility to surround hydrophilic and hydrophobic drugs together with genes and proteins.[20] High stress-absorbing property makes them a possible candidate for cartilage tissue engineering.

Polymeric nanoparticles like dendrimers, hyperbranched polymers, liposomes, polymeric micelles, nanogels, and core–shell polymeric particles have achieved great importance in drug delivery application as a result of their ability to entrap hydrophobic or hydrophilic drugs. Among these systems, dendrimers and hyperbranched polymeric nanoparticles are attractive due to their branched and spherical structure.[21] Owing to this distinctive nanostructure, they show a large number of functional groups on their surface that end in higher reactivity and loading potency. In addition, a range of bioactive agents like drugs, proteins, and genes are often encapsulated on the outer boundary of these nanoparticles. Among the manifold nanomaterials which will be incorporated into hydrogels, a range of chemical compound nanoparticles have additionally been used, primarily aiming to endow the final NCH with controlled drug release ability, together with the mechanical reinforcement. Dendritic nanoparticles are often used to reinforce the network within the hydrogel via covalent or noncovalent interactions with the compound chains. For example, polyamide amine dendritic nanoparticles were physically integrated inside collagen scaffolds to boost the structural integrity and mechanical stiffness. This could conjointly improve the human conjunctival fibroblast proliferation in collagen hydrogels. The applications of these hybrid hydrogels from dendrimers and hyperbranched nanoparticles are used within the pharmaceutical and biomedical area which needs a porous structure with controlled drug delivery properties. By changing the dendrimer concentration, the stiffness of the hydrogels as well as the degradation properties and also the hydration kinetics may be tailored. The nanocomposite is also helpful for cartilage tissue engineering applications attributable to the high stress-absorbing capability. To beat the difficulties of low drug loading and kinetic properties related to the low generation macromolecules that are utilized in tissue engineering applications, hydrogels from hyperbranched nanoparticles (generation 5) are fabricated. The physical properties of the hydrogels together with the microstructure and mechanical strength may be tuned by controlling the cross-linking density. These photo-cross-linkable hydrogels will encapsulate the hydrophobic drug within the inner cavities of the nanoparticles. A controlled drug release for more than a week may be achieved, which is difficult to achieve using conventional hydrogels made of linear polymers (Fig. 8.3).

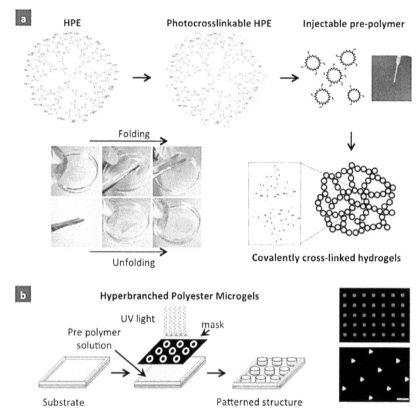

FIGURE 8.3 Microfabricated photocross-linkable hyperbranched poly (HPE) hydrogels made by cross-linking HPE macromers. (a) Precursor solution containing 50% HPE macromers showed low viscosity and can be injected using 22-ga needle. When subjected to ultraviolet radiation the prepolymer solution quickly formed covalently cross-linked network. The optical images representing the flexibility of 150-µm-thin methylene blue dyed HPE hydrogel sheet. The HPE hydrogel sheet can be easily handled without breaking. After folding, the hydrogel sheet easily unfolds and regains its original shape in the Dulbecco's phosphate buffered saline bath. (b) Schematic for developing patterned HPE hydrogel structures using photolithography. Fluorescent images of different micropatterned structures obtained using photolithography (scale bar = 500 µm). Adapted with permission from Reference [21]. ©2011 American Chemical Society.

8.2.3   INORGANIC NANOPARTICLES PRIMARILY BASED NCHs FOR MEDICINE APPLICATIONS

Researchers are developing future generation of advanced biomaterials by combining inorganic ceramic nanoparticles with natural or synthetic polymers. Similarly, the employment of inorganic nanomaterials for improving the hydrogel properties is additionally being reported.[22] Most of the inorganic nanomaterials are already present in our body and are necessary for the normal functioning, and there will not be any adverse impact. Some of them, like calcium and silicon, facilitate with preventing bone loss and skeletal development. Others, like nanoclays, improve the structural formation. A variety of bioactive nanoparticles, like hydroxy-apatite (nHA), synthetic silicate nanoparticles, bioactive glasses, silica, calcium phosphate, glass ceramic, β-wollastonite, etc., are reported for biomedical applications. From these materials, advanced biomaterials can be created by combining inorganic ceramic nanoparticles with natural or synthetic polymers. The inorganic nanoparticles contain minerals that are already present within the body that is critical for the normal func-tioning of human tissues. It also shows desirable biological responses, for instance, calcium, which is present in bone and plays an important role in bone development and maintenance. The presence of intracellular calcium and phosphate within the osteoblasts promote the deposition of the miner-alized matrix and stop bone loss. Similarly, silicon, another component present in ceramic nanoparticles, will play a key role in skeletal develop-ment. Considering these facts, incorporating such inorganic nanoparticles in polymeric hydrogels is predicted to supply bioactive characteristics to the network. Different sorts of NCHs are synthesized by incorporating ceramic nanoparticles inside the polymer matrix. For instance, nHA was incorporated within poly(ethylene glycol) (PEG) matrix to get extremely elastomeric NCHs. It enhances the mechanical strength and improves the physiological stability of the nanocomposite. Another vital material is nanoclay (synthetic salt nanoplatelets). It will induce osteogenic differ-entiation of human mesenchymal stem cells without the utilization of exogenous growth factors. It will trigger a series of events that follow the temporal pattern of osteogenic differentiation, bone-related matrix protein deposition (osteocalcin and osteopontin), followed by matrix mineraliza-tion. The distinctive bioactive properties of the nanoclay may be wont to create devices like injectable tissue repair matrices, bioactive fillers, or therapeutic agents helpful for triggering specific cellular responses toward bone-related tissue engineering approaches. The synthetic silicate nano-platelets produce mechanically robust and tissue-adhesive NCHs once they are mixed with linear and branched polymers. The physical and chem-ical properties of silicate-based nanocomposites result in the high surface interactions and also the noncovalent interactions because the polymer chains will reversibly sorb and desorb on the silicate surfaces. The addition of silicates will improve the elongation of the polymeric hydrogels attributable to the formation of physically cross-linked networks. This modified PEG network enhances the cell and tissue adhesive properties of the NCHs.

8.2.4   METAL AND METAL-OXIDE NANOPARTICLES BASED NCHs FOR BIOMEDICAL APPLICATIONS

The electrical conductivity is extremely vital for the hydrogels in forming functional tissue and to use as imaging agents, drug delivery systems, conductive scaffolds, sensors, etc. Once the electrically conductive metal and metal-oxide nanoparticles are incorporated within the hydrogel, the electrical conduction and antimicrobial property of the hydrogels are increased. Various sorts of metallic nanoparticles used to fabricate NCHs for medicine applications, including gold (Au), silver (Ag), and different metal nanoparticles, whereas metal-oxide nanoparticles include the iron oxide (Fe_3O_4, Fe_2O_3), titania (TiO_2), alumina, and zirconia. Once such conductive nanoparticles are embedded or entrapped inside a polymeric hydrogel network, they enhance the electrical conduction of the NCHs. In a very recent try, researchers incorporated Au nanowires within macroporous alginate hydrogels to boost the electrical conduction of the polymeric network (Fig. 8.4).[22] Such semiconductive scaffolds can be used to engineer tissues that need propagation of electrical signals to facilitate the formation of practical tissues. As an example, neonatal rat cardiomyocytes once seeded in the semiconductive colloidal gel network, contract synchronously once electrically stimulated.[22] Moreover, the addition of Au nanowire to the alginate hydrogel conjointly showed a better expression of the cardiac markers (such as troponin I and connexin 43) within the seeded cardiac cells in comparison to the alginate hydrogels.

Metal and metal-oxide nanoparticles possess desired physical and electrical properties (conductivity in Au nanorods), magnetic properties (iron oxides), and antimicrobial properties (Ag nanoparticles). Owing to these characteristics, NCHs containing metal or metal-oxide nanoparticles are extensively used as imaging agents, drug delivery systems, conductive scaffolds, switchable electronics, actuators, sensors, etc. The functionalization of the nanoparticle surfaces enhances the interactions between the polymer and also the nanoparticles. This improvement influences the physical, chemical, and biological properties of the NCHs. As an example, once the thiol-functionalized Au nanoparticles are allowed to cross-link with the polymeric network; there is a big increase within the stiffness is observed. Metallic nanoparticles possess high electrical and thermal conduction. Once these conductive nanoparticles are incorporated among a polymeric hydrogel network, they will enhance the electrical conduction of the NCHs. As an example, macroporous alginate hydrogels incorporated with Au nanowires shows electrical conduction within the polymeric

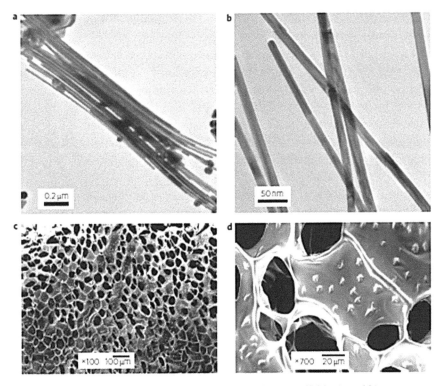

FIGURE 8.4 Incorporation of nanowires within alginate scaffolds. (a and b) Transmission electron microscopy images of a typical distribution of gold nanowires, which exhibited an average length of 1 mm and average diameter of 30 nm. (c and d) scanning electron microscopy image revealed that the nanowires (1 mg mL^{-1}) assembled within the pore walls of the scaffold into star-shaped structures with a total length scale of 5 mm. The assembled wires were distributed homogeneously within the matrix (c) at a distance of 5 mm from one another (d). Adapted with permission from Reference [23]. ©2013 Nature.

network. Such semiconductive scaffolds are often used to develop tissues that need propagation of electrical signals to facilitate the formation of functional tissue

The magnetic nanoparticles incorporated within hydrogel network will act with external magnetic fields. Once these hybrid hydrogels are subjected to external magnetic fields, the magnetic nanoparticle entrapped among the hydrogel network generate heat. The rise within the temperature of encompassing matrix higher than the lower critical solution temperature of the compound ends up in a coil-to-globule transition of the compound chains. Such sort of triggers will be used to release therapeutic agents or cells from the NCHs. Application of NCHs for biomedical biomedicine depends on the sort of physical and chemical biomedicine of the network. As an example, by incorporating magnetic nanoparticles among colloidal gel network, the nanocomposite network will remotely interact with external magnetic fields. In an exceedingly recent study, magnetic nanoparticles are covalently conjugated with biocompatible and thermoresponsive hydroxypropyl cellulose to get stimuli-responsive hydrogels.[23] Once these hybrid hydrogels are subjected to external magnetic fields, the magnetic nanoparticle entrapped within the colloidal gel network generate heat. The rise within the temperature of encompassing matrix on top of the lower essential resolution temperature of the compound ends up in a coil-to-globule transition of the compound chains. However, varied varieties of antimonial and metal chemical compound nanoparticles will be used to explore coming up with of NCHs with desired practicality. Besides using the metallic or metal-oxide-based NCHs for tissue engineering applications, they are conjointly explored for biosensing, diagnostic, and bioactuation applications.

8.3 CONCLUSION

NCHs are advanced biomaterials that may probably be used for numerous biomedical and pharmaceutical applications. The application of these hydrogel networks includes sensors, actuators, drug delivery, stem cell engineering, regenerative medicine, and different biomedical devices. Compared to traditional polymeric hydrogels, NCHs have superior physical, chemical, electrical, and biological properties. Electrically conductive scaffolds, mechanically stiff and extremely elastomeric network, cell and tissue adhesive matrices, injectable matrices, and controlled drug delivery depots are some examples of NCHs. The improved performance of the NCH network is especially attributed to the improved interactions between the polymer chains and nanoparticles. The NCHs are evaluated for various biomedical applications. The improved surface interactions between the nanoparticles and also the polymer chains build the fabric properties helpful for applications. However, the present situations lack the control over some essential options like stimuli responsiveness and biodegradation. To beat these challenges, alternate ways ought to be developed. Many properties will be incorporated to mimic structure and properties of native tissues, by combining multiple phases among a nano-composite colloidal gel network. As an example, aliphatic ester dendrimers conjugated with PEG spacer will physically act with inorganic nanoplate-lets (ceramic nanoparticles) to make physically cross-linked hydrogels. It conjointly possesses self-healing characteristics, high mechanical strength, and adhesive properties. Researchers had made up multicomponent NCHs from ferritin (iron core adorned with protein shell) and poly(vinyl alcohol) using electrospinning. The addition of ferritin considerably increased the mechanical stiffness and reduced creep throughout the cyclic deformation of the NCHs. Such properties were attributed to the presence of ferritin nanoparticles that act as nanoscale springs within the polyvinyl acetate matrix. These sorts of hybrid hydrogel networks will probably be used to fabricate scaffolds for tissue engineering applications, stimuli responsive matrix for drug delivery, or as actuators. These multicomponent systems are able to offer superior properties compared to one or two parts systems; however, the complexness involved in optimizing the specified composi-tion might have to be addressed in future. The NCHs have proven that microscale technologies can be used to entrap cells in predefined geom-etries and cellular microenvironment. These microfabrication technolo-gies enable better positioning of various sorts of cells among the hydrogel network to facilitate the formation of functional tissues. The utilization of microscale technologies in nanomaterials analysis can change us to grasp and address a number of the complexness related to engineering functional tissues.

KEYWORDS

- **hydrogels**
- **polymer**
- **nanomaterials**
- **composite**
- **nanomedicine**
- **nanodevices**

REFERENCES

1. Ahmed, E. M. Hydrogel: Preparation, Characterization, and Applications: A Review. *J. Adv. Res.* **2015,** *6,* 105–121.
2. Ullah, F.; Othman, M. B. H.; Javed, F.; Ahmad, Z.; Akil, H. M. Classification, Processing and Application of Hydrogels: A Review. *Mater. Sci. Eng. C* **2015,** *57,* 414–433.
3. Koetting, M. C.; Peters, J. T.; Steichen, S. D.; Peppas N. A. Stimulus-Responsive Hydrogels: Theory, Modern Advances, and Applications. *Mater. Sci. Eng., R* **2015,** *93,* 1–49.
4. Bahram, M.; Mohseni, N.; Moghtader, M. An Introduction to Hydrogels and Some Recent Applications. In *Emerging Concepts in Analysis and Applications of Hydrogels*; Majee, S. B., Ed.; InTech: Rijeka, 2016; Ch. 02.
5. Wheeldon, I.; Ahari, A. F.; Khademhosseini, A. Microengineering Hydrogels for Stem Cell Bioengineering and Tissue Regeneration. *JALA Charlottesv. Va.* **2010,** *15,* 440–448.
6. Singh, A.; Peppas, N. A. Hydrogels and Scaffolds for Immunomodulation. *Adv. Mater.* **2014,** *26,* 6530–6541.
7. Gooch, A.; Westenfelder, C. Modified Hydrogels to Enhance Cellular Therapy for AKI: A Translational Challenge. *J. Am. Soc. Nephrol.* **2016,** *27,* 2219–2221.
8. Ebara, M.; Kotsuchibashi, Y.; Uto, K.; Aoyagi, T.; Kim, Y.-J.; Narain, R., et al. *Smart Hydrogels. Smart Biomaterials*; Springer: Tokyo, Japan, 2014; pp 9–65.
9. Obradovic, B.; Stojkovska, J.; Jovanovic, Z.; Miskovic-Stankovic, V. Novel Alginate Based Nanocomposite Hydrogels with Incorporated Silver Nanoparticles. *J. Mater. Sci.: Mater. Med.* **2012,** *23,* 99–107.
10. Annabi, N.; Tamayol, A.; Uquillas, J. A.; Akbari, M.; Bertassoni, L. E.; Cha, C., et al. 25th Anniversary Article: Rational Design and Applications of Hydrogels in Regenerative Medicine. *Adv. Mater.* **2014,** *26,* 85–124.
11. Satarkar, N. S.; Biswal, D.; Hilt, J. Z. Hydrogel Nanocomposites: A Review of Applications as Remote Controlled Biomaterials. *Soft Matter* **2010,** *6,* 2364–2371.
12. Zhu, C.-H.; Lu, Y.; Peng, J.; Chen, J.-F.; Yu, S.-H. Photothermally Sensitive Poly(*N*-Isopropylacrylamide)/Graphene Oxide Nanocomposite Hydrogels as Remote Light-Controlled Liquid Microvalves. *Adv. Funct. Mater.* **2012,** *22,* 4017–4022.

13. Cha, C.; Shin, S. R.; Annabi, N.; Dokmeci, M. R.; Khademhosseini, A. Carbon-Based Nanomaterials: Multi-Functional Materials for Biomedical Engineering. *ACS Nano* **2013**, *7*, 2891–2897.

14. Li, C.; Mezzenga, R. Functionalization of Multiwalled Carbon Nanotubes and Their pH-Responsive Hydrogels with Amyloid Fibrils. *Langmuir* **2012**, *28*, 10142–10146.

15. Liu, J.; Chen, C.; He, C.; Zhao, J.; Yang, X.; Wang, H. Synthesis of Graphene Peroxide and Its Application in Fabricating Super Extensible and Highly Resilient Nanocomposite Hydrogels. *ACS Nano* **2012**, *6*, 8194–8202.

16. Shin, S. R.; Jung, S. M.; Zalabany, M.; Kim, K.; Zorlutuna, P.; Kim, S. B., et al. Carbon-Nanotube-Embedded Hydrogel Sheets for Engineering Cardiac Constructs and Bioactuators. *ACS Nano* **2013**, *7*, 2369–2380.

17. Cha, C.; Shin, S. R.; Annabi, N.; Dokmeci, M. R.; Khademhosseini, A. Carbon-Based Nanomaterials: Multifunctional Materials for Biomedical Engineering. *ACS Nano* **2013**, *7*, 2891–2897.

18. Biondi, M.; Borzacchiello, A.; Mayol, L.; Ambrosio, L. Nanoparticle-Integrated Hydrogels as Multifunctional Composite Materials for Biomedical Applications. *Gels* **2015**, *1*, 162.

19. Gillies, E. R.; Fréchet, J. M. J. Dendrimers and Dendritic Polymers in Drug Delivery. *Drug Discovery Today* **2005**, *10*, 35–43.

20. Zhang, H.; Patel, A.; Gaharwar, A. K.; Mihaila, S. M.; Iviglia, G.; Mukundan, S., et al. Hyperbranched Polyester Hydrogels with Controlled Drug Release and Cell Adhesion Properties. *Biomacromolecules* **2013**, *14*, 1299–1310.

21. Oral, E.; Peppas, N. A. Responsive and Recognitive Hydrogels Using Star Polymers. *J. Biomed. Mater. Res. Part A* **2004**, *68A*, 439–447.

22. Dvir, T.; Timko, B. P.; Brigham M. D.; Naik, S. R.; Karajanagi, S. S.; Levy, O., et al. Nanowired Three Dimensional Cardiac Patches. *Nat. Nanotechnol.* **2011**, *6*, 720–725.

23. Gaharwar, A. K.; Peppas, N. A.; Khademhosseini, A. Nanocomposite Hydrogels for Biomedical Applications. *Biotechnol. Bioeng.* **2014**, *111*, 441–453.

CHAPTER 9

DEVELOPMENT OF BIOACTIVE MULTIFUNCTIONAL INORGANIC–ORGANIC HYBRID RESIN WITH POLYMERIZABLE METHACRYLATE GROUPS FOR BIOMEDICAL APPLICATIONS

VIBHA C. and LIZYMOL P. P.[*]

Division of Dental Products, Biomedical Technology Wing, Sree Chitra Tirunal Institute for Medical Sciences and Technology, Poojappura, Thiruvananthapuram 695012, India

[*]Corresponding author. E-mail: lizymol@rediffmail.com

ABSTRACT

The implication of biodegradable polymers in the field of biomedical applications urges its innovation. In the hierarchy of synthetic biomaterials, acrylic bone cement has its uniqueness make the usage of its brands highly demanding in arthroplasty even now. Most of the acrylic bone–cement brands are based on poly(methyl methacrylate), but they are overwhelmed with many drawbacks. Several research efforts and reviews were cited to overcome these drawbacks by implementing a large number of formulations. The objective of the present work was to present a critical review which summarizes the main advances and drawbacks in the field of bone cement and to develop a novel material, which can be a potential candidate in biomedical applications. In the present study, we discussed about the mechanical properties and bioactivity of the new composite, which we have developed for orthopedic applications.

9.1 INTRODUCTION

Bone is a unique tissue that gives a mechanical support to the body. It is a major reservoir of calcium minerals and bone marrow from which blood cells originate. Fracture in the bone weakens its mechanical support. A bone fracture can be the result of high force impact or stress, or trivial injury as a result of certain medical conditions that weaken the bones, such as osteoporosis, bone cancer, or osteogenesis imperfecta, etc. Acrylic bone cement [poly(methyl methacrylate) (PMMA)] is the most commonly used nonmetallic implant material in orthopedics to fix cemented prostheses to the hosting bone over 60 years of age. Dr. Gluck described the use of ivory ball-and-socket joints which were especially useful in the treatment of diseases of the hip joint in 1890. These joints were stabilized in the bone with cement and composed of colophony, pumice powder, and plaster.[1] While in 1951, Dr. Haboush used a self-curing acrylic dental cement to secure a total hip replacement.[2] During that time, PMMA cements were used primarily in dentistry to fabricate partial dentures, orthodontic retainers, artificial teeth, denture repair resins, and an all-acrylic dental restorative.[3,4] However, its application flourish in other areas like bone cements, contact and intraocular lens, screw fixation in bone, filler for bone cavities and skull defects, vertebrae stabilization, etc., due to its biocompatibility, reliability, relative ease of manipulation, and low toxicity.[5] Its use as a grouting material to stabilize hip prostheses was pioneered by Sir John Charnley at Wrightington Hospital, Wigan, United Kingdom in the late 1950s, who borrowed it from dentistry.[6–8] Bone cements are self-curing systems generally dispensed as powder and liquid phases, which are packaged separately and are mixed prior to use and introduced in the intramedullary canal in a dough state which subsequently undergoes in situ polymerization to yield the hardened cement mass. The solid phase of bone cement is mainly composed of spherical particles (beads) of PMMA or acrylic copolymers containing ethyl acrylate, methyl acrylate, or methylmethacrylate (MMA)-styrene copolymers. Average molecular weights of the beads are reported to be in the range from 44,000 to 1,980,000.[9] The solid phase also includes the initiator, benzoyl peroxide (BPO), in a concentration ranging from 0.75 to 2.5 wt%, which is physically mixed with the beads. Another component of this phase is the radiopaque agent, either barium sulfate or zirconium dioxide, that are normally incorporated as an X-ray contrast agent to allow X-ray follow-up analysis of the prosthesis. The concentration of the radiopaque agent is generally around 10 wt% on the basis of the solid component, and the average diameter of the radiopaque particles is around 5 μm.[10] The liquid phase of the cement formulation consists of MMA monomer at a concentration of 95 wt%. In order to initiate the polymerization at room temperature, a tertiary aromatic amine which acts as the activator especially N,N-dimethyl-4-toluidine (DMT) is added in the liquid phase in a concentration range of 0.89–2.7 wt% in conventional bone cement. Finally, an inhibitor is added to the liquid phase to avoid any premature polymerization that may occur in the presence of heat or light during storage. The inhibitor belongs to the family of quinines which acts as a radical scavenger forming radicals that are stabilized by resonance. The solid:liquid ratio of acrylic bone cement is very sensitive for the cement's setting parameters which need to be optimized. Care must be taken as manipulation of this ratio may have an adverse effect on excessive polymerization shrinkage or under-polymerization.[10] While mixing the two phases, the monomer begins to solvate the surface of the prepolymerized beads producing a paste-like consistency. The viscosity increases with time and then attains a rigid state. This setting process has been classified into four phases, namely, the (1) mixing phase, (2) waiting phase, (3) working phase, and (4) hardening phase. When both phases are mixed, the reaction between the initiator and the activator starts giving rise to primary free radicals, which initiate the free radical polymerization of the monomer. In the mixing step, the wetting process predominates, whereas during the waiting phase, the radical polymerization commences causing the viscosity to increase. During the working phase, the polymerization progresses result in a reduction in the mobility of the growing macromolecular species and a sudden increase in the viscosity and heat generation. In the setting step, the cement hardens with the subsequent cessation of the polymerization due to vitrification of the material. The bone cement is subjected to high mechanical stress in the body. The international standard ISO 5833 describes the methods for determining compressive strength (CS), bending strength, and bending modulus. The main tasks of the cement layer are to resist and transfer the loads between the natural and synthetic coupled materials, function as a mechanical buffer, reduce the stress concentrations, and absorbing mechanical shock.[11] If the external stress factors are greater than the inherent strength of the cement, a break will result. For this reason, it is extremely important that the bone cement is mechanically stable, possessing a substantial degree of strength and toughness. The usage of PMMA assures optimal implant stability which should be guaranteed for the entire implant life. However, the conventional bone cement is beset with a number of drawbacks such as chemical necrosis of bone, evoking inflammatory periprosthetic tissue responses by interact with the surrounding tissues, modulus mismatch, polymerization shrinkage, and high polymerization exotherm.[12] Bone cement is formed by mixing the prepolymerized PMMA beads with methylmethacrylate monomer liquid. The polymerization proceeds as a result of a redox reaction, which occurs between the BPO present in the polymer powder and the *N,N*-dimethyl-*p*-toluidine (DMPT) included in the monomer liquid. BPO decomposes to produce free radicals; it reacts with monomer to generate an active core. The reaction is exothermic, and temperatures can exceed 100°C. The polymerization proceeds through an addition reaction which takes place in three distinct steps: initiation, propagation, and termination. These processes do not occur sequentially but are instantaneous, and according to their particular rates that vary with time from the commencement of mixing.[13] Hasenwinkel[14] pointed out that porosity within the bulk of the cement can also have adverse effects which act as stress risers and initiating sites for crack formation. Topoleski et al.[15] reported that pores in bone cement act as nucleation sites for microcracks which strengthen Hasenwinkel's opinion. He and his coworkers developed novel high-viscosity two-solution acrylic bone cement. Two separate pore-free solutions are prepared by predissolving PMMA powder into MMA monomer and adding the initiator to one solution and the activator to the other. The two solutions can be stored for prolonged periods before being mixed and delivered to the surgical site where they polymerize through the same reaction mechanism as powder/liquid bone cements. While comparing with conventional cement, the amount of the liquid monomer in two-solution cement is greater. It has been already reported by Lewis[12] that the unreacted monomer (MMA) in the bone cement in vivo implicated in chemical necrosis of the bone, and it also can affect the material properties of the cement. Many researchers have developed bone cements as an alternative for the conventionally prepared bone cement by mixing powder and liquid by modifying the activator, for example, 4,*N,N*-(diethylamino) phenethanol, 4,4-(dimethylamino) phenyl acetic acid, 4-dimethylamino benzyl methacrylate, 4-dimethylamino benzyl alcohol, 4,4-dimethylamino benzydrol, 4-*N,N*-dimethylamino-4-benzyl laurate, 4-*N,N*-dimethylamino-4-benzyl oleate, etc., and monomers, that is, diethyl amino ethyl methacrylate, 4-methacryloyloxybenzoic acid, 4-diethylamino benzyl methacrylate, etc., but it is still beset with the poor interface between the cement and the bone.[16] The mechanical properties of the cemented prosthesis system may be improved by increasing the mechanical properties of the acrylic bone cement. The mechanical properties of acrylic bone cements can be increased by the addition of reinforcements. The reinforcements used have been fibers made of Kevlar,[17,18] polyethylene,[19] carbon,[20] hydroxyapatite (HAP),[21,22] bone mineral,[23] PMMA,[24] titanium,[25] and stainless steel[26] as well as particles of glass,[27] alumina,[28] and ABS.[29] The bonding between the reinforcements and the acrylic bone cement is extremely poor due to failure at the reinforcement–cement interface except PMMA fibers.[26,30,31] The adhesion between the acrylic matrix and the reinforcement can be improved through the use of a coupling agent (silane-bonding agents) which will increase the mechanical properties of acrylic bone cements.[32,33] Silane-bonding agents improve bonding by the formation of strong covalent bonds in between the acrylic matrix and reinforcements[33–36] which act as a bridge between the reinforcement and the acrylic matrix. They bond covalently to the reinforcement on one end, while the organic moiety on the other end forms covalent bonds with the methylmethacrylate. These new bone cements have improved mechanical properties but can't overcome the problems beset with conventional PMMA bone cement like poor adhesion to bone, poor bioactivity, and lack of osteoconductivity.[37–39] Oonishi suggests that using of bioactive bone cement is a solution to overcome this problem at the interface and categorize bone cement into three, that is, all-bioactive bone cement, surface-bioactive cement, and interface-bioactive bone cement.[40] Bioactive materials have the capability of bioactivity (bone bonding) when implanted in bone defects. Bioactive bone cement is designed to produce a sufficient interlocking between cement and the surrounding cancellous bone. Thus, bioactivity should be induced into the bone cement. Bioactivity in bone cement can exhibit by incorporating silanol groups and calcium ions. This type of modification makes the cement to have the potential to form a layer of bone-like HAP on its surface through a reaction between the material and surrounding body fluid results in osteoconduction.[41,42] It has been previously reported that in order to achieve better fixation of orthopedic implants, that is, for providing bioactivity bioactive ceramics like HAP,[43] glass–ceramics A-W, titanium dioxide,[44–46] and bisphenol-a-glycidyl methacrylate (bis-GMA)-based resin can be added to the powder of the

PMMA-based bone cement.[47–51] Osborn and Newesely found out that among different calcium phosphate ceramics, HAP is identical with the original bone mineral.[52] So, it can be reinforced with PMMA filler part to improve its bioactivity. This investigation was performed by a group of Italian researchers at Bolgana University.[53–56] Chu et al. conducted its in vitro study and compared its mechanical properties and bioactivity. In terms of bioactivity, HA is mostly preferred as a reinforcing particle in bioactive bone cement even though its mechanical properties are not ideal, with a relatively low tensile strength of 50–70 MPa, a Young's modulus in the range 35–120 GPa and a low fracture toughness, 0.5–1.5 $MN/m^{3/2}$.

In vitro apatite formation on implants is induced by modifying organic polymer with silanol groups in combination with calcium salts enables the polymer to show bioactivity. Kawai et al.[57] modified carboxyl groups of polyamide with different amounts of silanol groups and studied its apatite formation ability in 1.5 simulated body fluid (SBF) which contain $CaCl_2$. Increased concentration of silicate ions provides the vicinity for the apatite nucleation. It has been reported in earlier studies that silanol groups are highly effective for the induction of apatite formation. Deposition of apatite on polyamide films containing carboxyl (–COOH) groups when exposed to 1.5 SBF which contain calcium chloride was reported earlier by Miyazaki et al.[58] which paved the way for the modification of –COOH group of polyamide. In spite of –COOH group, sulfonic groups can also trigger apatite nucleation in a body environment. Miyazaki et al.[59] prepared organic–inorganic hybrids from hydroxyethylmethacrylate (HEMA) in presence of vinylsulfonic acid sodium salt and calcium chloride ($CaCl_2$) and assessed bioactivities of the hybrids in vitro by examining the apatite formation in SBF. Induction of nucleation of apatite on polymers by sulfonic groups can also be attained through covalent coupling. Leonor et al.[60] sulfonated substrates of polycaprolactam (Nylon 6), high molecular weight polyethylene (HMWPE), polyethylene terephthalate, and ethylene-vinyl alcohol copolymer by soaking it in sulfuric acid (H_2SO_4) or chlorosulfonic acid ($ClSO_3H$) with different concentrations which is then soaked in saturated solution of calcium hydroxide ($Ca(OH)_2$) for incorporating calcium ions. They further soaked the treated substrates in a SBF for apatite formation. From this point of view, Yang et al.[61] prepared an organic–inorganic hybrid material by incorporating PMMA structure units covalently into a SiO_2 glass network. The interfacial adhesion between the particles and the filler can be improved by establishing covalent chemical bonding between both materials by using a silane coupling agent.[62] The purposes of using coupling agent are (1) to improve the mechanical properties of composites by improving the interfacial bonding strength and (2) to prevent the surface dissolution of fibers. The use of silane coupling agents in mineral filled dental resins has been employed since the 1960s as a method to react with the oxide surface of the inorganic particles and to copolymerize with organic monomers. Plueddemann[35] reported that appropriate silane coupling agents were chosen for a variety of mineral fillers to improve the mechanical properties of composites. Venhoven et al.[63] reported that addition of 3-(trimethoxysilyl) propyl methacrylate treated glass filler with bis-GMA reduces the wear rates. Jones et al.[64] also reported that silane-treated bioactive glass filler was found to have a significant effect on the elastic modulus of bis-GMA based composites. Normally silane treated glass–ceramic or quartz particles might have an encouraging effect on the stiffness, strength, and wear resistance and other aspects of the composites. Behiri et al. reported that while reinforcing 3-(trimethoxysilyl)propyl methacrylate (γ-MPS)-treated HA particles, poly(ethylmethacrylate)-based cements have enhanced tensile modulus and yield strength compared with untreated fillers, but its radiopacity is poor. Abboud et al.[65] proposed that the radiopacity of acrylic bone cements can be improved by the incorporation of alumina (Al_2O_3) particles treated with γ-MPS. Sugino et al.[66] replicate the experiment by taking calcium acetate instead of calcium chloride. They compared the compare the handling, mechanical, and histological properties of the modified bone cement with those of the conventional cement. The specimens were implanted into the femur of nine 2-year-old female beagle dogs weighing 8–11 kg. They concluded that the modified specimens exhibited higher bonding strength and osteoconduction, which the conventional PMMA bone cement lacked, providing stable fixation for a long period after implantation. Sugino et al.[67] investigated the apatite-forming ability of water soluble calcium salts such as calcium acetate and calcium hydroxide reinforced PMMA-based bone cement by soaking it in SBF for 7 days. They found out that apatite formation would be effective not only for bioactivity but also for decreasing the reduction of mechanical strength. The incorporation of bioactive fillers of diminutive particle size and high molecular weight in PMMA powder favorably affected the bioactivity of the PMMA-based cements.[68,69] It was confirmed by the study performed by Goto et al.[70] They evaluated the mechanical properties and osteoconductivity of titania-containing cements by performing mechanical tests, examining affinity indices in rat tibiae, and histologically examining the bone–cement interface. Some metallic materials such as stainless steel, Co–Cr–Mo alloys, titanium, and its alloys have been widely used for dental and orthopedic implants under load-bearing conditions. These metals themselves cannot directly bond to living bone. So far, various surface modifications based on physical coating with ceramic materials with bone-bonding ability have been endeavoring to afford these metals with bone-bonding ability. Electrochemical deposition is a useful method for coating calcium phosphate on metal substrates with complicated shapes, such as dental and orthopedic implants, and the deposition time in this method is usually within few hours.[71]

9.2 CROSS-LINKING AGENTS

The extent to which monomers react to form a polymer during polymerization and the degree of conversion of the reactive species are expected to have an important effect on the physicomechanical properties of bone cement. Incorporation of a cross-linking agent into the liquid monomer phase causes an insoluble cross-linked network to form during the course of polymerization which improves many of the physical properties, such as modulus of elasticity, heat distortion, solvent resistance, shrinkage, stiffness, hardness, and glass transition temperature. Commonly used cross-linking agents are ethylene glycol dimethacrylate (EGDMA), triethyleneglycoldimethacrylate (TEGDMA), and poly(ethyleneglycoldimethacrylate) (PEGDMA, Mn = 400). Incorporation of a cross-linking agent to the bone cement would improve the toughness of the cement by providing anchor points for the chains, and these anchor points will restrain excessive movement and maintain the properties of the chain in the network. Deb et al. conducted a comparative study on the mechanical properties of bone–cement formulations in which the liquid phase has been modified by incorporation of three different cross-linking agents with different side chain lengths—EGDMA, TEGDMA, and PEGDMA.[72] Vallo et al. also studied the mechanical behavior of self-curing acrylic cements when it is reinforced with cross-linking agents like TEGDMA and PEGDMA and reported that the influence of reinforcing agent accounts for superior flexural strength compared with the other conventional cements.[73]

The gaps at these interfaces occur as result of the shrinkage of the cement when its polymerization can be minimized by improving the adhesion of the cement to both the bone and the implant. The improved hydrolysis resistance of the implant-bone–cement interface may diminish the incidence of osteolysis and, consequently, reduce the potential for aseptic loosening of the prosthesis. It has been reported that cement comprising PMMA beads, tributyl borane as the initiator of the polymerization reaction, and 4-trimethacryloyloxyrthyl trimellitate (4-META) and MMA as the polymers in the liquid monomer (METAMMA cement) may reduce the incidence of osteolysis[74] and a cement in which between 5% and 20% of 3-methacryloxyproyl-trimethoxysilane (3-MPS) were added to the liquid monomer (MPS cement).[75] The tensile bonding strength and the interfacial shear strength of bone cement with the METAMMA are found to be 100–200% higher than PMMA cement. The compressive and flexural strengths of the 3-MPS cement were not significantly different from those of PMMA cement. However, when a Ti implant was grouted in a 3-MPS cement bed, hydrolysis-resisting covalent bonding formed between the cement and the implant has the interfacial strength increased by 20 times compared to PMMA cement.

BIS-GMA is a monomer of high viscosity, usually necessitating a diluent monomer, which cross-links upon polymerization producing a brittle polymer. It possesses similar water uptake to PMMA but lower exotherm and shrinkage upon curing. Since no polymer bead is present in the formulation, a higher percentage of bioactive filler particles can be added, although this does lead to a "sticky" mixture which is difficult to handle.

9.3 HYDROXYAPATITE

HAP $(Ca_{10}(PO_4)_6(OH)_2)$ is the most suitable material for repairing the bone defects because of its excellent biocompatibility. HA particles also have been used as a filler material to construct HA polymer composites.[76] Such composites combine the osteoconductivity of HA with the easy processing ability of polymers to provide wide variety of mechanical properties to polymers, and completely degradable HA/polymer composites can be made with a biodegradable polymer matrix. These biodegradable composites have the ability to stimulate new bone growth and progressively degrade, thus facilitating the load to relocate slowly from the material to the newly grown bone.[77] The most commonly used bone cement is composed mainly of PMMA, which is an amorphous polymer, brittle and hard at its glass transition temperature (105–120°C).[78] Bone cement is formed by combining the prepolymerized PMMA beads with methylmethacrylate monomer liquid. The polymerization proceeds as a result of a redox reaction, which occurs between the BPO present in the polymer powder and the DMPT included in the monomer liquid. BPO decomposes to produce free radicals, which reacts with the monomer to create an active center. The reaction is exothermic, and temperatures can exceed 100°C. Also, the polymerization proceeds via an addition reaction that takes place in three distinctive steps: initiation, propagation, and termination. All these processes do not occur sequentially but are simultaneous, and it is only their respective rates that change relatively with time from the onset of mixing.[13] The addition of bioactive HA is potentially advantageous for forming a stable bone/implant interface, which act as "flaws" in the continuous polymer matrix but it limits the effect on toughening mechanisms.[79] HAP have no chemical bonding at the interface.[80,81] It can be benefited by using acrylics bis-GMA/TEG-DMA by silane coupling to HAP.[82–87] However, the migration of individual particles from the implant site before tissue in growth causes inconvenience, difficult to fill the surface of irregular bone defect and reconstruction. In recent years, numerous bioactive bone cements have been developed to overcome this problem. This shortcoming of HAP was overcome by employing biodegradable HAP-particulate composites and composites based on PMMA, polyethylene polyacrylic acid, chitosan, gelatin, etc. are reported.[88,89] The mechanical properties of HAP reinforced poly(ethylmethacrylate) bone cement under the influence of a silane coupling agent was studied by Harper et al. It has been reported that the fatigue resistance of the reinforced PEMA cement was increased when a silane coupling agent was present upon the HA particle surface.[33]

9.4 GLASS-IONOMER CEMENT

Glass-ionomer cements (GICs) have been used as bioactive cements which have shown better adhesion, higher CS, less shrinkage, and lower exotherm, as compared to PMMA.[90] But it exhibits lower mechanical strengths compared to their counterparts due to ionic cross-links in conventional

GICs which are brittle in nature. However, resin-modified GICs demonstrate higher strengths because they have polymerizable covalent bonds for cross-linking in addition to ionic bonds.[91] 2-HEMA is commonly used as a comonomer in resin-modified GICs, which may cause potential cytotoxicity to the surrounding tissue if it is not completely polymerized.[92] Higher mechanical strengths exhibited by GIC are due to the presence of aluminum which is a major filler component in GIC. However, aluminum ions in the composition of GIC are cytotoxic.[93] Therefore, both aluminum and HEMA would not be suitable components in GICs. Osteoconductivity and osteogenesis of zinc cation (Zn^{2+}) have been studied by Ishikaw et al.[94] Hill et al founded that zinc oxide (ZnO) was able to form salt-bridges between zinc cations and carboxyl anions by preparing bone cement by blending the mixture of ZnO/HAP with polyacrylic acid aqueous solution.[95] Xie et al. evaluated the mechanical strengths and other properties of cements prepared from sintered zinc–calcium–silicate phosphate glass and formulated it with polyalkenoate having a pendant methacrylate group as well as acryloyl beta-alanine (act as comonomer).[96] An appetite and wollastonite containing glass–ceramic (AW glass–ceramic) developed at Kyoto University[97–100] is one of the most promising bioactive glass–ceramic materials and used clinically for hip arthoplasy.[101]

Despite these developmental advances, bone cement still faces the problems such as polymerization shrinkage and wears resistance. In order to diminish polymerization shrinkage and to improve wear resistance and biocompatibility organically, modified ceramic materials are used as monomer matrix in dental restoratives.[102] These organically modified ceramic materials are inorganic–organic hybrid materials with molecules containing a metal core bonded to reactive organic groups. The concept is based on sol–gel type reaction by combining properties of organic polymers with properties of glass like materials to generate new/synergistic properties which are already well known for the synthesis of ceramics. In this process, additional organic network or cross-linking is formed above the inorganic network. Because of the formation of covalent bonds and the elimination of macroscopic interfaces between polymers and inorganic components, physical properties of the organic–inorganic hybrid sol–gel materials could be designed and controlled by varying the nature and composition of both the polymer and the inorganic components.[103,104] It had wide range of applications such as wear resistant coating in microelectronics, microoptics, electrooptics, and photonics, as matrices for dental composites and as scaffolds for tissue engineering.[105] Yang et al. prepared an epoxy–SiO_2 hybrid sol–gel material by incorporating epoxy structure units covalently into a SiO_2 glass network via the sol–gel approach and reported that the mechanical properties of the new bone cement are superior to those of commercial bone cements.[106] While comparing with conventional organic oligomers, organically modified ceramic hybrid materials are thermally stable.[107] It was reported that the type of inorganic material incorporated in organically modified ceramic resin has considerable influence on its thermal characteristics. Its properties like polymer shrinkage, depth of cure, and in vitro cytotoxic behavior overweighed than that of dimethacrylate and tetramethacrylate resins.[108]

Bioactive properties of fillers are made use in most of the studies to impart bioactivity to dental restoratives. Inorganic organic hybrid resins containing alkoxides or mixture of alkoxides of silicone, aluminum, calcium, and titanium were developed by Lizymol.[108] In this study, silicone, aluminum, calcium, and titanium are selected because of their reported osteoconductive and biocompatibility characteristics, which can enhance remineralization. This inspired us to develop a choice of inorganic–organic hybrid resins containing alkoxides/mixture of alkoxides of calcium, magnesium, zinc, strontium, etc. Among these, photocured composites developed from zinc-containing inorganic–organic hybrid resin exhibited 22.15% reduction in *Escherichia coli* ATCC 25922 compared to bis-GMA after 1 h exposure which proved that it can impart antimicrobial activity.[109] Figure 9.1 shows the schematic representation of synthesis of inorganic–organic hybrid resin containing inorganic content. All the synthesized resins showed good thermal stability, low polymerization shrinkage, good mechanical properties, and good cytocompatibility with enhanced bioactivity.[110] One of the synthesized resins containing calcium got bioactivity, which is compared with control bis-GMA-based composite shown in Figure 9.1.

Apatite formation on novel composite after 14 days evidenced that the material is bioactive in nature. Controlled bis-GMA-based composite does not exhibit any indication of apatite growth. The mechanical property of the novel composite satisfies the criteria as stipulated by ISO 5833:2002 (Table 9.1). As per the international standard, the material should possess CS of minimum of 70 MPa. The polymerization shrinkage of the composite is also low compared to bis-GMA-based composites. Good mechanical properties indicate good monomer conversion and effective bonding

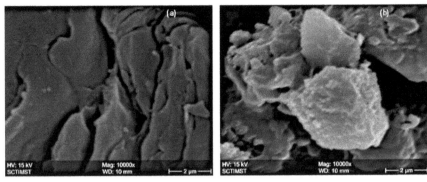

FIGURE 9.1   SEM images showing the comparison of bioactivity of (a) control and (b) calcium containing inorganic–organic hybrid resin based composite stored in simulated body fluid after 14 days.

between the inorganic filler and the resin matrix. Low polymerization shrinkage of the composite evidenced the availability of multifunctional monomers.

TABLE 9.1   Properties of Novel Composite Based on Inorganic–Organic Hybrid Resins.

Sl. no.	Properties	Values
1	Compressive strength (MPa)	80±2.36
2	Polymerization shrinkage (%)	1.73±0.095

9.5   CONCLUSION

Several techniques have been employed in order to reduce polymerization shrinkage, enhanced physicomechanical properties, bioactivity, antimicrobial property, radiopacity, biocompatibility, and the antimicrobial activity of composites employed in biomedical applications. An ideal biomaterial should meet all the above criteria. However, in the conventional composites, one or two properties can be achieved at a time, whereas the inorganic–organic hybrid resins can be tailored to get combination of desirable properties. The new material exhibited low polymerization shrinkage with good mechanical properties and enhanced bioactivity which makes it a potential candidate in biomedical applications.

ACKNOWLEDGMENTS

Financial support from Kerala State Council for Science, Technology and Environment, Government of Kerala, India is gratefully acknowledged. We thank Head BMT Wing and Director Sree Chitra Tirunal Institute for Medical Sciences and Technology for extending the facilities.

KEYWORDS

- acrylic bone cement
- arthroplasty

- poly(methyl methacrylate)
- inorganic–organic hybrid resins

REFERENCES

1. Lu, J. Orthopedic Bone Cements. In *Biomechanics and Biomaterials in Orthopedics*; Poitout, D.G., Ed.; Springer: New York, 2004; p 86.
2. Jianxi, L. Orthopeadic Bone Cement. In *Biomechanics and Biomaterials in Orthopedics*, 2nd ed.; Poitout, D.G., Ed.; Springer: New York, 2016; pp 10, 123.
3. Vallo, C. I.; Schroeder, W. F. Properties of Acrylic Bone Cements Formulated with Bis-GMA. *J. Biomed. Mater. Res. B: Appl. Biomater.* **2005,** *74*(2), 676–685.
4. Alla, R. K.; Raghavendra Swamy, K. N.; Vyas, R.; Konakanchi, A. Conventional and Contemporary Polymers for the Fabrication of Denture Prosthesis: Part I—Overview, Composition and Properties. *Int. J. Appl. Dent. Sci.* **2015,** *1*(4), 82–89.
5. Aherwar, A.; Singh, A. K.; Patnak, A. Current and Future Biocompatibility Aspects of Biomaterials for Hip Prosthesis. *AIMS Bioeng.* **2015,** *3*(1), 23–43.
6. Webb, J. C.; Spencer, R. F. The Role of Polymethylmethacrylate Bone Cement in Modern Orthopaedic Surgery. *J. Bone Joints Surg.* **2007,** *89 B*, 851–857.
7. Smith, L. D. C. Glues, Cements and Adhesives in Medicine and Dentistry. *BioMed. Eng.* **1973,** *8*, 108–115.
8. Chamley. J. *Acrylic Cement in Orthopedic Surgery*; Williams and Wilkins: Baltimore, 1970.
9. Lasa, V. Poly(Methylmethacrylate) Bone Cement: Chemical Composition and Chemistry. In *Orthopaedic Bone Cements*; Sanjukta, D., Ed.; CRC: Boca Raton, 2008; pp 183–200.
10. Kuhn, K.-D. *Bone Cements*; Springer: New York, 2000; pp 246–247.
11. Dunne, N. Mechanical Properties of Bone Cements. In *Orthopaedic Bone Cements*; Sanjukta, D., Ed. CRC: Boca Raton, 2008; pp 240–255.

12. Vaishya, R.; Chauhan, M.; Vaish, A.; Bone Cement. *J. Clin. Orthop. Trauma* **2013**, *4*, 157–163.

13. Rehman, I.; Harper, E. J.; Bonfield, W. In Situ Analysis of the Degree of Polymerization of Bone Cement by Using FT-Raman Spectroscopy. *Biomaterials* **1996**, *17*, 1615–1619.

14. Hasenwinkel, J. M.; Lautenschlager, E. P.; Wixson, R. L.; Gilbert, J. L. A Novel High-Viscosity, Two-Solution Acrylic Bone Cement: Effect of Chemical Composition on Properties. *J. Biomed. Mater. Res.* **1999**, *47*(1), 36–45.

15. Topoleski, L. D. T.; Ducheyne, P.; Cuckler, J. M. Microstructural Pathway of Fracture in Poly(Methyl Methacrylate) Bone Cement. *Biomaterials* **1993**, *14*, 1165–1172.

16. Harper, E. J. Bioactive Bone Cements. *Review Paper, Proc Instn Mech Engrs* Vol 212 Part H, H03697 © IMechE, 1998.

17. Pourdeyhimi, B.; Wagner, H. D.; Schwartz, P. A Comparison of Mechanical Properties of Discontinuous Kevlar 29 Fiber Reinforced Bone and Dental Cements. *J. Mater. Sci.* **1986**, *21*, 4468–4474.

18. Ulcay, Y.; Agarwal, V.; Pourdeyhimi, B. In *Elastic Properties of 3D Randomly Distributed Spectra 900 and Kevlar 49 Discontinuous Fiber Reinforced Composites*, Proceedings of the 14th Annual Energy Sources Technology Conference and Exhibition, Houston, ASME, 1991; pp 41–49.

19. Pourdeyhimi, B.; Wagner, H. D. Elastic and Ultimate Properties of Acrylic Bone Cement Reinforced with Ultra High Molecular Weight Polyethylene Fibers. *J. Biomed. Mater. Res.* **1989**, *23*, 63–80.

20. Friis, E. A.; Kumar, B.; Cooke, F. W.; Yasuda, H. K. In *Fracture Toughness of Surface-Treated Carbon Fiber Reinforced Composite Bone Cement*, Fifth World Biomaterials Congress, 1996; p 913.

21. Harper, E. J.; Behiri, J. C.; Bonfield, W. Flexural and Fatigue Properties of a Bone Cement Based Upon Polyethylmethacrylate and Hydroxyapatite. *J. Mater. Sci. Mater. Med.* **1995**, *6*, 799–803.

22. Kobayashi, M.; Nakamura, T.; Okada, Y.; Fukumoto, A.; Furukawa, T.; Kato, H.; Kokubo, T.; Kikutani, T. Bioactive Bone Cement: Comparison of Apatite and Wollastonite Containing Glass–Ceramic, Hydroxyapatite, and Beta-Tricalcium Phosphate Fillers on Bone-Bonding Strength. *J. Biomed. Mater. Res.* **1998**, *42*(2), 223–237.

23. Kim, Y. S.; Kang, Y. H.; Kim, J. K.; Park, J. B. The Effect of Bone Mineral Particles on the Porosity of Bone Cement. *Biomed. Mater. Eng.* **1994**, *4*, 37–46.

24. Gilbert, J. L.; Net, S. S.; Lauthenschlager, E. P. Self-Reinforced Composite Poly(Methylmethacrylate): Static and Fatigue Properties. *Biomaterials* **1995**, *16*, 1043–1055.

25. Topoleski, L. D. T.; Ducheyne, P.; Cuckler, J. M. The Fracture Toughness of Titanium-Fiber-Reinforced Bone Cement. *J. Biomed. Mater. Res.* **1992**, *26*, 1599–1617.

26. Kotha, S.P.; Li, C.; Schmid, S. R.; Mason, J. J. Fracture Toughness of Steel Fiber Reinforced Bone Cement. *J. Biomed. Mater. Res.* **2004**, *70*(3), 514–521.

27. Vallo, C. I. Influence of Filler Content on Static Properties of Glass-Reinforced Bone Cement. *J. Biomed. Mater. Res.* **2000**, *53*, 717–727.

28. Abboud, M.; Casaubieilh, L.; Morvan, F.; Fontanille, M.; Duguet, E. PMMA-Based Composite Materials with Reactive Ceramic Fillers. IV. Radiopacifying Particles

Embedded in PMMA Beads for Acrylic Bone Cements. *J. Biomed. Mater. Res.* **2000,** *53*, 728–736.

29. Vila, M. M.; Ginebra, M. P.; Gil, F. J.; Planell, J. A. Effect of Porosity and Environment on the Mechanical Behavior of Acrylic Bone Cement Modified with Acrylonitrile-Butadiene-Styrene Particles. Part I. Fracture Toughness. *J. Biomed. Mater. Res. Appl. Biomater.* **1999,** *48*, 121–127.

30. Kim, H. Y.; Yasuda, H. K. Improvement of Fatigue Properties of Poly(Methyl Methacrylate) Bone Cement by Means of Plasma Surface Treatment of Fillers. *J. Biomed. Mater. Res.* **1999,** *48*(2), 135–142.

31. Hild, D. N.; Schwartz, P. Plasma-Treated Ultra-High-Strength Polyethylene Fibers Improved the Fracture-Toughness of Poly(Methyl Methacrylate). *J. Mater. Sci. Mater. Med.* **1993,** *4*(5), 481–493.

32. Ohashi, K. L.; Yerby, S. A.; Dauskardt, R. H. Effects of an Adhesion Promoter on the Debond Resistance of a Metal-Polymethylmethacrylate Interface. *J. Biomed. Mater. Res.* **2001,** *54*(3), 419–427.

33. Harper, E. J.; Braden, M.; Bonfield, W. Mechanical Properties of Hydroxyapatite Reinforced Poly(Ethylmethacrylate) Bone Cement After Immersion in Physiological Solution: Influence of a Silane Coupling Agent. *J. Mater. Sci. Mater. Med.* **2000,** *11*(8), 491–497.

34. Yerby, S. A.; Paal, A. F.; Young, P. M.; Beaupre, G. S.; Ohashi, K. L.; Goodman, S. The Effect of a Silane Coupling Agent on the Bond Strength of Bone Cement and Cobalt-Chrome Alloy. *J. Biomed. Mater. Res.* **2000,** *49*(1), 127–133.

35. Plueddemann, E. P. *Silane Coupling Agents*; 2nd ed. Plenum Press: New York, 1991.

36. Freeman, M. A. R.; Bradley, G. W.; Ravell, P. A. Observation upon the Interface Between Bone and Polymethylmethacrylate Cement. *J. Bone Joint Surg.* **1982,** *64B*, 489–493.

37. Saha, S.; Pal, S. Mechanical Properties of Bone Cement: A Review. *J. Biomed. Mater. Res.* **1984,** *18*, 435–462.

38. Malony, W. J.; Jasty, M.; Rosenberg, A.; Harris, W. H. Bone Lysis in Well-Fixed Cemented Femoral Components. *J. Bone Joint Surg.* **1990,** *72B*, 966–970.

39. Vorndran, E.; Spohn, N.; Nies, B.; Rößler, S.; Storch, S.; Gbureck, U. Mechanical Properties and Drug Release Behavior of Bioactivated PMMA Cements. *J. Biomater. Appl.* **2010,** *26*(5), 581–594.

40. Oonishi, H. Bioactive Bone Cements. In *The Bone Biomaterial Interface*; Davies, J.E., Ed.; University of Toronto Press: Toronto, Canada, 1991; pp 321–333.

41. Kokubo, T.; Kushitani, H.; Sakka, S.; Kitsugi, T.; Yamamuro, T. Solutions Able to Reproduce In Vivo Surface-Structure Changes in Bioactive Glass–Ceramic A-W. *J. Biomed. Mater. Res.* **1990,** *24*, 721–734.

42. Miyazaki, T.; Ohtsuki, C.; Kyomoto, M.; Tanihara, M.; Mori, A.; Kuramoto, K.-I. Bioactive PMMA Bone Cement Prepared by Modification with Methacryloxypropyltrimethoxysilane and Calcium Chloride. *J. Biomed. Mater. Res. A.* **2003,** *67*(4), 1417–1423.

43. Chu, K. T.; Oshidac, Y.; Hancock, E. B.; Kowolik, M. J.; Barco, T.; Zunt, S. L.; Hydroxyapatite/PMMA Composites as Bone Cements. *Bio-Med. Mater. Eng.* **2004,** *14*, 87–105.

44. Heikkaila, J. T.; Aho, A. J.; Kangasniemi, I.; Yliurpo, A. *Biomaterials* **1996**, *17*, 1755.

45. Shinzato, S.; Kobayashi, M.; Mousa, W. F.; Kaminuma, M.; Neo, M.; Kitamura, Y.; Kokubo, T.; Nakamura, T. *J. Biomed. Mater. Res.* **2000**, *51*, 258.

46. Goto, K.; Tamura, J.; Shinzato, S.; Fujibayashi, S.; Hashimoto, M.; Kawashita, M.; Kokubo, T.; Nakamura, T. *Biomaterials* **2005**, *26*, 6496.

47. Kawanabe, K.; Tamura, J.; Yamamuro, T.; Nakamura, T.; Kokubo, T.; Yoshihara, S. A New Bioactive Bone Cement Consisting of BIS-GMA Resin and Bioactive Glass Powder. *J. Appl. Biomater.* **1993**, *4*, 135–141.

48. Tamura, J.; Kawanabe, K.; Yamamuro, T., et al. Bioactive Bone Cement: The Effect of Amounts of Glass Powder and Histological Changes with Time. *J. Biomed. Mater. Res.* **1995**, *29*, 551–559.

49. Tamura, J.; Kawanabe, K.; Kobayashi, M., et al. Mechanical and Biological Properties of Two Types of Bioactive Bone Cements Containing MgO–CaO–SiO_2–P_2O_5–CaF_2 Glass and Glass–Ceramic Powder. *J. Biomed. Mater. Res.* **1996**, *30*, 85–94.

50. Tamura, J.; Kitsugi, T.; Iida, H., et al. Bone Bonding Ability of Bioactive Bone Cements. *Clin. Orthop.* **1997**, *343*, 183–191.

51. Fujita, H.; Nakamura, T.; Tamura, J., et al. Bioactive Bone Cement: Effect of the Amount of Glass–Ceramic Powder on Bone Bonding Strength. *J. Biomed. Mater. Res.* **1998**, *40*, 145–152.

52. Osborn, J. F.; Newesely, H. The Material Science of Calcium Phosphate Ceramics. *Biomaterials* **1980**, *1*, 108–111.

53. Olmi, R.; Moroni, A.; Castaldini, A.; Romagnoli, R. Hydroxyapatite Alloyed with Bone Cement: Physical and Biological Characterisation. In *Ceramics in Surgery;* Vincenzini, P., Ed.; Elsevier Scientific: Amsterdam, Netherlands, 1983; pp 91–96.

54. Castaldini, A.; Cavallini, A.; Moroni, A.; Olmi, R. Young's Modulus of Hydroxyapatite Mixed Bone Cement. In *Biomaterials and Biomechanics;* Ducheyne, P., Van der Perre, G., Aubert, A. E., Eds.; Elsevier Scientific: Amsterdam, Netherlands, 1984; pp 427–432.

55. Castaldini, A.; Cavallini, A. Setting Properties of Bone Cement with Added Synthetic Hydroxyapatite. *Biomaterials* **1985**, *6*, 55–60.

56. Castaldini, A.; Cavallini, A. Creep Behaviour of Composite Bone Cement. In *Biological and Biomedical Performance of Biomaterials;* Christel, P., Meunier, A. Lee, A.J.C., Eds.; Elsevier Scientific: Amsterdam, Netherlands, 1986; pp 525–530.

57. Kawai, T.; Ohtsuki, C.; Kamitakahara, M.; Hosoya, K.; Tanihara, M.; Miyazaki, T.; Sakaguchi, Y.; Konagaya, S. In Vitro Apatite Formation on Polyamide Containing Carboxyl Groups Modified with Silanol Groups. *J. Mater. Sci.: Mater. Med.* **2007**, *18*, 1037–1042.

58. Miyazaki, T.; Ohtsuki, C.; Akioka, Y.; Tanihara, M.; Nakao, J.; Sakaguchi, Y.; Konagaya, S. *J. Mater. Sci. Mater. Med.* **2003**, *14*, 569.

59. Miyazaki, T.; Imamura, M.; Ishida, E.; Ashizuka, M.; Ohtsuki, C.; Apatite Formation Abilities and Mechanical Properties of Hydroxyethylmethacrylate-Based Organic–Inorganic Hybrids Incorporated with Sulfonic Groups and Calcium Ions. *J. Mater. Sci.: Mater. Med.* **2009**, *20*, 157–161.

60. Leonor, I. B.; Kim, H. M.; Balas, F.; Kawashita, M.; Reis, R. L.; Kokubo, T.; Nakamura, T. Functionalization of Different Polymers with Sulfonic Groups as a Way to

Coat them With a Biomimetic Apatite Layer. *J. Mater. Sci.: Mater. Med.* **2007**, *18*, 1923–1930.

61. Yang, J.-M.; Lu, C.-S.; Hsu, Y.-G.; Shih, C.-H. Mechanical Properties of Acrylic Bone Cement Containing PMMA-SiO$_2$ Hybrid Sol–Gel Material. *J. Biomed. Mater. Res.* **1997**, *38*(2), 143–154.

62. Behiri, J. C.; Braden, M.; Khorosani, D.; Wiwattanadate, D.; Bonfield, W. Advanced Bone Cement for Long Term Orthopaedic Implantation. *Bioceramics* **1991**, *4*, 301–307.

63. Venhoven, B. A. M.; de Gee, A. J.; Werner, A.; Davidson, C. L. Silane Treatment of Filler and Composite Blending in a One Step Procedure for Dental Restoratives. *Biomaterials* **1994**, *15*, 1152–1156.

64. Jones, D. W.; Rizkalla, A. S. Characterization of Experimental Composite Biomaterials. *J. Biomed. Mater. Res. (Appl. Biomater.)* **1996**, *33*, 89–100.

65. Abboud, M.; Vol, S.; Duguet, E.; Fontanille, M. PMMA-Based Composite Materials with Reactive Ceramic Fillers Part III: Radiopacifying Particle-Reinforced Bone Cements. *J. Mater. Sci.: Mater. Med.* **2000**, *11*, 295–300.

66. Sugino, A.; Ohtsuki, C.; Miyazaki, T. In Vivo Response of Bioactive PMMA-based Bone Cement Modified with Alkoxysilane and Calcium Acetate. *J. Biomater. Appl.* **2008**, *23*, 213–228.

67. Sugino, A.; Miyazaki, T.; Kawachi, G.; Kikuta, K.; Ohtsuki, C. Relationship between Apatite-Forming Ability and Mechanical Properties of Bioactive PMMA-Based Bone Cement Modified with Calcium Salts and Alkoxysilane. *J. Mater. Sci.: Mater. Med.* **2008**, *19*, 1399–1405.

68. Shinzato, S., et al. Bioactive Polymethyl Methacrylate-Based Bone Cement: Comparison of Glass Beads, Apatite- and Wollastonite-Containing Glass–Ceramic, and Hydroxyapatite Fillers on Mechanical and Biological Properties. *J. Biomed. Mater. Res.* **2000**, *51*(2), 258–272.

69. Shinzato, S.; Nakamura, T.; Kokubo, T.; Kitamura, Y. Bioactive Bone Cement: Effect of Filler Size on Mechanical Properties and Osteoconductivity. *J. Biomed. Mater. Res.* **2001**, *56*, 452–458.

70. Goto, K.; Tamura, J.; Shinzato, S.; Fujibayashi, S.; Hashimoto, M.; Kawashita, M.; Kokubo, T.; Nakamura, T. Bioactive Bone Cements Containing Nano-Sized Titania Particles for Use as Bone Substitutes. *Biomaterials* **2005**, *26*, 6496–6505.

71. Kawashita, M.; Itoh, S.; Miyamoto, K.; Takaoka, G. H. Apatite Formation on Titanium Substrates by Electrochemical Deposition in Metastable Calcium Phosphate Solution. *J. Mater. Sci.: Mater. Med.* **2008**, *19*, 137–142.

72. Deb, S.; Vazquez, B.; Bonfield, W. Effect of Crosslinking Agents on Acrylic Bone Cements Based on Poly(Methylmethacrylate). *J. Biomed. Mater. Res.* **1997**, *37*, 465–473.

73. Vallo, C. I.; Abraham, G. A.; Cuadrado, T. R.; San Roma, J. Influence of Cross-Linked PMMA Beads on the Mechanical Behavior of Self-Curing Acrylic Cements. *J. Biomed. Mater. Res. B Appl. Biomater.* **2004**, *70B*, 407–416.

74. Sakai, T.; Morita, S.; Shinomiya, K.-I.; Watanabe, A.; Nakabayashi, N.; Ishihara, K. In Vivo Evaluation of the Bond Strength of Adhesive 4-META/MMA-TBB Bone Cement Underweight Bearing Conditions. *J. Biomed. Mater. Res. A* **2000**, *52*, 128–134.

75. Gbureck, U.; Grübel, S.; Thull, R.; Barralet, J. E. Modified PMMA Cements for a Hydrolysis Resistant Metal–Polymer Interface in Orthopaedic Applications. *Acta Biomater.* **2005,** *1,* 671–676.

76. Labella, R.; Braden, M.; De, S. Novel Hydroxyapatite-Based Dental Composites. *Biomaterials* **1994,** *15,* 1197–2000.

77. Liu, Q.; deWijn, J. R.; van Blitterswijk, C. A. Covalent Bonding of PMMA, PBMA, and Poly (HEMA) to Hydroxyapatite Particles. *J. Biomed. Mater. Res.* **1998,** *40,* 257–263.

78. Lee, R. R.; Ogiso, M.; Watanabe, A.; Ishihara, K. Examination of Hydroxyapatite Filled 4-META/MMA-TBB Adhesive Bone Cement In Vitro and In Vivo Environment. *J. Biomed. Mater. Res. Appl. Biomater.* **1997,** *38,* 11–16.

79. Kurtz, S. M.; Devine, J. N. PEEK Biomaterials in Trauma, Orthopedic, and Spinal Implants. *Biomaterials* **2007,** *28,* 4845–4869.

80. Deb, S., et al. Hydroxyapatite-Polyethylene Composites: Effect of Grafting and Surface Treatment of Hydroxyapatite. *J. Mater. Sci. Mater. Med.* **1996,** *7*(4), 191–193.

81. Wang, M.; Deb, S.; Bonfield, W. Chemically Coupled Hydroxyapatite-Polyethylene Composites: Processing and Characterisation. *Mater. Lett.* **2000,** *44*(6), 119–124.

82. Lewis, G. Alternative Acrylic Bone Cement Formulations for Cemented Arthroplasties: Present Status, Key Issues, and Future Prospects. *J. Biomed. Mater. Res.* **2007,** *84B,* 301–319.

83. Saito, M., et al. Experimental Studies on a New Bioactive Bone Cement: Hydroxyapatite Composite Resin. *Biomaterials* **1994,** *15*(2), 156–160.

84. Labella, R.; Braden, M.; Deb, S. Novel Hydroxyapatite-Based Dental Composites. *Biomaterials* **1994,** *15*(15), 1197–1200.

85. Kobayashi, M., et al. Bioactive Bone Cement: Comparison of Apatite and Wollastonite Containing Glass–Ceramic, Hydroxyapatite, and β-Tricalcium Phosphate Fillers on Bone Bonding Strength. *J. Biomed. Mater. Res.* **1998,** *42*(2), 223–237.

86. Kobayashi, M. et al. Effect of Bioactive Filler Content on Mechanical Properties and Osteoconductivity of Bioactive Bone Cement. *J. Biomed. Mater. Res.* **1999,** *46*(4), 447–457.

87. Santos, C., et al. Hydroxyapatite as a Filler for Dental Composite Materials: Mechanical Properties and In Vitro Bioactivity of Composites. *J. Mater. Sci. Mater. Med.* **2001,** *12,* 565–573.

88. Kargupta, K.; Rao, P.; Kumar, A. *J. Appl. Polymer Sci.* **1993,** *49,* 1309–1317.

89. Balamurugan, A.; Kannan, S.; Selvaraj, V.; Rajeswari, S. Development and Spectral Characterization of Poly(Methyl Methacrylate)/Hydroxyapatite Composite for Biomedical Applications. *Trends Biomater. Artif. Organs* **2004,** *18*(1), 41–45.

90. Craig, R.G. Ed. Restorative Dental Materials. 10th ed. Mosby-Year Book, Inc.: St. Louis, 1997.

91. Wilson, A. D. Resin-Modified Glass-Ionomer Cement. *Int. J. Prosthodont.* **1990,** *3,* 425–429.

92. Oliva, A.; Della Ragione, F.; Salerno, A.; Riccio, V.; Tartaro, G.; Cozzolino, A.; D'Amato, S.; Pontoni, G.; Zappia, V. Biocompatibility Studies on Glass Ionomer Cements by Primary Cultures of Human Osteoblasts. *Biomaterials* **1996,** *17*(13), 1351–1356.

93. Renard, J. L.; Felten, D.; Bequet, D. Post-Otoneurosurgery Aluminium Encephalopathy. *Lancet* **1994**, *344*(8914), 63–64.

94. Ishikaw, K.; Miyamoto, Y.; Yuasa, T.; Ito, A.; Nagayama, M.; Suzuki, K. Fabrication of Zn Containing Apatite Cement and Its Initial Evaluation Using Human Osteoblastic Cells. *Biomaterials* **2002**, *23*(2), 423–428.

95. Hill, R. G.; Kenny, S.; Fennell, B. In *The Influence of Poly (Acrylic Acid) Molar Mass and Concentration on Polyalkenoate Bone Cements*, Proceedings of the 25th Annual Meeting of the Society for Biomaterials, Kamuela, Hawaii, USA, 2000, p 1114.

96. Xie, D.; Feng, D.; Chung, I.-D.; Eberhardt, A. W. A Hybrid Zinc–Calcium–Silicate Polyalkenoate Bone Cement. *Biomaterials* **2003**, *24*, 2749–2757.

97. Kitsugi, T.; Yamamuro, T.; Nakamura, T. Bone-Bonding Behavior of Three Kinds of Apatite-Containing Glass–Ceramic. *J. Biomed. Mater. Res.* **1986**, *20*, 1295–1307.

98. Kokubo, T.; Ito, S.; Shigematsu, M.; Sakka, S.; Yamamuro, T. Mechanical Properties of a New Type of Apatite-Containing Glass–Ceramic for Prosthetic Application. *J. Mater. Sci.* **1985**, *20*, 2001–2004.

99. Kokubo, T.; Ito, S.; Sakka, S.; Yamamuro, T. Formation of a Highstrength Bioactive Glass–Ceramic in the System $MgO–CaO–SiO_2–P_2O_5$. *J. Mater. Sci.* **1986**, *21*, 536–540

100. Nakamura, T.; Yamamuro, T.; Higashi, S.; Kokubo, T.; Ito, S. A New Glass–Ceramic for Bone Replacement: Evaluation of Its Bonding Ability to Bone Tissue. *J. Biomed. Mater. Res.* **1985**, *19*, 685–698.

101. Yamamuro, T.; Shikata, J.; Okumura, H.; Kitsugi, T.; Kakutani, Y.; Matsui, T.; Kokubo, T. Replacement of Lumber Vertebrae of Sheep with Ceramic Prosthesis. *J. Bone Joint Surg. Br.* **1990**, *72-B*, 889–893.

102. Hass, K.-H. Hybrid Inorganic–Organic Polymers Based on Organically Modified Si-Alkoxides. *Adv. Eng. Mater.* **2000**, *2*, 571.

103. Yang, J. M.; Chen, H. S.; Hsu, Y. G.; Lin, F. H.; Chang, Y. H. Organic–Inorganic Hybrid Sol–Gel Materials: 2. Application for Dental Composites. *Angew. Makromol. Chem.* **1997**, *251*, 61–72.

104. Yang, J. M.; Lu, C. S.; Hsu, Y. G.; Shih, C. H. Mechanical Properties of Acrylic Bone Cement Containing $PMMA-SiO_2$ Hybrid Sol–Gel Material. *J. Biomed. Mater. Res. (Appl. Biomater.)* **1997**, *38*, 143–154.

105. Ovsianikov, A.; Schlie, S.; Ngezahayo, A.; Haverich, A.; Chichkov, B. N. Shrinkage of Microstructures Produced by Two-Photon Polymerization of Zr-Based Hybrid Photosensitive Materials. *J. Tissue Eng. Reg. Med.* **2007**, *1*, 443.

106. Yang, J. M.; Shih, C. H.; Chang, C.-N.; Lin, F. H.; Jiang, J. M.; Hsu, Y. G.; Su, W. Y.; See, L. C. Preparation of Epoxy-SiO_2 Hybrid Sol–Gel Material for Bone Cement. *J. Biomed. Mater. Res. A* **2002**, *64*(1), 138–146.

107. Lizymol, P. P. Thermal Studies: A Comparison of the Thermal Properties of Different Oligomers by Thermogravimetric Techniques. *J. Appl. Polym. Sci.* **2004**, *93*, 977–985.

108. Lizymol, P. P. Studies on Shrinkage, Depth of Cure, and Cytotoxic Behavior of Novel Organically Modified Ceramic Based Dental Restorative Resins. J. Appl. Polym. Sci. **2010**, *116*, 2645–2650.

109. Vibha, C.; Lizymol, P. P. Effect of Inorganic Content on Thermal Stability and Antimicrobial Properties of Inorganic–Organic Hybrid Dimethacrylate Resins. *Int. J. Sci. Res.* **2015,** *4*(8), 59–60.

110. Vibha, C.; Lizymol, P. P. Multifunctional Inorganic–Organic Hybrid Resins with Polymerizable Methacrylate Groups for Biomedical Applications; Effects of Synthesis Parameters on Polymerisation Shrinkage and Molecular Weight. *Adv. Mater. Lett.* **2016,** *7*(4), 289–295.

CHAPTER 10

THE MULTIPLE FACETS OF MESENCHYMAL STEM CELLS IN MODULATING TUMOR CELLS' PROLIFERATION AND PROGRESSION

RAJESH RAMASAMY* and VAHID HOSSEINPOUR SARMADI

Stem Cell & Immunity Research Group, Immunology Laboratory, Department of Pathology, Faculty of Medicine and Health Sciences, Universiti Putra Malaysia, 43400 Seri Kembangan, Selangor, Malaysia

Corresponding author. E-mail: rajesh@upm.edu.my

ABSTRACT

Cancer has become the global health, social, and economic burden that direly needs a revolutionized novel treatment for a broad spectrum of tumors. Among the innovative and promising therapeutic modalities, the use of mesenchymal stem cells (MSCs) as a tool for delivering an anti-tumorigenic activity via its inherent antiproliferative capability or transportation of proteins or genes that suppress tumor has created a new way to circumvent cancers. MSCs are one of the most studied adult stem cells in the field of regenerative medicine, gene therapy, and immunomodulation due to its unique biological characteristics. Upon an in vivo administration, MSCs are able to migrate and home to the tumor site and exert either stimulatory or inhibitory effects on tumor cell growth, invasion, and metastasis via regulating angiogenesis, altering immune surveillance, modifying signaling pathways, and regulating apoptosis. Nonetheless, the mechanisms involved in the reported inhibition or stimulation are still elusive. Therefore, a better understanding of the biological consequences of MSCs–tumor interaction, prior to a successful MSCs based therapy, should be warranted. In this chapter, a number of previous findings associated with mutual interplays between MSCs and tumor cells are summarized and highlighted.

10.1   INTRODUCTION

Cancer, the second leading cause of death after cardiovascular diseases, is a generic term for a group of diseases characterized by uncontrolled growth and spread of abnormal cells to other organs. According to the World Health Organization predictions, the number of cancers (excluding nonmelanoma skin cancer) will increase approximately to 17 million, and 10 million patients are projected to perish due to cancer in the year 2020.[1] Despite a considerable improvement in diagnosis, prevention, and treatment of cancer, unresponsiveness of cancer to the conventional treatments such as chemotherapy, immunotherapy, and radiotherapy has been the main obstacles to cancer treatment. In fact, the lifespan of most patients who are undergoing such treatments is severely affected by therapy-related life-threatening complications or recurrence of metastasis. Therefore, it is deemed necessary to explore the effective treatment strategies that specifically target tumor cells. Although most of the studies in the past decades were focused on cancer cells, recent studies have shown that the tumor microenvironment has a major impact on cancer cell growth and metastasis.[2] The tumor microenvironment is composed of vascular endothelial cells (ECs), pericytes, different types of inflammatory cells associated with the immune system,[3] stromal cells such as tumor-associated fibroblasts (TAFs) and mesenchymal cells.[4]

The recent studies have shown that the development and progression of tumor cells are dependent on the mutual interactions of tumor cells and stromal cells through direct or indirect manners.[5–7] Several studies have outlined that mesenchymal stem cells (MSCs) could potentially migrate and engraft into the tumor sites and promote or inhibit tumor cells proliferation and metastasis.[8–10]

10.2   MESENCHYMAL STEM CELLS

MSCs were initially generated and characterized from the bone marrow (BM) compartment by Friedenstein et al. in 1974.[11] In general, MSCs are nonhematopoietic, multipotent adult stem cells that adhere to the plastic surface as spindle-shaped cells.[12] Besides the self-renewal, MSCs as well differentiates into various mesodermal linage-derived mature cells such as osteoblast, adipocytes, and chondrocytes but also cells of ectoderm and endoderm lineages.[13–15] Although BM is still regarded as the main source of MSCs, yet MSCs have been generated from almost all tissues, namely, adipose, umbilical cord and blood, placenta, dental pulp, and fetal tissues.[16–21] One of the major hurdles in identifying MSCs has been the lack of a unique surface marker for MSCs. Hence, three criteria have been proposed by International Society for Cellular Therapy to define MSCs minimally:

1. adherence to the plastic surface under standard culture conditions
2. the capability of differentiating into mesenchymal lineages: fat, bone, and cartilage
3. expression of CD73, CD105, CD90 and lack of CD11b, CD14, CD34, CD45, and human leukocyte antigen—antigen D-related surface markers.[22]

10.3 INTERACTION BETWEEN MSCs AND TUMOR CELLS

The tropic nature of tumor microenvironment produces and secretes different cytokines and growth mediators that can attract various cell types such as MSCs into the tumor site. The migration of MSCs into tumor microenvironment has been witnessed in the most of the tumor models such as breast, colon, pancreatic, and brain cancers.[23,24] Although there is no dispute on MSCs homing and migration capabilities toward the tumor microenvironment, the post interaction effect of MSCs and tumor cells is controversial. The accumulating scientific evidence has recorded the dualistic effects of MSCs on tumor cells such a double-edged sword, either suppressing or promoting tumor growth and invasion.[19,25,26]

10.3.1 MSCs SUPPORT TUMOR CELLS PROLIFERATION

Several studies have shown that MSCs can support the in vitro and in vivo growth of tumor cells.[27–29] This tumor-supporting activities of MSCs can be further deciphered by the following mechanisms.

10.3.1.1   SUPPRESSION OF IMMUNE RESPONSE

The immune system plays a vital part in detecting and eliminating tumor cells through a complex cytotoxicity activity exerted by either T or natural killer (NK) cells. Numerous studies have demonstrated that the immunosuppressive nature of MSCs affects the proliferation, activation, and effector functions of immune cells.[30–34] However, the immunoregulatory mechanism/s of MSCs have not been entirely elucidated. In this regard, several studies have reported that a direct cell–cell contact between MSCs and immune cells is mandatory to execute the inhibitory functions[35,36], whereas others have shown that the immunosuppressive activity of MSCs is attributable to the soluble molecules secreted by MSCs, such as hepatocyte growth factor (HGF),[37] interleukin-1 (IL-1), IL-6,[38] IL-10,[39] transforming growth factor β (TGF-β),[40] interferon-γ (IFN-γ), tumor necrosis factor α (TNF-α),[41] prostaglandin E2,[42] nitric oxide (NO),[43] and indoleamine 2,3-dioxygenase.[44] Regardless of the mode or mechanisms of immunosuppression, studies on animal models had revealed that MSCs indirectly promote the development of tumors by suppressing the gatekeepers of cancer, namely, the immune cells.[45] Moreover, MSCs as well could alter the conditions or status of the immune cells by inflicting changes in various parameters. For instance, MSCs promote the carcinomatous pancreas via polarizing macrophage to express IL-6 and IL-10 which are known to favor tumor growth.[46]

10.3.1.2   STIMULATION OF ANGIOGENESIS

In tumor microenvironment, the vasculature provides nutrients and oxygen for tumor cells' survival and function and also disposes wastes from the tumor sites. In addition, angiogenesis is also crucial for cancer cells to roam around the body to home into a new site in the body, where nutrients and space are not limited.

MSCs are sensitive to the inflammatory mediators where their migration is dictated by various proinflammatory cytokines and chemokines. Since cancer is considered as chronic inflammation, thus the possibility of MSCs either from BM or other transplanted sources to home at the tumor niche is a high likelihood. It could be feasible that the inflammatory cytokines and hypoxia conditions trigger MSCs to secrete the proangiogenic factors such as vascular endothelial growth factor (VEGF), HGF, angiopoietin-1, platelet-derived growth factor (PDGF), and basic fibroblast growth factors (FGF).[47–49] It has been known that VEGF, one of the important proangiogenic molecules produced by MSCs, stimulates the EC and elevates the density of tumor vessels in breast and prostate tumors.[50] In addition, Liu et al. have shown that inflammatory molecules such IFN-γ and TNF-α could highly elevate the angiogenesis through VEGF which were produced by MSCs.[47] In parallel, MSCs can promote angiogenesis by differentiating into TAF, ECs, smooth muscle cells (SMCs), and pericytes.[51–53] Lee et al. showed that MSCs in response to TGF-β could transdifferentiate toward SMCs and trigger angiogenesis via a prostaglandin $F_{2\alpha}$.[54] Furthermore, several studies have demonstrated that VEGF, PDGF, and NO are able to induce the transdifferentiation of MSCs into ECs, hence participate in angiogenesis.[55–57]

10.3.1.3   INHIBITION OF APOPTOSIS

Several studies have revealed that MSCs induce cancer proliferation by exploiting the antiapoptotic mechanism within the cancer cells. Apoptosis is a well-regulated mechanism controlled by anti and proapoptotic proteins, could be induced by intracellular or extracellular signals such as first apoptosis signal (FAS) receptor and FAS ligand, TNF-α, caspase, DNA-damaging agents, or mitochondrial activation and p53. In this regard, Castells et al. showed that conditioned medium of MSCs (soluble factor/s) induced an antiapoptotic effect on ovarian cancer cells by preventing the caspases 3 and 7 activity and activating the PI3K/Akt pathway signaling.[58] Furthermore, another study found that murine and human MSCs induced the antiapoptotic effect in chronic lymphocytic leukemia via cell-to-cell contact manner by increasing the level of myeloid cell leukemia-1 and poly ADP ribose polymerase protein.[59] In addition, Wu et al. demonstrated that MSCs promote colorectal cancer (HCT116 cells) proliferation via decreasing the protein expression level of Bax and p53; apoptosis-related proteins and increasing the protein expression level of antiapoptotic protein, B-cell lymphoma 2 (BCL-2).[60]

10.3.1.4   INDUCTION OF EPITHELIAL-MESENCHYMAL TRANSITION

Under the influence of certain reversible biological processes and biochemical molecules, epithelial cells are able to transdifferentiate phenotypically into MSCs. These transdifferentiated cells display MSCs features, and this process is called EMT or epithelial-mesenchymal transition.[61,62] Previous studies have demonstrated that MSCs secrete soluble factors such as FGF, HGF, PDGF, and TGF-β molecules which can act as EMT-inducing agents.[37,40] During the EMT process, the epithelial cells lose its cell-to-cell contact and morphologically resemble MSCs.[63] Although EMT is a critical process during embryonic development and wound healing[64]; however, an aberrant EMT could possibly induce cancer cell metastasis, invasion and increase resistance to apoptosis.[65,66] In line with these findings, Mele et al. have confirmed that MSCs derived from EMT process favor the occurrence of an aggressive human colorectal cancer through TGF-β.[66] Furthermore, Strong et al. showed that leptin produced by MSCs supports the breast cancer cell growth and progression which may result from increased EMT and metastasis genes (IL-6, MMP-2, and SERPINE1) expression.[67] Additionally, exosomes derived from MSCs promote proliferation, tumorigenesis, and migration of nasopharyngeal carcinoma cells (NPCs) by FGF19-mediated activation of the FGFR4 signaling cascade and consequently induce EMT.[68]

10.3.2 MSCs INHIBIT TUMOR CELL PROLIFERATION

Many studies have indicated that MSCs from different origins and species can inhibit tumor cells proliferation either by a direct cell-to-cell contact or secretion of soluble factor/s at various in vivo and in vitro models.[69–72] Amongst the capability of homing to the tumor site, initiation of cell cycle arrest, induction of apoptosis, antiangiogenesis activity and suppression of WNT and PI3K/Akt signaling are reported as mechanisms that been recruited by MSCs to suppress tumor growth.[24,73–75]

10.3.2.1 HOMING CAPABILITY OF MSCs

Similar to the immune cells, MSCs as well migrate and home to the tumor niche and inflammatory regions which make them a beneficial agent in delivering a particular treatment such as gene therapy and drug delivery to the specific tissues.[76–78] Genetically modified MSCs that express a high level of IL-15 gene has shown to inhibit the pancreatic tumor cell growth by attracting cytotoxic T lymphocytes and NK cells in mice.[79] Likewise, several scientific reports have shown that genetically engineered MSCs can specifically migrate to various tumors and locally secrete therapeutic proteins such as IFN-γ,[80] IFN-β,[81] IL-24,[82] IL-18,[83] and TNF-related apoptosis-inducing ligand (TRAIL).[84] Additionally, Xie et al. showed that IFN-β-induced MSCs inhibit the proliferation of hepatocellular carcinoma by decreased expression of cyclin D1 in cell cycle through inhibiting AKT/FOXO3a pathway.[85] Moreover, Grisendi et al. reported that TRAIL-transduced-MSCs induced the antitumor activity of MSCs by triggering significant apoptosis and exerting robust antiangiogenic functions on xenograft models of Ewing's Sarcoma.[86] These findings support the notion that the homing capacity of MSCs can be modified genetically toward delivering specific antitumor proteins and molecules, hence could serve as a new approach for the targeted antitumor therapy.

10.3.2.2   *INDUCTION OF CELL CYCLE ARREST AND APOPTOSIS*

MSCs are capable of suppressing tumor cells' proliferation either by inducing apoptosis and/or cell cycle arrest through modulating the expression of proteins such as p53, p21, BAX, BCL-2, and various caspases.[87–89] In this regard, it has been shown that MSCs inhibit lymphoid origin hematopoietic tumor cell proliferation by arresting the tumor cells in G_0/G_1 phase with no sign of apoptosis.[90] In addition, Fonseka et al. showed that MSCs suppress the proliferation of K562 cells, human erythromyeloblastoid leukemia cell line, by predominantly accumulating the tumor cells in the G_0/G_1 phase of cell cycle.[75]

In line with this, MSCs limited the cell growth of ovarian carcinoma (TOV-112D), breast adenocarcinoma (MDA-MB-231), and osteosarcoma (MG-63) cells by arresting them in metaphase stage of G_2/M phase. This cell cycle arrest was accompanied by induction of apoptosis where BAX (proapoptotic) and BCL-2 (antiapoptotic) were upregulated and downregulated, respectively, in all three cancer cell lines.[10] Additionally, Tang et al. showed that the inhibitory effect of MSCs on hepatocellular carcinoma cells (HCCs) (HepG2) is due to induction of apoptosis through downregulation of BCL-2, survivin, and α-fetoprotein.[91] Furthermore, there is evidence indicating that MSCs inhibit cancer stem cells derived from primary human breast cancer cells, MCF-7 and MDA-MB-231 cell lines by induction of apoptosis and reduction of PI3K and AKT proteins.[92] Moreover, research by Dasari et al. suggested that MSCs inhibit human glioblastoma tumor cells growth by induction of apoptosis via downregulation x-linked inhibitor of apoptosis protein.[93]

Besides, it has been shown that microvesicles secreted by MSCs attenuate T24 cell line, a bladder tumor by halting the tumor cells in G_0/G_1 phase by mean of p21 overexpression. Similarly, MSCs as well induce apoptosis in T24 cell line by initiating the phosphorylation of Akt and upregulation of cleaved caspase 3 at in vivo and in vitro.[87] Moreover, almost a similar result was shown by Dzobo et al. who illustrated that MSCs inhibit WHCO1 (esophageal cancer) and MDA-MB 231 (breast cancer) cell proliferation either by arresting cells in the G_2 phase of the cell cycle via induction p21 protein and by inducing apoptosis via activation of caspases 3 and 9.[94]

10.3.2.3 WNT AND PI3K/AKT SIGNALING PATHWAY

Numerous studies have suggested that WNT and PI3K/Akt signaling pathways play a major role in cell proliferation, invasion, migration, and angiogenesis.[95,96] In line with this, Qiao et al. illustrated that MSCs suppress tumorigenesis in H7402 and HepG2 cell lines (human hepatoma cell) by targeting the WNT signaling pathway in in vivo and in vitro models.[97] Moreover, it has been reported that HCCs were suppressed by MSCs in rats model due to downregulation of WNT signaling pathway by targeting β-catenin, cyclin D, surviving and proliferating cell nuclear antigen genes expression in HCC.[98] Furthermore, the same finding was reported by Hou et al. who documented that the inhibitory effect of MSCs on HepG2 cells (human hepatoma cell) might be due to downregulation of WNT signaling pathway. They found that Dkk-1 secreted from MSCs downregulates the expression of c-Myc, BCL-2, surviving and β-catenin which are WNT signaling pathway-related factors and consequently inhibit the proliferation of HepG2 cells.[99] Additionally, Liu et al. investigated the possible role of Akt and WNT pathways in MSCs exerted antitumor activity. This study revealed that MSCs-mediated inhibition on intrahepatic cholangiocarcinoma cell growth is due to molecular crosstalk between two signaling pathways, the Wnt/β-catenin and PI3K/Akt signaling pathways, with glycogen synthase kinase 3β as the key enzyme bridging these signaling pathways.[100] In accordance with these findings, it has been stated that MSCs elevate apoptosis in prostate cancer (PC-3 cell) and attenuate the proliferation of cancer cells through downregulation of PI3K/AKT signaling pathway.[72] In 2015, Yulyana et al. discovered that the conditioned media derived from MSCs (CM-MSCs) might provide a useful tool as a treatment for HCC. This study showed that CM-MSCs impose a profound inhibitory effect on HCC proliferation by reducing the level of activated PI3K/AKT.[101]

10.3.2.4   ANGIOGENESIS

The nutrition and oxygen provided by vasculatures are vital for the survival and function of tumor cells. Consequently, evacuation of metabolic wastes and carbon dioxide by vasculatures are also essential for cells. Therefore, any mechanisms that disrupt the angiogenesis in the tumor microenvironment would be a great attempt to control tumor cell proliferation and metastasis. Several studies have delineated the role of MSCs in coercing the angiogenesis process at various levels of different tumor models.[102,103] It has been reported that the antitumor effect of MSCs on human glioma tumor cells (ΔGli36) could be the result of downregulation of proangiogenic soluble factors such as FGF-2, IGF-1, PDGF-BB, and IL-1b which their downregulation leads to suppression of angiogenesis in ΔGli36 cells.[104] In addition, Lee et al. showed that MSCs-derived exosomes suppress mouse breast cancer cell proliferation by an antiangiogenic mechanism in tumor cells via suppression of VEGF at both in vivo and in vitro.[9] Furthermore, it has been recently illustrated that MSCs upregulate the secretion of tissue inhibitor of metalloproteinase-1, as an antiangiogenic molecule, by inflammatory-associated cytokine in inflammatory disease.[105] Likewise, Yang et al. documented that the engineered MSCs to secrete pigment epithelium-derived factor, an antiangiogenic molecule, might inhibit tumor angiogenesis and, consequently, suppress the cell growth and tumor metastasis of murine colorectal peritoneal carcinomatosis.[106]

10.4   CONCLUSION

Since several decades, MSCs have attracted much attention in the field of regenerative medicine as potential therapeutic agents for the treatment of a broad range of diseases such as cancer. It has been shown that MSCs contain several therapeutic potentials such as regenerating or repairing damaged tissues,[107] ameliorating graft-versus-host disease,[108] tools for gene therapy,[84] and affecting the tumor growth.[70] Although considerable progress toward the identification of the various effect of MSCs on tumor cell proliferation was made but what causes them to exert an inhibitory or promoting effect on tumor cells remain a puzzle to be solved. MSCs migrate and infiltrate the tumor microenvironment in order to promote tumor cell growth.[76,109] Several mechanisms, particularly induction of angiogenesis and antiapoptosis have been instrumental to the MSCs to favor tumor cells through inducing proliferation, metastasis, and invasion. On the other hand, MSCs as well inhibit the tumor cell proliferation by promoting cell cycle arrest, induction of apoptosis, blocking angiogenesis and suppressing WNT and PI3K/Akt signaling. There are many factors and parameters obscure the delineation of MSCs' role in tumor growth. The source of MSCs (individual donor variability and origins) type of tumors such as solid or nonsolid, tumor microenvironment and its secretion, the amount of secretion and the intrinsic signaling pathway that govern cell proliferation, cell cycle, invasion, and migration in various tumor cells might contribute to this discrepancy. Despite the difficulty of anticipating the outcome of MSCs–tumor cells interaction, the engineered MSCs could deliver a favorable antitumor activity by migrating into the tumor site and systematically express the antitumor cytokines or proteins. However, a need is felt to understand better the biological consequences of MSC–tumor interactions before MSCs based therapies can truly be successful in the clinical setting. The specificity of MSCs exerted antitumor activity is an important point in clinical implications since different types of tumors evolve and exploit the host mechanism to escape from the immune system.

KEYWORDS

- **mesenchymal stem cells**
- **tumor cells**
- **microenvironment**
- **tumor-supporting activity**
- **antitumorigenic activity**

REFERENCES

1. GLOBOCAN. *International Agency for Research on Cancer 2012.* http://globocan. iarc.fr/Pages/burden_sel.aspx (accessed Oct 21, 2016).
2. Calon, A., et al. Dependency of Colorectal Cancer on a TGF-Beta-Driven Program in Stromal Cells for Metastasis Initiation. *Cancer Cell* **2012,** *22*(5), 571–584.
3. Gorgun, G. T., et al. Tumor-Promoting Immune-Suppressive Myeloid-Derived Suppressor Cells in the Multiple Myeloma Microenvironment in Humans. *Blood* **2013,** *121*(15), 2975–2987.
4. D'Souza, N., et al. MSC and Tumors: Homing, Differentiation, and Secretion Influence Therapeutic Potential. *Adv. Biochem. Eng. Biotechnol.* **2013,** *130*, 209–266.
5. Mbeunkui, F.; Johann, D. J., Jr. Cancer and the Tumor Microenvironment: A Review of an Essential Relationship. *Cancer. Chemother. Pharmacol.* **2009,** *63*(4), 571–582.
6. Whiteside, T. L. The Tumor Microenvironment and Its Role in Promoting Tumor Growth. *Oncogene* **2008,** *27*(45), 5904–5912.
7. Nomoto-Kojima, N., et al. Interaction Between Adipose Tissue Stromal Cells and Gastric Cancer Cells In Vitro. *Cell Tissue Res.* **2011,** *344*(2), 287–298.
8. Shinagawa, K., et al. Mesenchymal Stem Cells Enhance Growth and Metastasis of Colon Cancer. *Int. J. Cancer* **2010,** *127*(10), 2323–2333.
9. Lee, J. K., et al. Exosomes Derived from Mesenchymal Stem Cells Suppress Angiogenesis by Down-Regulating VEGF Expression in Breast Cancer Cells. *PLoS One* **2013,** *8*(12), e84256.
10. Gauthaman, K., et al. Human Umbilical Cord Wharton's Jelly Stem Cell (hWJSC) Extracts Inhibit Cancer Cell Growth in vitro. *J. Cell. Biochem.* **2012,** *113*(6), 2027–2039.
11. Friedenstein, A. J., et al. Precursors for Fibroblasts in Different Populations of Hematopoietic Cells as Detected by the in vitro Colony Assay Method. *Exp. Hematol.* **1974,** *2*(2), 83–92.
12. Pevsner-Fischer, M.; Levin, S.; Zipori, D. The Origins of Mesenchymal Stromal Cell Heterogeneity. *Stem Cell Rev.* **2011,** *7*(3), 560–568.
13. Wang, G., et al. Adult Stem Cells from Bone Marrow Stroma Differentiate into Airway Epithelial Cells: Potential Therapy for Cystic Fibrosis. *Proc. Natl. Acad. Sci. U.S.A.* **2005,** *102*(1), 186–191.
14. Tropel, P., et al. Functional Neuronal Differentiation of Bone Marrow-Derived Mesenchymal Stem Cells. *Stem Cells* **2006,** *24*(12), 2868–2876.
15. Pittenger, M. F., et al. Multilineage Potential of Adult Human Mesenchymal Stem Cells. *Science* **1999,** *284*(5411), 143–147.
16. Joshi, M., et al. Fetal Liver-Derived Mesenchymal Stromal Cells Augment Engraftment of Transplanted Hepatocytes. *Cytotherapy* **2012,** *14*(6), 657–669.
17. Ponnaiyan, D.; Bhat, K. M.; Bhat, G. S. Comparison of Immuno-Phenotypes of Stem Cells from Human Dental Pulp and Periodontal Ligament. *Int. J. Immunopathol. Pharmacol.* **2012,** *25*(1), 127–134.
18. Ishige, I., et al. Comparison of Mesenchymal Stem Cells Derived from Arterial, Venous, and Wharton's Jelly Explants of Human Umbilical Cord. *Int. J. Hematol.* **2009,** *90*(2), 261–269.

19. Gruber, H. E., et al. Human Adipose-Derived Mesenchymal Stem Cells: Direction to a Phenotype Sharing Similarities with the Disc, Gene Expression Profiling, and Coculture with Human Annulus Cells. *Tissue Eng. Part A* **2010,** *16*(9), 2843–2860.

20. Erices, A.; Conget, P.; Minguell, J. J. Mesenchymal Progenitor Cells in Human Umbilical Cord Blood. *Br. J. Haematol.* **2000,** *109*(1), 235–242.

21. In't Anker, P. S., et al. Isolation of Mesenchymal Stem Cells of Fetal or Maternal Origin from Human Placenta. *Stem Cells* **2004,** *22*(7), 1338–1345.

22. Dominici, M., et al. Minimal Criteria for Defining Multipotent Mesenchymal Stromal Cells. The International Society for Cellular Therapy Position Statement. *Cytotherapy* **2006,** *8*(4), 315–317.

23. Hagenhoff, A., et al. Harnessing Mesenchymal Stem Cell Homing as an Anticancer Therapy. *Expert Opin. Biol. Ther.* **2016,** *16*(9), 1079–1092.

24. De Becker, A.; Riet, I. V. Homing and Migration of Mesenchymal Stromal Cells: How to Improve the Efficacy of Cell Therapy? *World J. Stem Cells* **2016,** *8*(3), 73–87.

25. Abd-Allah, S. H., et al. Effect of Bone Marrow-Derived Mesenchymal Stromal Cells on Hepatoma. *Cytotherapy* **2014,** *16*(9), 1197–1206.

26. Yan, X. L., et al. Hepatocellular Carcinoma-Associated Mesenchymal Stem Cells Promote Hepatocarcinoma Progression: Role of the S100A4-miR155-SOCS1-MMP9 Axis. *Hepatology* **2013,** *57*(6), 2274–2286.

27. Waterman, R. S.; Henkle, S. L.; Betancourt, A. M. Mesenchymal Stem Cell 1 (MSC1)-Based Therapy Attenuates Tumor Growth Whereas MSC2-Treatment Promotes Tumor Growth and Metastasis. *PLoS One* **2012,** *7*(9), e45590.

28. Yang, X., et al. Human Umbilical Cord Mesenchymal Stem Cells Promote Carcinoma Growth and Lymph Node Metastasis When Co-Injected with Esophageal Carcinoma Cells in Nude Mice. *Cancer Cell Int.* **2014,** *14*(1), 93.

29. Li, T., et al. Umbilical Cord-Derived Mesenchymal Stem Cells Promote Proliferation and Migration in MCF-7 and MDA-MB-231 Breast Cancer Cells Through Activation of the ERK Pathway. *Oncol. Rep.* **2015,** *34*(3), 1469–1477.

30. English, K.; French, A.; Wood, K. J. Mesenchymal Stromal Cells: Facilitators of Successful Transplantation? *Cell Stem Cell.* **2010,** *7*(4), 431–442.

31. Vellasamy, S., et al. Mesenchymal Stem Cells of Human Placenta and Umbilical Cord Suppress T-Cell Proliferation at G0 Phase of Cell Cycle. *Cell Biol. Int.* **2013,** *37*(3), 250–256.

32. El Omar, R., et al. Umbilical Cord Mesenchymal Stem Cells: The New Gold Standard for Mesenchymal Stem Cell-Based Therapies? *Tissue Eng. Part B Rev.* **2014,** *20*(5), 523–544.

33. Zhou, C., et al. Immunomodulatory Effect of Human Umbilical Cord Wharton's Jelly-Derived Mesenchymal Stem Cells on Lymphocytes. *Cell Immunol.* **2011,** *272*(1), 33–38.

34. Deuse, T., et al. Immunogenicity and Immunomodulatory Properties of Umbilical Cord Lining Mesenchymal Stem Cells. *Cell Transplant.* **2011,** *20*(5), 655–667.

35. Chao, K. C.; Yang, H. T.; Chen, M. W. Human Umbilical Cord Mesenchymal Stem Cells Suppress Breast Cancer Tumourigenesis through Direct Cell–Cell Contact and Internalization. *J. Cell Mol. Med.* **2012,** *16*(8), 1803–1815.

36. Ciccocioppo, R., et al. Ex Vivo Immunosuppressive Effects of Mesenchymal Stem Cells on Crohn's Disease Mucosal T Cells Are Largely Dependent on Indoleamine 2,3-Dioxygenase Activity and Cell–Cell Contact. *Stem Cell Res. Ther.* **2015,** *6,* 137.

37. Chen, P. M., et al. Induction of Immunomodulatory Monocytes by Human Mesenchymal Stem Cell-Derived Hepatocyte Growth Factor through ERK1/2. *J. Leukoc. Biol.* **2014,** *96*(2), 295–303.

38. Munir, H., et al. Comparative Ability of Mesenchymal Stromal Cells from Different Tissues to Limit Neutrophil Recruitment to Inflamed Endothelium. *PLoS One* **2016,** *11*(5), e0155161.

39. Quaranta, P., et al. Human Wharton's Jelly-Derived Mesenchymal Stromal Cells Engineered to Secrete Epstein–Barr Virus Interleukin-10 Show Enhanced Immunosuppressive Properties. *Cytotherapy* **2016,** *18*(2), 205–218.

40. Kim, H. S., et al. Human Umbilical Cord Blood Mesenchymal Stem Cell-Derived PGE2 and TGF-Beta1 Alleviate Atopic Dermatitis by Reducing Mast Cell Degranulation. *Stem Cells* **2015,** *33*(4), 1254–1266.

41. Prasanna, S. J., et al. Pro-Inflammatory Cytokines, IFNGamma and TNFAlpha, Influence Immune Properties of Human Bone Marrow and Wharton Jelly Mesenchymal Stem Cells Differentially. *PLoS One* **2010,** *5*(2), e9016.

42. Duffy, M. M., et al. Mesenchymal Stem Cell Inhibition of T-Helper 17 Cell-Differentiation is Triggered by Cell–Cell Contact and Mediated by Prostaglandin E2 via the EP4 Receptor. *Eur. J. Immunol.* **2011,** *41*(10), 2840–2851.

43. Ren, G., et al. Mesenchymal Stem Cell-Mediated Immunosuppression Occurs via Concerted Action of Chemokines and Nitric Oxide. *Cell Stem Cell* **2008,** *2*(2), 141–150.

44. Zimmermann, J. A.; Hettiaratchi, M. H.; McDevitt, T. C. Enhanced Immunosuppression of T Cells by Sustained Presentation of Bioactive Interferon-Gamma within Three-Dimensional Mesenchymal Stem Cell Constructs. *Stem Cells Transl. Med.* **2016,** *6*(1), 223–237.

45. Nakamura, K., et al. Antitumor Effect of Genetically Engineered Mesenchymal Stem Cells in a Rat Glioma Model. *Gene Ther* **2004,** *11*(14), 1155–1164.

46. Mathew, E., et al. Mesenchymal Stem Cells Promote Pancreatic Tumor Growth by Inducing Alternative Polarization of Macrophages. *Neoplasia* **2016,** *18*(3), 142–151.

47. Liu, Y., et al. Effects of Inflammatory Factors on Mesenchymal Stem Cells and Their Role in the Promotion of Tumor Angiogenesis in Colon Cancer. *J. Biol. Chem.* **2011,** *286*(28), 25007–25015.

48. Feng, B.; Chen, L. Review of Mesenchymal Stem Cells and Tumors: Executioner or Coconspirator? *Cancer Biother. Radiopharm.* **2009,** *24*(6), 717–721.

49. Tan, C. Y., et al. Mesenchymal Stem Cell-Derived Exosomes Promote Hepatic Regeneration in Drug-Induced Liver Injury Models. *Stem Cell Res. Ther.* **2014,** *5*(3), 76.

50. Zhang, T., et al. Bone Marrow-Derived Mesenchymal Stem Cells Promote Growth and Angiogenesis of Breast and Prostate Tumors. *Stem Cell Res. Ther.* **2013,** *4*(3), 70.

51. Silva, G. V., et al. Mesenchymal Stem Cells Differentiate into an Endothelial Phenotype, Enhance Vascular Density, and Improve Heart Function in a Canine Chronic Ischemia Model. *Circulation* **2005,** *111*(2), 150–156.

52. Armulik, A.; Genove, G.; Betsholtz, C. Pericytes: Developmental, Physiological, and Pathological Perspectives, Problems, and Promises. *Dev. Cell* **2011,** *21*(2), 193–215.

53. Spaeth, E. L., et al. Mesenchymal Stem Cell Transition to Tumor-Associated Fibroblasts Contributes to Fibrovascular Network Expansion and Tumor Progression. *PLoS One* **2009**, *4*(4), e4992.

54. Lee, M. J., et al. Macrophages Regulate Smooth Muscle Differentiation of Mesenchymal Stem Cells via a Prostaglandin F(2)Alpha-Mediated Paracrine Mechanism. *Arterioscler. Thromb. Vasc. Biol.* **2012**, *32*(11), 2733–2740.

55. Xu, Y., et al. Efficient Commitment to Functional CD34+ Progenitor Cells from Human Bone Marrow Mesenchymal Stem-Cell-Derived Induced Pluripotent Stem Cells. *PLoS One* **2012**, *7*(4), e34321.

56. Lin, H., et al. Adenoviral Expression of Vascular Endothelial Growth Factor Splice Variants Differentially Regulate Bone Marrow-Derived Mesenchymal Stem Cells. *J. Cell Physiol.* **2008**, *216*(2), 458–468.

57. Lin, R. Z., et al. Equal Modulation of Endothelial Cell Function by Four Distinct Tissue-Specific Mesenchymal Stem Cells. *Angiogenesis* **2012**, *15*(3), 443–455.

58. Castells, M., et al. Microenvironment Mesenchymal Cells Protect Ovarian Cancer Cell Lines from Apoptosis by Inhibiting XIAP Inactivation. *Cell Death Dis.* **2013**, *4*, e887.

59. Kurtova, A. V., et al. Diverse Marrow Stromal Cells Protect CLL Cells from Spontaneous and Drug-Induced Apoptosis: Development of a Reliable and Reproducible System to Assess Stromal Cell Adhesion-Mediated Drug Resistance. *Blood* **2009**, *114*(20), 4441–4450.

60. Wu, X. B., et al. Mesenchymal Stem Cells Promote Colorectal Cancer Progression through AMPK/mTOR-Mediated NF-KappaB Activation. *Sci. Rep.* **2016**, *6*, 21420.

61. Grunert, S.; Jechlinger, M.; Beug, H. Diverse Cellular and Molecular Mechanisms Contribute to Epithelial Plasticity and Metastasis. *Nat. Rev. Mol. Cell Biol.* **2003**, *4*(8), 657–665.

62. Jechlinger, M., et al. Expression Profiling of Epithelial Plasticity in Tumor Progression. *Oncogene* **2003**, *22*(46), 7155–7169.

63. Klymkowsky, M. W.; Savagner, P. Epithelial-Mesenchymal Transition: A Cancer Researcher's Conceptual Friend and Foe. *Am. J. Pathol.* **2009**, *174*(5), 1588–1593.

64. Cannito, S., et al. Epithelial-Mesenchymal Transition: From Molecular Mechanisms, Redox Regulation to Implications in Human Health and Disease. *Antioxid. Redox Signal.* **2010**, *12*(12), 1383–1430.

65. Fan, F., et al. Overexpression of Snail Induces Epithelial-Mesenchymal Transition and a Cancer Stem Cell-Like Phenotype in Human Colorectal Cancer Cells. *Cancer Med.* **2012**, *1*(1), 5–16.

66. Mele, V., et al. Mesenchymal Stromal Cells Induce Epithelial-to-Mesenchymal Transition in Human Colorectal Cancer Cells through the Expression of Surface-Bound TGF-Beta. *Int. J. Cancer* **2014**, *134*(11), 2583–2594.

67. Strong, A. L., et al. Leptin Produced by Obese Adipose Stromal/Stem Cells Enhances Proliferation and Metastasis of Estrogen Receptor Positive Breast Cancers. *Breast Cancer Res.* **2015**, *17*, 112.

68. Shi, S., et al. Mesenchymal Stem Cell-Derived Exosomes Facilitate Nasopharyngeal Carcinoma Progression. *Am. J. Cancer Res.* **2016**, *6*(2), 459–472.

69. Ayatollahi, M.; Talaei-Khozani, T.; Razmkhah, M. Growth Suppression Effect of Human Mesenchymal Stem Cells from Bone Marrow, Adipose Tissue, and Wharton's Jelly of Umbilical Cord on PBMCs. *Iran J. Basic Med. Sci.* **2016**, *19*(2), 145–153.

70. Ohta, N., et al. Human Umbilical Cord Matrix Mesenchymal Stem Cells Suppress the Growth of Breast Cancer by Expression of Tumor Suppressor Genes. *PLoS One* **2015**, *10*(5), e0123756.

71. Hendijani, F.; Javanmard, S. H.; Sadeghi-aliabadi, H. Human Wharton's Jelly Mesenchymal Stem Cell Secretome Display Antiproliferative Effect on Leukemia Cell Line and Produce Additive Cytotoxic Effect in Combination with Doxorubicin. *Tissue Cell* **2015**, *47*(3), 229–234.

72. Han, I., et al. Umbilical Cord Tissue-Derived Mesenchymal Stem Cells Induce Apoptosis in PC-3 Prostate Cancer Cells through Activation of JNK and Downregulation of PI3K/AKT Signaling. *Stem Cell Res. Ther.* **2014**, *5*(2), 54.

73. Rhee, K. J.; Lee, J. I.; Eom, Y. W. Mesenchymal Stem Cell-Mediated Effects of Tumor Support or Suppression. *Int. J. Mol. Sci.* **2015**, *16*(12), 30015–30033.

74. Hass, R.; Otte, A. Mesenchymal Stem Cells as All-Round Supporters in a Normal and Neoplastic Microenvironment. *Cell Commun. Signal* **2012**, *10*(1), 26.

75. Fonseka, M., et al. Human Umbilical Cord Blood-Derived Mesenchymal Stem Cells (hUCB-MSC) Inhibit the Proliferation of K562 (Human Erythromyeloblastoid Leukaemic Cell Line). *Cell Biol. Int.* **2012**, *36*(9), 793–801.

76. Auffinger, B., et al. Drug-Loaded Nanoparticle Systems and Adult Stem Cells: A Potential Marriage for the Treatment of Malignant Glioma? *Oncotarget* **2013**, *4*(3), 378–396.

77. Durinikova, E.; Kucerova, L.; Matuskova, M. Mesenchymal Stromal Cells Retrovirally Transduced with Prodrug-Converting Genes are Suitable Vehicles for Cancer Gene Therapy. *Acta Virol.* **2014**, *58*(1), 1–13.

78. Park, J. S., et al. Engineering Mesenchymal Stem Cells for Regenerative Medicine and Drug Delivery. *Methods* **2015**, *84*, 3–16.

79. Jing, W., et al. Human Umbilical Cord Blood-Derived Mesenchymal Stem Cells Producing IL15 Eradicate Established Pancreatic Tumor in Syngeneic Mice. *Mol. Cancer Ther.* **2014**, *13*(8), 2127–2137.

80. Yang, X., et al. IFN-Gamma-Secreting-Mesenchymal Stem Cells Exert an Antitumor Effect In Vivo via the TRAIL Pathway. *J. Immunol. Res.* **2014**, *2014*, 318098.

81. Ryu, H., et al. Adipose Tissue-Derived Mesenchymal Stem Cells Cultured at High Density Express IFN-Beta and Suppress the Growth of MCF-7 Human Breast Cancer Cells. *Cancer Lett.* **2014**, *352*(2), 220–227.

82. Zhang, X., et al. Experimental Therapy for Lung Cancer: Umbilical Cord-Derived Mesenchymal Stem Cell-Mediated Interleukin-24 Delivery. *Curr. Cancer Drug Targets* **2013**, *13*(1), 92–102.

83. Liu, X., et al. Mesenchymal Stem Cells Expressing Interleukin-18 Suppress Breast Cancer Cells In Vitro. *Exp. Ther. Med.* **2015**, *9*(4), 1192–1200.

84. Yan, C., et al. Suppression of Orthotopically Implanted Hepatocarcinoma in Mice by Umbilical Cord-Derived Mesenchymal Stem Cells with sTRAIL Gene Expression Driven by AFP Promoter. *Biomaterials* **2014**, *35*(9), 3035–3043.

85. Xie, C., et al. Interferon-Beta Gene-Modified Human Bone Marrow Mesenchymal Stem Cells Attenuate Hepatocellular Carcinoma through Inhibiting AKT/FOXO3a Pathway. *Br. J. Cancer* **2013**, *109*(5), 1198–1205.

86. Grisendi, G., et al. Mesenchymal Progenitors Expressing TRAIL Induce Apoptosis in Sarcomas. *Stem Cells* **2015**, *33*(3), 859–869.

87. Wu, S., et al. Microvesicles Derived from Human Umbilical Cord Wharton's Jelly Mesenchymal Stem Cells Attenuate Bladder Tumor Cell Growth In Vitro and In Vivo. *PLoS One* **2013**, *8*(4), e61366.

88. Lim, P. K., et al. Gap Junction-Mediated Import of microRNA from Bone Marrow Stromal Cells Can Elicit Cell Cycle Quiescence in Breast Cancer Cells. *Cancer Res.* **2011**, *71*(5), 1550–1560.

89. Ramsey, R., et al. Mesenchymal Stem Cells Inhibit Dendritic Cell Differentiation and Function by Preventing Entry into the Cell Cycle. *Transplantation* **2007**, *83*(1), 71–76.

90. Sarmadi, V. H., et al. Mesenchymal Stem Cells Inhibit Proliferation of Lymphoid Origin Haematopoietic Tumour Cells by Inducing Cell Cycle Arrest. *Med. J. Malays.* **2010**, *65*(3), 209–214.

91. Tang, Y. M., et al. Umbilical Cord-Derived Mesenchymal Stem Cells Inhibit Growth and Promote Apoptosis of HepG2 Cells. *Mol. Med. Rep.* **2016**, *14*(3), 2717–2724.

92. Ma, Y., et al. The in vitro and in vivo Effects of Human Umbilical Cord Mesenchymal Stem Cells on the Growth of Breast Cancer Cells. *Breast Cancer Res. Treat.* **2012**, *133*(2), 473–485.

93. Dasari, V. R., et al. Cord Blood Stem Cell-Mediated Induction of Apoptosis in Glioma Downregulates X-Linked Inhibitor of Apoptosis Protein (XIAP). *PLoS One* **2010**, *5*(7), e11813.

94. Dzobo, K., et al. Wharton's Jelly-Derived Mesenchymal Stromal Cells and Fibroblast-Derived Extracellular Matrix Synergistically Activate Apoptosis in a p21-Dependent Mechanism in WHCO1 and MDA MB 231 Cancer Cells in vitro. *Stem Cells Int.* **2016**, *2016*, 4842134.

95. Gilbert, S. F. *Developmental Biology*, 9th ed.; Sinauer Associates: Sunderland, MA, 2010.

96. Manning, B. D.; Cantley, L. C. AKT/PKB Signaling: Navigating Downstream. *Cell* **2007**, *129*(7), 1261–1274.

97. Qiao, L., et al. Suppression of Tumorigenesis by Human Mesenchymal Stem Cells in a Hepatoma Model. *Cell Res.* **2008**, *18*(4), 500–507.

98. Abdel Aziz, M. T., et al. Efficacy of Mesenchymal Stem Cells in Suppression of Hepatocarcinorigenesis in Rats: Possible Role of Wnt Signaling. *J. Exp. Clin. Cancer Res.* **2011**, *30*, 49.

99. Hou, L., et al. Inhibitory Effect and Mechanism of Mesenchymal Stem Cells on Liver Cancer Cells. *Tumour Biol.* **2014**, *35*(2), 1239–1250.

100. Liu, J., et al. Suppression of Cholangiocarcinoma Cell Growth by Human Umbilical Cord Mesenchymal Stem Cells: A Possible Role of Wnt and Akt Signaling. *PLoS One* **2013**, *8*(4), e62844.

101. Yulyana, Y., et al. Paracrine Factors of Human Fetal MSCs Inhibit Liver Cancer Growth through Reduced Activation of IGF-1R/PI3K/Akt Signaling. *Mol. Ther.* **2015**, *23*(4), 746–756.

102. Otsu, K., et al. Concentration-Dependent Inhibition of Angiogenesis by Mesenchymal Stem Cells. *Blood* **2009,** *113*(18), 4197–4205.

103. Ronca, R., et al. Long Pentraxin-3 as an Epithelial-Stromal Fibroblast Growth Factor-Targeting Inhibitor in Prostate Cancer. *J. Pathol.* **2013,** *230*(2), 228–238.

104. Ho, I. A., et al. Human Bone Marrow-Derived Mesenchymal Stem Cells Suppress Human Glioma Growth through Inhibition of Angiogenesis. *Stem Cells* **2013,** *31*(1), 146–155.

105. Zanotti, L., et al. Mouse Mesenchymal Stem Cells Inhibit High Endothelial Cell Activation and Lymphocyte Homing to Lymph Nodes by Releasing TIMP-1. *Leukemia* **2016,** *30*(5), 1143–1154.

106. Yang, L., et al. Mesenchymal Stem Cells Engineered to Secrete Pigment Epithelium-Derived Factor Inhibit Tumor Metastasis and the Formation of Malignant Ascites in a Murine Colorectal Peritoneal Carcinomatosis Model. *Hum. Gene Ther.* **2016,** *27*(3), 267–277.

107. Zhao, Y., et al. Exosomes Derived from Human Umbilical Cord Mesenchymal Stem Cells Relieve Acute Myocardial Ischemic Injury. *Stem Cells Int.* **2015,** *2015*, 761643.

108. Wu, K. H., et al. Effective Treatment of Severe Steroid-Resistant Acute Graft-Versus-Host Disease with Umbilical Cord-Derived Mesenchymal Stem Cells. *Transplantation* **2011,** *91*(12), 1412–1416.

109. Kidd, S., et al. Direct Evidence of Mesenchymal Stem Cell Tropism for Tumor and Wounding Microenvironments Using In Vivo Bioluminescent Imaging. *Stem Cells* **2009,** *27*(10), 2614–2623.

CHAPTER 11

WATER PURIFICATION USING NANOPARTICLES

NITHEESHA SHAJI[1], SUNIJA SUKUMARAN[1],
PARVATHY PRASAD[1], JIYA JOSE[1*], SABU THOMAS[1,2],
and NANDAKUMAR KALARIKKAL[3]

[1]*International and Inter University Centre for Nanoscience and Nanotechnology, Mahatma Gandhi University, Kottayam 686560, Kerala, India*

[2]*School of Chemical Sciences, Mahatma Gandhi University, Kottayam 686560, Kerala, India*

[3]*School of Pure and Applied Physics, Centre for Nanoscience and Nanotechnology, Mahatma Gandhi University, Kottayam 686560, Kerala, India*

Corresponding author. E-mail: jiyajose@gmail.com

ABSTRACT

Water is the most important asset of human civilization, and clean water supply is a basic human necessity. Due to the population growth and climate change, challenges in meeting global water needs require novel water treatment technologies in order to ensure the drinking water supply and reduce global water pollution. To overcome this situation, highly advanced nanotechnology can be adapted to develop new opportunities in technological developments for advanced water and wastewater technology processes. This chapter is an overview of recent advances in nanotechnologies for water and wastewater treatment processes, including nanobased materials, such as nanoadsorbents, nanometals, nanomembranes, and photocatalysts. These materials have favorable properties when compared with conventional processes reported.

11.1 INTRODUCTION

Clean water is essential to human health. The world is facing challenges to overcome the rising demands of clean water because of the available supplies of freshwater are draining due to (1) extended droughts, (2) population growth, (3) more stringent health based regulations, and (4) competing demands from a variety of users.[1] Water which is free of toxic chemicals and pathogens is essential to human health. Water that has been adversely affected in quality by anthropogenic influence is called wastewater. Wastewater can originate from a combination of domestic, industrial, commercial, or agricultural activities, surface runoff or storm water, and from sewer inflow or infiltration. Several chemical and biological contaminants have endangered the quality of drinking water. The contents of wastewater widely varied. Wastewater may contain pathogens such as bacteria, viruses, prions and parasitic worms, nonpathogenic bacteria, organic particles, soluble organic materials, inorganic particles, soluble inorganic materials, gases, emulsions, toxins, pharmaceuticals and hormones, and other hazardous substances. There are many processes that can be used to clean up wastewaters based on the type and degree of contamination.[2]

A number of techniques are used for treatment of water which includes chemical and physical agent such as chlorine and its derivatives, ultraviolet light, boiling, low-frequency ultrasonic irradiation, distillation, reverse osmosis, water sediment filters (fiber and ceramic)-activated carbon, solid block, pitcher and faucet-mount filters, bottled water, ion exchange water softener, ozonization, activated alumina "Altered" water. Halogens such as chlorine (Cl) and bromine (Br) are widely used as antibacterial agents because of the high toxicity and vapor pressure in pure form, the direct use of halogens as bactericides has many problems. NH_4^+ is the most common cation in water affecting human and animal health. In drinking water, ammonia removal is very important to prevent oxygen depletion and algae bloom and due to its extreme toxicity to most fish species. To reduce the effects of NH_4^+, it can be replaced with biologically acceptable cations, like Na^+, K^+, or Ca^{2+} in the zeolite. During the past few decades, several investigations have been carried out concerning the use of synthetic and natural zeolites, polymer films, and metal ions (Ag^+, Cu^{2+}, Zn^{2+}, Hg^{2+}, Ti^{3+}, Ni^{2+}, Co^{2+}) as bactericides for water disinfection.[3]

In the area of water purification, nanotechnology offers the possibility of an efficient removal of pollutants and germs. Nanotechnology deals with the manipulation of matter at size scales of less than 100 nm. Because of their high reactivity due to the large surface-to-volume ratio, it holds the promise of creating new materials and devices which take advantage of unique phenomena realized at those length scales. Nanomaterials have unique size-dependent properties related to their high-specific surface area (fast dissolution, high reactivity, strong sorption) and discontinuous properties (such as superparamagnetism, localized surface plasmon resonance, and quantum confinement effect). These specific nanobased characteristics allow the development of novel high-tech materials for more efficient water and wastewater treatment processes, namely, membranes, adsorption materials, nanocatalysts, functionalized surfaces, coatings, and reagents. Nanoparticles are expected to play a crucial role in water purification. In case of materials selection and design for water purification, the environmental fate and toxicity of materials are crucial issues. With no doubt we can say that nanotechnology is better than other techniques used in water treatment.

The adsorption technique is one of the most frequently studied and industrially adopted techniques.[4] Activated carbon from various sources such as coconut coir, jute stick, rice husk, etc. is the most popular of the adsorbents. The treatment of water includes adsorption methods using specific ion exchangers and combination of adsorption with catalytic treatment methods, redox processes, and magnetic processes. The application of carbon nanotube (CNT) clusters for adsorption also reported.[5] The unique property of this cluster is in the removal of bacteria from water by adsorption method. The surface-to-volume ratio increases drastically with the reduction of the size of the particle from bulk to nanodimensions. It leads to the availability of higher numbers of atoms/molecules on the surface for adsorption of contaminants. Accordingly, the surface energy available with each particle also increases significantly. This provides many advantages to drinking water purification:

- Effective contaminant removal even at low concentrations.
  Less waste is generated posttreatment as less quantity of nanomaterial will be required with respect to its bulk form. This happens since more adsorbent atoms/molecules are present per unit mass of the adsorbent and thus are actively utilized for adsorption.
- Novel reactions can be accomplished at the nanoscale due to an increase in the number of surface atoms (i.e., surface energy),

which is not possible with the analogous bulk material. An example of one such process is the use of noble metal nanoparticles for the degradation of pesticides which cannot be done by noble metals in their bulk form.[6]

11.2 ADSORPTION

Adsorption is a surface phenomenon in which the adhesion of atoms, ions, or molecules from a gas, liquid, or dissolved solid to a surface. This process creates a film of the adsorbate on the surface of the adsorbent. Adsorption is the capability of all solid substances to attract to their surfaces, molecules of gases, or solutions with which they are in close contact. Solids that are used to adsorb gases or dissolved substances are called adsorbents, and the adsorbed molecules are usually referred to collectively as the adsorbate.[7] Due to their high specific surface area of nanoadsorbents, they show a considerably higher rate of adsorption for organic compounds compared with granular or powdered activated carbon. They have great potential for novel, more efficient, and faster decontamination processes aimed at removal of organic and inorganic pollutants like heavy metals and micropollutants. Adsorption is usually employed as a polishing phase to eliminate organic and inorganic contaminants in water and wastewater treatment. For conventional adsorbents, efficiency is commonly restricted by the surface area or active sites, the lack of selectivity, and the adsorption kinetics. Nanoadsorbents offer significant improvement with their extremely high specific surface area and associated sorption sites, short intraparticle diffusion distance, and tunable pore size and surface chemistry. Current research activities mainly focus on the following types of nanoadsorbents:

- Carbon-based nanoadsorbents, that is, CNTs
- Polymeric nanoadsorbents
- Zeolites
- Metal-based nanosorbents.

11.2.1 CARBON-BASED NANOADSORBENTS

CNTs are allotropes of carbon with a cylinder-shaped nanostructure. CNTs are categorized as single-walled nanotubes and multiwalled nanotubes, depending on their manufacturing process. CNTs having a high specific surface area, it possesses highly assessable adsorption sites and also an adjustable surface chemistry. Due to their hydrophobic surface, CNTs have to be stabilized in aqueous suspension in order to avoid aggregation that reduces the active surface. They can be used for adsorption of persistent contaminants as well as to preconcentrate and detect contaminants.[8] Metal ions are adsorbed by CNTs through electrostatic attraction and chemical bonding. CNTs showed higher efficiency than activated carbon on adsorption of various organic chemicals. Its high adsorption capacity mainly arises from the large specific surface area and the diverse contaminant–CNT interactions. In case of individual CNTs, their external surfaces are the available surface area for adsorption. In the aqueous phase, due to the hydrophobicity of their graphitic surface CNTs form loose bundles or aggregates, reducing the effective surface area. On the other hand, CNT aggregates contain interstitial spaces and grooves, which are high adsorption energy sites for organic molecules.[9] Although activated carbon possesses comparable measured specific surface area as CNT bundles, it contains a significant number of micropores inaccessible to bulky organic molecules such as many antibiotics and pharmaceuticals.[10] Thus, CNTs have much higher adsorption capacity for some bulky organic molecules because of their larger pores in bundles and more accessible sorption sites.

Oxidized CNTs possess high adsorption capability for metal ions with fast kinetics. The surface functional groups such as carboxyl, hydroxyl, and phenol of CNTs are the major adsorption sites for metal ions through electrostatic attraction and chemical bonding. Surface oxidation can significantly enhance the adsorption capacity of CNTs. Several studies show that CNTs are better adsorbents than activated carbon for heavy metals and the adsorption kinetics is fast on CNTs due to the highly accessible adsorption sites and the short intraparticle diffusion distance.[11]

Conventional desalination methods are energy-consuming and technically demanding, whereas adsorption-based techniques are simple and easy to use for water purification devices, yet their capacity to remove salts is limited. Researchers developed plasma-modified ultralong CNTs that feature an ultrahigh specific adsorption capacity for salt that is two orders of magnitude higher when compared with conventional carbon-based water treatment systems. These ultralong CNTs can be implemented in multifunctional membranes that are able to remove not only salt but also organic and metal contaminants. Next-generation drinking water purification devices equipped with these novel CNTs are expected to have superior desalination, disinfection, and filtration properties.[12] Recently, a team of US researchers developed a sponge made of pure CNTs with a dash of boron that shows a remarkable ability to absorb oil from water. The oil can be deposited in the sponge for future retrieval or burned off so the sponge can be reused. If they succeed in generating large sheets, the sponge material can be applied in removing oil spills for oil remediation.[13]

11.2.2   POLYMERIC NANOADSORBENTS

Dendrimers are highly branched, star-shaped macromolecules with nanometer-scale dimensions. Dendrimers are defined by three components: a central core, an interior dendritic structure (the branches), and an exterior surface with functional surface groups. Polymeric nanoadsorbents such as dendrimers are utilizable for removing organics and heavy metals. Organic compounds can be adsorbed by the interior hydrophobic shells, whereas heavy metals can be adsorbed by the tailored exterior branches.[14] Diallo et al. integrated dendrimers in an ultrafiltration device in order to remove copper from water.[15] Nearly all copper ions were recovered by use of this combined dendrimer-ultrafiltration system. The adsorbent is regenerated simply through a pH shift. However, due to the complex multistage synthesis of dendrimers, up until now there are no commercial suppliers, except for some recently founded companies in the People's Republic of China.

Sadeghi-Kiakhani et al. produced a highly efficient bioadsorbent for the removal of anionic compounds such as dye from textile wastewater by preparing a combined chitosan-dendrimer nanostructure.[16] The bioadsorbent is biodegradable, biocompatible, and nontoxic. They achieve removal rates of certain dyes up to 99%.

11.2.3   ZEOLITES

Zeolites are microporous, aluminosilicate minerals commonly used as commercial adsorbents and catalysts. Zeolites in combination with silver atoms have been known since the early 1980s.[17] Nanoparticles such as silver ions can be embedded in zeolites due to its porous structure. The embedded nanoparticles are released from the zeolite matrix by exchange with other cations in solution. Egger et al. compared various materials containing nanosilver, including zeolites. When used for sanitary purposes, the silver attacks microbes and inhibits their growth.[19] A small amount of silver ions is released from the metallic surface when placed in contact with liquids. The success of this composition in water disinfection was shown once again by Petrik et al.[20] The Water Research Commission Report No KV 297/12 shows the innovative use of zeolites as an adsorbing platform for silver nanoparticles as a source of silver ions for a disinfectant.[21] Another possibility is applying zeolites themselves as nanoparticles. Jung et al. used nanozeolites in sequencing batch reactors for wastewater treatment.[22] With regard to this application field, nanometals and zeolites benefit from their cost-effectiveness and compatibility with existing water treatment systems since they can be implemented in pellets and beads for fixed absorbers.

11.2.4 METAL-BASED NANOSORBENTS

Nanoparticles have two significant properties that make them particularly attractive as sorbents. They have much larger surface areas than bulk particles and also be functionalized with various chemical groups to increase their affinity toward target compounds. Nanoscale metal oxides are promising alternatives to activated carbon and effective adsorbents to eliminate heavy metals and radionuclides. It is having a high specific surface area, a short intraparticle diffusion distance and is compressible without a significant reduction of surface area. Some of these nanoscale metal oxides are superparamagnetic, which facilitates separation and recovery by a low-gradient magnetic field. They can be employed for adsorptive media filters and slurry reactors.[23] Nanoiron hydroxide [α-FeO(OH)] is a robust abrasion-resistant adsorbent with a large specific surface area that enables adsorption of arsenic from waste and drinking water.[24] ArsenXnp is a commercially available hybrid ion exchange medium comprising iron oxide nanoparticles and polymers and is highly efficient in removing arsenic and requires little back wash. Usually, for industrial use, nanometals and nanometal oxides are compressed into porous pellets or used in powders. Metal oxides such as iron oxide, titanium dioxide, and alumina are effective, low cost and are used as adsorbents for heavy metals and radionuclides. The sorption is generally controlled by complexation among dissolved metals and the oxygen in metal oxides. It is a two-step process, fast adsorption of metal ions on the exterior surface, followed by the rate-limiting intraparticle diffusion along the microporewalls. Because of the higher specific surface area, shorter intraparticle diffusion distance, and larger number of surface reaction sites such as corners, edges, vacancies in their nanoscale counterparts have higher adsorption capacity and faster kinetics.

In addition to high adsorption capacity, some iron oxide nanoparticles, such as nanomaghemite and nanomagnetite, are superparamagnetic. Magnetism is highly volume dependent as it arises from the collective interaction of atomic magnetic dipoles. These magnetic nanoparticles can be either used directly as adsorbents or as the core material in a core–shell nanoparticle structure where the shell provides the desired function while the magnetic core realizes magnetic separation. Metal oxide nanocrystals can be compressed into porous. Metal-based nanomaterials have been explored to remove a variety of heavy metals such as arsenic, lead, mercury, copper, cadmium, chromium, nickel, and have shown great potential to outcompete activated carbon.[25] Several metal oxide nanomaterials including nanosized magnetite and TiO_2 have shown arsenic adsorption performance greater than activated carbon.[26]

Nanosilver has been used in the photodevelopment process since the late 1800s and has been registered with the Environmental Protection Agency for use in swimming pool algaecides since 1954 and drinking water filters since the 1970s. Although nanosilver exhibits a strong and broad-spectrum antimicrobial activity, it has hardly any harmful effects in humans. It is already applied water disinfection systems and antibiofouling surfaces. Nano-titanium dioxide (TiO_2) has high chemical stability and low human toxicity, at a cheap price is utilizable in disinfection and decontamination processes. The antimicrobial effect of nanosilver is based on the continuous release of silver ions. After a certain operation period, depending on the thickness and composition of the nanosilver layers, the coating or the complete device has to be renewed, leading to significant replacement costs. However, TiO_2 needs energy-consuming ultraviolet lamps for activation, but nanosilver kills bacteria with no need of additional energy-consuming devices. That makes nanosilver a favorable disinfectant for remote areas.

Magnetic nanoparticles are a class of nanoparticle that can be manipulated using magnetic fields. Such particles commonly consist of two components, a magnetic material such as iron, nickel, and cobalt and a chemical component that has functionality. Magnetic nanoparticles can be injected directly into the contaminated ground, and loaded particles can be removed simply through a magnetic field. Since magnetic nanoparticles have been established in medical applications and are approved by the US Food and Drug Administration as a contrast agent for magnetic resonance imaging, comprehensive in vitro/in vivo toxicity studies have already been carried out. Thus, an extensive database for the toxicity of magnetic nanoparticles is provided that helps to develop and apply magnetic nanoparticles with minimal toxicity for water and wastewater processes. Another benefit of magnetic nanoparticles in terms of the removal and fate of nanoparticles in an aqueous environment is the fact that they can be simply recovered by applying a magnetic field.

Nano zerovalent iron (nZVI) is emerging as a new option for the treatment of contaminated soil and groundwater. Due to their small size, the particles are very reactive and can be used for in situ treatment. nZVI effectively reduces chlorinated organic contaminants such as trichloroethylene, perchloroethylene, polychlorinated biphenyls, 1,1,1-trichloroethane, pesticides, solvents, and also inorganic anions like perchlorate. It can even be used to recover/remove dissolved metals from solution. The nZVI particles have limited mobility and the lifetime. Therefore, some modifications of nZVI are studied, tested, and commercialized. The most important examples are surface-modified nZVI, emulsified nZVI, bimetallic nZVI (higher reactivity), and nZVI on carbon support (better distribution within the soil).[27]

Nanoadsorbents can be readily integrated into existing treatment processes in slurry reactors or absorbers. Application of nanoadsorbents in powder form into slurry reactors can be highly efficient since all surfaces of the adsorbents are utilized and the mixing greatly facilitates the mass transfer. However, an additional separation unit is required to recover the nanoparticles. Nanosized adsorbents can also be used in fixed or fluidized absorbers in the form of pellets/beads or porous granules loaded with nanoadsorbents. Applications of nanoadsorbents for arsenic removal have been commercialized, and their performance and cost have been compared to other commercial adsorbents in pilot tests.

11.3   MEMBRANES AND MEMBRANE PROCESSES

To remove undesired constituents from water is the basic goal of water treatment. Membranes provide a physical barrier for such constituents based on their pore size and molecule size, allowing use of unconventional water sources. As the key component of water treatment and reuse, they provide high level of automation, require less land and chemical use, and the modular configuration allows flexible design.[28] Membrane separation processes are rapidly advancing applications for water and wastewater treatment. Membrane technology is well established in the water and wastewater area as a reliable and largely automated process. Novel research activities with regard to nanotechnology focus on improving selectivity and flux efficiency.

11.3.1   NANOFILTRATION MEMBRANES

Nanofiltration is one of the membrane filtration techniques and can be defined as a pressure-driven process wherein molecules and particles less than 0.5–1 nm are rejected by the membrane. Nanofiltration membranes are characterized by a unique charge-based repulsion mechanism allowing the separation of various ions.[29] They are mostly applied for the reduction of hardness, color, odor, and heavy metal ions from groundwater. The conversion of sea water into potable water (desalination) is another prosperous field of application since comparable desalination technologies are very cost intensive.

11.3.2   NANOCOMPOSITE MEMBRANES

Nanocomposite membranes are considered as a new set of filtration materials consists of mixed matrix membranes and surface-functionalized membranes. In case of mixed matrix membranes, nanofillers added in a matrix material. Generally, the nanofillers are inorganic in nature and embedded in a polymeric or inorganic oxide matrix. These nanofillers possess a larger specific surface area leading to a higher surface-to-mass ratio.[30] Metal oxide nanoparticles (Al_2O_3, TiO_2) can help to increase the mechanical and thermal stability as well as permeate flux of polymeric membranes. The incorporation of zeolites improves the hydrophilicity of membranes resulting in raised water permeability. Thin-film nanocomposite membranes are semipermeable membranes with a selective layer on the upper surface that is commonly applied to reverse osmosis. The surface of the hydrophobic ordered mesoporous carbons is treated by applying atmospheric pressure plasma to achieve hydrophilized ordered mesoporous carbons with higher solubility. To increase the pure water permeability of membrane, a small proportion of hydrophilized ordered mesoporous carbon is used to rise the hydrophilicity of the membrane surface.

11.3.3   SELF-ASSEMBLING MEMBRANES

Some research activity is aiming to nanostructure new membrane materials, for gas permeation, by means of self-assembly of block copolymer membranes. Self-assembly is defined as the "autonomous organization of components into patterns or structures without human intervention."[31] The structure of these membranes can be systematically controlled by the process parameters, so that the targeted membranes feature specific tailor-made characteristics. High-density cylindrical nanopores can be formed that way to be useful for micro/nanofluidic devices and also for water filtration. Such ultrafiltration membranes provide enhanced selectivity and permeate efficiency. However, due to difficulties with adaption of the techniques to large-scale membrane areas, until now self-assembling membranes have produced only in small quantities in the laboratory.

11.3.4   NANOFIBER MEMBRANES

Electrospinning is a fiber production method which uses electric force to draw charged threads of polymer solutions or polymer melts up to fiber diameters in the order of some hundred nanometers. It shares characteristics of both electrospraying and conventional solution dry spinning of fibers. Electrospinning produces fibers by drawing very fine different materials like polymers or ceramics into an electric field. These so-called nanofibers are in the range of nanometers and are often used in separation and filtration processes.[32] The technical process ensures a high specific surface porosity and thus a high surface-to-mass ratio. Another benefit of nanofibers is their capacity to be tailored to specifics such as membrane thickness and an interconnected open pore structure. The electrospinning process is a fully developed technology in air treatment, but nearly unknown in water and wastewater treatment. Therefore, it involves a very high research potential.

NanoCeram® (Argonide Corporation, Sanford, FL, USA) is one of the already patented nanofiber filters, which is an electropositive filter medium that is implemented in a filter cartridge. NanoCeram is a small diameter fiber, with a high surface area, that can be produced in kilogram quantities by sol–gel reaction. The product is a white, free-flowing powder consisting of fibers approximately 2 nm in diameter and tens to hundreds of nanometers in length that collects in aggregates. Embedded into glass and cellulose nonwoven sheets, a filtration material is created that is comparable with ultrafiltration but has higher flow rates that attract dirt, bacteria, viruses, and proteins using an electrostatic effect. NanoCeram can be applied in a prefiltration mode for ultrapure water systems, to produce laboratory or process water, in commercial/industrial water treatment, as a microbiological sampler, or as a stand-alone filtration device.[33] The main advantage of nanofiber materials according to biofilm immobilization is their comparability with the dimensions of micro-organisms, the surface morphology, and biocompatibility. The micro-organisms benefit from protection against shear forces and the toxic effects of the surrounding environment. Thus, bionanofibers are ideal substrates for formation of microbial biofilms, resulting in accelerated biodegradation due to a higher rate of carrier ingrowth. Common nanofibers are made of polymers like polyurethane, polylactic acid, and polyethylene oxide.

11.3.5 AQUAPORIN-BASED MEMBRANES

Aquaporins are integral membrane proteins that serve as channels in the transfer of water and in some cases small solutes across the membrane. From the structural analysis of the molecule, the presence of a pore at the center of each aquaporin molecule is revealed.[34] Under certain conditions, they form highly selective water channels that are able to reject most ionic molecules. The combination of high water permeability and selective rejection make them an ideal material for creating novel high flux biomimetic membranes. Aquaporins are incorporated in vesicles to stabilize them. Since the membranes based on these vesicles are too weak for their intended technical applications, like osmosis, therefore they are embedded in a polymeric matrix or deposited onto polymeric substrates such as nanofiltration membranes. Aquaporin Inside™ (Aquaporin A/S, Copenhagen, Denmark) is the first commercially available biomimetic membrane with aquaporins embedded. This kind of membrane is able to withstand pressures up to 10 bar and allows a water flux. 100 L/(h m^2) for example required for brackish water desalination.[35] In order to enter the water and wastewater market, aquaporin-based membranes have to be competitive with conventional membranes in terms of stability and useful life. Currently, there exists no such membrane that can permanently withstand the operating pressures of reverse osmosis, the harsh cleaning conditions (high temperature, acidic, and alkaline cleaners), and fouling-based corrosion.

11.4 PHOTOCATALYSIS

Photocatalysis is the acceleration of a photoreaction in the presence of a catalyst. Photocatalysis is an advanced oxidation process that is employed in the field of water and wastewater treatment, in particular for oxidative elimination of micropollutants and microbial pathogens.[36] Most organic pollutants can be degraded by heterogeneous photocatalysis. Due to its high availability, low toxicity, cost efficiency, and well-known material properties, TiO_2 is widely utilized as a photocatalyst. When TiO_2 is irradiated by ultraviolet light with an appropriate wavelength in the range of 200–400 nm, electrons will be photoexcited and moved into the conduction band. As a result of photonic excitation, electron–hole pairs are created, leading to a complex chain of oxidative–reductive reactions. Hence, the biodegradability of heavily decomposable substances can be increased in a pretreatment step. Principally, persistent compounds like antibiotics or other micropollutants can be photocatalytically eliminated in polishing processes, such as tertiary clarification steps in municipal wastewater treatment plants. However, as the ultraviolet radiation is only about 5% that of sunlight, the photon efficiency is quite low, limiting use on an industrial scale.

Organic contaminants from the groundwater can be removed by the impregnation of adsorbents with photoactive catalysts. These adsorbents attract contaminating organic atoms/molecules like tetrachloroethylene to them. The impregnated photoactive catalyst speeds-up the degradation of the organic contaminants. Adsorbents are placed in packed beds for 18 h, which would attract and degrade the organic compounds. To remove the remaining attached organics, the used adsorbents would then be placed in regeneration fluid, by passing hot water countercurrent to the flow of water during the adsorption process to speed up the reaction. The regeneration fluid then gets passed through the fixed beds of silica gel photocatalysts to remove and decompose the rest of the organics left. Through the use of fixed bed reactors, the regeneration of adsorbents can help increase the efficiency.

In recent years, many research groups have investigated the combination of separation and catalytic processes (using a membrane photocatalytic reactor) in order to both purify the water and retain the catalytic particles. When using highly efficient nanoparticles, an appropriate filtration system like nanofiltration has to be implemented in order to ensure complete rejection of possibly harmful nanoparticles. Thus, an extensive and relatively costly installation technology is necessary, including high pressure pumps.

A solution for rejection of photocatalytic nanoparticles is their immobilization on defined materials by use of suitable coating processes, such as physical or chemical vapor deposition, as well as wet chemical coating processes. When using a microfilter material as a substrate, a beneficial multibarrier effect comprising mechanical filtration and chemical decontamination is obtained. Dirt particles and larger microorganisms are rejected by microfiltration membranes at the same time that viruses, spores, and contaminants are chemically eliminated and degraded. In recent years, some advanced oxidation processes were successfully introduced to the water technology market. For example, Purific Water established a combined water treatment process comprising photocatalysis and ceramic membrane filtration, with a capacity of more than 4 million cubic meters per day in particular for the degradation of volatile organic compounds and 1.4 dioxine for groundwater remediation.[37]

Commercialization of nanoengineered materials for water and wastewater technology strongly depends on their impact on the aqueous environment. Numerous studies including toxicity tests, life cycle analysis, technology assessment, and pathways and dispersal of nanoparticles in water bodies have been carried out in order to evaluate the health risks of nanomaterials. The results of these studies have led to a better understanding of the behavior of nanoparticles such as CNTs, TiO_2, and silver nanoparticles in aqueous systems; thus, stakeholders from administration, politics, and industry are supported to create new laws and regulations or modify present ones. However, many studies have yielded contradictory results, since no general standards and conditions for experimental tests and measurements have been determined, which slows down the necessary decision processes.[38]

11.5   CONCLUSION

To meet the global safe water need, we need water purification tools that provide high-quality drinking water by removing micropollutants, and other contaminants. Unique properties of nanoparticles are suitable for developing rapid water-treatment technology. Here, different advances in nanotechnology are discussed in this chapter, some of them are in laboratory research stage, some reached to pilot testing, and some are commercial. Nanoadsorbents, nanomembranes, and nanophotocatalysts are most promising among these technologies. Nanoparticles have the ability to integrate various properties, resulting in multifunctional systems such as nanocomposite membranes that enable both particle retention and elimination of contaminants, which is one of the most important advantages of nanomaterials when compared with conventional water technologies. Further due to their unique characteristics such as high reaction rate, nanomaterials facilitate higher process efficiency.

Ecofriendly waste management methods are required to avoid hazards and toxicities of recovered pollutants and exhausted nanoparticles. The future of the nanoparticles in water treatment is rather progressive, but it needs combined efforts of academic and industrial resources to materialize a fast, economical, and feasible water-treatment technology.

KEYWORDS

- **nanotechnology**
- **water technology**
- **nanoadsorbents**
- **nanometals**
- **nanomembranes**
- **photocatalysis**

REFERENCES

1. U.S. Bureau of Reclamation and Sandia National Laboratories. Desalination and Water Purification Technology Roadmap a Report of the Executive Committee Water Purification, 2003.

2. World Health Organization. *Guidelines for Drinking-Water Quality*; WHO: Geneva, 1996; Vol. 2.

3. Jung, J. Y.; Chung, Y. C.; Shin, H. S.; Son, D. H. Enhanced Ammonia Nitrogen Removal Using Consistent Biological Regeneration Ammonium Exchange of Zeolite in Modified SBR Process. *Water Res.* **2004,** *38*(Jan (2)), 347–354.

4. Jiuhui, Q. U. Research Progress of Novel Adsorption Process in Water Purification: A Review. *J. Environ. Sci.* **2008,** *20*(1), 1–13.

5. Jin-Woo, K.; Hyung-Mo, M.; Steve, T.; Hyun-Ho, L. In *Highly Effective Bacterial Removal System Using Carbon Nanotube Clusters*, Proceedings of 2009 4th IEEE International Conference on Nano/Micro Engineered and Molecular systems, 2009. Available from: http://doiieeecomputersociety.org.10.1109/NEMS.2009.5068756.

6. Pradeep, T.; Anshup. Noble Metal Nanoparticle for Water Purification: A Critical Review. *Thin Solid Films* **2009,** *517*, 6441–6478.

7. Encyclopaedia Britannica. Adsorption. http://www.britannica.com/EBchecked /topic/6565/adsorption (March 10, 2017).

8. Pan, B.; Xing, B. S. Adsorption Mechanisms of Organic Chemicals on Carbon Nanotubes. *Environ. Sci. Technol.* **2008,** *42*, 9005–9013.

9. Yang, K.; Xing, B. S. Adsorption of Organic Materials by Carbon Nanomaterials in Aqueous Phase: Polanyi Theory and Its Application. *Chem. Rev.* **2010,** *110*(10), 5989–6008.

10. Ji, L. L.; Chen, W.; Duan, L.; Zhu, D. Q. Mechanism for Strong Adsorption of Tetracycline to Carbon Nanotubes: A Comparative Study Using Activated Carbon and Graphite as Adsorbents. *Environ. Sci. Technol.* **2009,** *43*(7), 2322–2327.

11. Rao, G. P.; Lu, C.; Su, F. Sorption of Divalent Metal Ions from Aqueous Solution by Carbon Nanotubes: A Review. *Sep. Purif. Technol.* **2007,** *58*(1), 224–231.

12. Yan, H. Y.; Han, Z. J.; Yu, S. F.; Pey, K. L.; Ostrikov, K.; Karnik, R. Carbon Nanotubes Membranes with Ultrahigh Specific Adsorption Capacity for Water Desalination and Purification. *Nat. Commun.* **2013,** *4*, 2220.

13. Hashim, D. P.; Narayanan, N. T.; Romo-Herrera, J. M. Covalently Bonded Three Dimensional Carbon Nanotube Solid via Boron Induced Nanojunction. *Sci. Rep.* **2012,** *2*, 363.

14. Hajeh, M.; Laurent, S.; Dastafkan, K. Nanoadsorbents: Classification, Preparation, and Applications (with Emphasis on Aqueous Media). *Chem. Rev.* **2013,** *113*, S7728–S7768.

15. Diallo, M. S.; Christie, S.; Swaminathan, P.; Johnson, J. H.; Goddard, W. A. Dendrimer Enhanced Ultrafiltration. 1. Recovery of Cu(II) from Aqueous Solutions Using PAMAM Dendrimers with Ethylenediamine Core and Terminal NH2 Groups. *Environ. Sci. Technol.* **2005,** *39*, S1366–S1377.

16. Sadeghi-Kiakhani, M.; Mokhtar Arami, M.; Gharanjig, K. Dye Removal from Colored-Textile Wastewater Using Chitosan-PPI Dendrimer Hybrid as a Biopolymer:

Optimization, Kinetic, and Isotherm Studies. *J. Appl. Polym. Sci.* **2013**, *127*, 2607–2619.

17. Baker, M. D.; Ozin, G. A.; Godber, J. Far-Infrared Studies of Silver Atoms, Silver Ions, and Silver Clusters in Zeolites A and Y. *J. Phys. Chem.* **1985**, *89*, 305–311.

18. Egger, S.; Lehmann, R. P.; Height, M. J.; Loessner, M. J.; Schuppler, M. Antimicrobial Properties of a Novel Silver–Silica Nanocomposite Material. *Appl. Environ. Microbiol.* **2009**, *75*, 2973–2976.

19. Nagy, A.; Harrison, A.; Sabbani, S.; Munson, R. S., Jr; Dutta, P. K.; Waldman, W. J. Silver Nanoparticles Embedded in Zeolite Membranes: Release of Silver Ions and Mechanism of Antibacterial Action. *Int. J. Nanomedicine* **2011**, *6*, 1833–1852.

20. Petrik, L.; Missengue, R.; Fatoba, M.; Tuffin, M.; Sachs, J. Silver/Zeolite Nano Composite-Based Clay Filters for Water Disinfection. *Report to the Water Research Commission*. No KV 297/12, 2014. http://www.ircwash.org/resources/silver-zeolite-nano-composite-based-clay-filters-water-disinfection (March 10, 2017).

21. Tiwari, D. K.; Behari, J.; Sen, P. Application of Nanoparticles in Waste Water Treatment. *World Appl. Sci. J.* **2008**, *3*, 417–433.

22. Jung, J. Y.; Chung, Y. C.; Shin, H. S.; Son, D. H. Enhanced Ammonia Nitrogen Removal Using Consistent Biological Regeneration and Ammonium Exchange of Zeolite in Modified SBR Process. *Water Res.* **2004**, *38*, 347–354.

23. Qu, X.; Alvarez, P. J.; Li, Q. Application of Nanotechnology in Water and Waste Water Treatment. *Water Res.* **2013**, *47*, 3931–3946.

24. Aredes, S.; Klein, B.; Pacolik, M. Removal of Arsenic from Water Using Natural Iron Oxide Minerals. *J. Cleaner Prod.* **2012**, *29–33*, 208–213.

25. Sharma, Y. C.; Srivastava, V.; Singh, V. K.; Kaul, S. N.; Weng, C. H. Nano-Adsorbents for the Removal of Metallic Pollutants from Water and Wastewater. *Environ. Technol.* **2009**, *30*(6), 583–609.

26. Deliyanni, E. A.; Bakoyannakis, D. N.; Zouboulis, A. I.; Matis, K. A. Sorption of As(V) Ions by Akaganeite-Type Nanocrystals. *Chemosphere* **2003**, *50*(1), 155–163.

27. Muller, N. C.; Nowack, B. Nano Zero Valent Iron—The Solution for Water and Soil Remediation? *Report of Observatory Nano*, 2010. www.observatorynano.edu.

28. Qu, X. L.; Brame, J.; Li, Q.; Alvarez, J. J. P. Nanotechnology for a Safe and Sustainable Water Supply: Enabling Integrated Water Treatment and Reuse. *Acc. Chem. Res.* **2013**, *46*(3), 834–843.

29. Sharma, V.; Sharma, A. Nanotechnology: An Emerging Future Trend in Wastewater Treatment with Its Innovative Products and Processes. *Int. J. Enhanc. Res. Sci. Technol. Eng.* **2012**, *1*, 121–128.

30. Feng, C.; Khulbe, K. C.; Matsuura, T.; Tabe, S.; Ismail, A. F. Preparation and Characterization of Electro-Spun Nanofiber Membranes and Their Possible Applications in Water Treatment. *Sep. Purif. Technol.* **2013**, *102*, 118–135.

31. Whitesides, G. M.; Grzybowski, B. Self-Assembly at All Scales. *Science* **2002**, *295*, 2418–2421.

32. Ramakrishna, S.; Fujihara, K.; Teo, W. E.; Yong, T.; Ma, Z. W.; Ramaseshan, R. Electrospun Nanofibers: Solving Global Issues. *Mater. Today* **2006**, *9*, 40–50.

33. Karim, M. R.; Rhodes, E. R.; Brinkman, N.; Wymer, L.; Shay Fout, G. New Electropositive Filter for Concentrating Enteroviruses and Noroviruses from Large Volumes of Water. *Appl. Environ. Microbiol.* **2009**, *75*, 2393–2399.

34. Takata, K.; Matsuzaki, T.; Tajka, Y. Aquaporins: Water Channel Proteins of the Cell Membrane. *Prog. Histochem. Cytochem.* **2004,** *39*(1), 1–83.

35. Tang, C. Y.; Zhao, Y.; Wang, R.; Hélix-Nielsen, C.; Fane, A. G. Desalination by Biomimetic Aquaporin Membranes: Review of Status and Prospects. *Desalination* **2013,** *308*, 34–40.

36. Friedmann, D.; Mendiveb, C.; Bahnemann, D. TiO$_2$ for Water Treatment: Parameters Affecting the Kinetics and Mechanisms of Photocatalysis. *Appl. Catal. B Environ.* **2010,** *99*, 398–406.

37. [No authors listed]. Purifics' Photo-Cat Cleans up Contaminated Groundwater. *Membr. Technol.* **2011,** *9*, 6.

38. Gehrke, I.; Geiser, A.; Schulz, A. S. Innovations in Nanotechnology for Water Treatment. *Nanotechnol. Sci. Appl.* **2015,** *8*, 1–17.

CHAPTER 12

DESIGN AND IN VITRO CHARACTERIZATION OF ETHOSOMES CONTAINING DICLOFENAC SODIUM FOR TRANSDERMAL DELIVERY

JAWAHAR NATARAJAN[1,2*] and
VEERA VENKATA SATYANARAYANA REDDY KARRI[1,2]

[1]*Department of Pharmaceutics, JSS College of Pharmacy, Ootacamund, Rocklands, Udhagamandalam 643001, Tamil Nadu, India*

[2]*JSS Academy of Higher Education and Research, Mysuru, Karnataka, India*

Corresponding author. E-mail: jawahar.n@jssuni.edu.in

ABSTRACT

Many drugs of topical delivery suffer from poor penetration through skin because of their poor solubility. During the past decades, there has been wide interest in exploring new techniques to increase drug absorption through skin. Ethosomes which are novel permeation-enhancing lipid vesicles embodying high concentration (20–45%) of ethanol have the capability of carrying away the poorly soluble drugs into various layers of the skin. Hence, in this study, a model drug, diclofenac sodium, was used to study the effect of ethosomes on topical skin penetration. The compatibility of Phospholipon 90H and diclofenac sodium was studied using Fourier-transform infrared spectroscopy (FTIR) and was found with no incompatibilities. In vitro permeation studies were performed using

Franz diffusion cell to compare the drug permeation and retention across pig skin, of ethosomal systems against hydroethanolic solution of drug and phospholipid and marketed Omnigel®. The ethosomal system showed the highest drug permeation (18.76%) after 12 h of study, more than 6% higher than that of the marketed formulation. The ehosomal system also showed the highest skin retention of diclofenac (3.62 µg/cm²) compared to the hydroethanolic solution and the marketed product. This study confirms the use of ethosomes as a better therapeutic tool for the topical delivery of poorly permeable drugs.

12.1   INTRODUCTION

Although the skin as a route for drug delivery can offer many advantages, including avoidance of first-pass metabolism, lower fluctuations in plasma drug levels, targeting of the active ingredient for a local effect and good patient compliance, the barrier nature of skin makes it difficult for most drugs to penetrate into and permeate through it.[1] Many drugs of topical delivery suffer from poor penetration because of their poor solubility. During the past decades, there has been wide interest in exploring new techniques to increase drug absorption through skin. Topical delivery of drugs by lipid vesicles has evoked a considerable interest. Liposomes are lipid vesicles that fully enclose an aqueous volume. Lipid molecules are usually phospholipids with or without some additives. Cholesterol may be included to improve bilayers characteristics of liposomes, increasing microviscosity of the bilayers, reducing permeability of the membrane to water soluble molecules, stabilizing the membrane, and increasing rigidity of the vesicles.[2,3]

Recently, it became evident that in most cases, classic liposomes are of little or no value as carriers for transdermal drug delivery as they do not deeply penetrate skin but rather remain confined to upper layers of the stratum corneum. Intensive research led to the introduction and development, over the past 15 years, of a new class of lipid vesicles, the highly deformable (elastic or ultraflexible) liposomes. Although liposomal formulations containing up to 10% ethanol and up to 15% propylene glycol were previously described due to the interdigitation effect of ethanol on lipid bilayers, it was commonly believed that vesicles cannot coexist with high concentrations ethanol.[4–7] The use of high ethanol content was first described by Touitou et al. for ethosomes.[8] Ethosomes which are novel permeation-enhancing lipid vesicles embodying high concentration (20–45%) of ethanol. Hence, in this present study, an attempt has been made to increase the topical penetration of a poorly soluble drug diclofenac sodium by using ethosomes.

12.2 EXPERIMENTAL

Diclofenac sodium (IP) and Ibuprofen (IP) were procured from Ranbaxy, India. Phospholipon® 90H and hydrogenated phosphatidylcholine from soybean were obtained from Lipoid GmbH, Germany. Dipotassium hydrogen phosphate dihydrate and sodium chloride were purchased from S.D. Fine Chemicals, India.

12.2.1 COMPATIBILITY STUDY

FTIR can be used to investigate and predict any physicochemical interactions or incompatibilities between different components in a formulation by matching of IR spectrum peaks of pure ingredients and physical mixture of the drug and excipients. Therefore, it can be applied to the selection of suitable and compatible excipients. While selecting ingredients, we chose excipients which are stable, compatible, and cosmetically and therapeutically acceptable.

12.2.2 DETERMINATION OF LIMIT OF DETECTION AND LIMIT OF QUANTIFICATION

A volume of 1 L of 0.1 M phosphate buffer (pH 5.8) was prepared by dissolving 1.19 g of dipotassium hydrogen phosphate dihydrate and 8.25 g of potassium dihydrogen phosphate in water and diluted to 1000 mL. The final pH was adjusted to 5.8 by addition of 0.1 M HCl and/or 0.1 M NaOH drop wise. The prepared buffer can be stored at 4°C for 1 month for further use. The parameters limit of detection (LOD) and limit of quantification (LOQ) were determined using the signal-to-noise ratio by comparing results of the test of samples with known concentrations of analyte to blank samples. The analyte concentration that produced a signal-to-noise ratio of 3:1 was accepted as the LOD. The LOQ was identified as the lowest drug concentration of the standard curve that could be quantified with acceptable accuracy and precision.

12.2.3 DEVELOPMENT OF CALIBRATION CURVE

A stock solution of diclofenac sodium was prepared by dissolving 50 mg of diclofenac sodium in 50 mL of methanol (HPLC grade) to give a stock solution of 1 mg/mL. For the determination of diclofenac in entrapment efficiency (EE) study, the stock solution was diluted with methanol to prepare standard solutions of 0.1, 0.25, 0.5, 0.75, 1.0, 1.25, and 1.5 µg/mL. To a 1 mL of 0.3 µg/mL solution of internal standard (Internal standard), ibuprofen, was added to 9 mL of each solution before injection. For in vitro drug permeation study, standard solutions of 0.001, 0.005, 0.01, 0.025, 0.05, 0.075, and 0.1 µg/mL were prepared, and 1 mL of 0.03 µg/mL IS is added to 9 mL of standard solution before injection. A calibration curve was obtained by plotting the relative peak area of diclofenac with that of IS against the concentration of the standard solutions.

HPCL CONDITIONS

Column	: Phenomenex RP C_{18} column (250 mm × 4.6 mm I.D., 5 µm)
Mobile phase	: Methanol and phosphate buffer (pH 5.8) (700:300)
Pressure	: 3000 psi
Temperature	: 25°C (ambient temperature)
Injection volume	: 20 µL
Flow rate	: 1 mL/min
Injection port	: Rheodyne manual injection with 50 µL loop
UV detection	: 239 nm for ibuprofen; 277 nm for diclofenac
Retention time	: 7.8 min for ibuprofen; 6.5 min for diclofenac
LOD	: 0.0005 µg/mL
LOQ	: 0.0008 µg/mL

12.2.4   METHOD OF PREPARATION OF ETHOSOMAL VESICLES

Ethosome colloidal suspensions were prepared as reported by Touitou et al. with some modifications.[8] The ethosomal systems investigated here were composed of 2–4% w/v Phospholipon 90H (PL), 20–40% v/v ethanol, 1% w/v diclofenac sodium, and water to 100% v/v. Phospholipid and drug were dissolved in ethanol. *Mili-pore* water was added slowly in a fine stream (flow rate ~1.5 mL/min) using a peristaltic pump with constant mixing at 850 ± 15 rpm on magnetic stirrer in a well-sealed container. Mixing was continued for an additional 15 min at 1200 ± 15 rpm. The system was maintained at $30 \pm 2°C$ throughout the preparation and was then left to cool at room temperature. The preparations were further extruded through a filter of pore size <0.45 µm to reduce the vesicular size distribution. The final ethosomal systems were stored at 4°C until required for characterization and in vitro studies (Table 12.1).

TABLE 12.1   Composition of Ethosomal Formulations.

Formulation code	Conc. of ethanol (% v/v)	Conc. of PL (% w/v)	Conc. of water (% v/v)
$F1_A$	20.0	2.0	78.0
$F1_B$	30.0	2.0	68.0
$F1_C$	40.0	2.0	58.0
$F2_A$	20.0	3.0	77.0
$F2_B$	30.0	3.0	67.0
$F2_C$	40.0	3.0	57.0
$F3_A$	20.0	4.0	76.0

PL, Phospholipon 90H.

12.2.5   CHARACTERIZATION STUDIES

12.2.5.1   ENTRAPMENT EFFICIENCY

EE of diclofenac sodium ethosomal vesicles was determined by centrifugation method. The vesicles were separated in a high-speed cooling centrifuge at 20,000 rpm for 90 min in the temperature maintained at 4°C. The sediment and supernatant liquids were separated; amount of drug in the supernatant was determined by HPLC. For quantification of EE, 10 mL of supernatant was pipette out and diluted to 100 mL with methanol, and 1 mL of 0.3 µg/mL of IS is added and analyzed in HPLC. From this, the EE was determined by the following equation:

$$\text{Entrapment efficiency (EE \%)} = \frac{(T-C)}{T} 100$$

where T is the total amount of drug in system (supernatant + sediment) and C is the amount of drug in supernatant only.

12.2.5.2 VESICLE SIZE DETERMINATION

The mean vesicle size and polydisperse index and the zeta potential of the resulting ethosomal vesicles with high EE were determined by photon correlation spectroscopy using a Zetasizer. Light scattering was monitored at 25°C at an angle of 90°.

12.2.5.3 SCANNING ELECTRON MICROSCOPY

The size and shape of the vesicles were observed in the scanning electron microscopy (SEM). One drop of ethosomal suspension was mounted on a clear glass stub. It was then air dried and gold coated using sodium auro thiomalate to visualize under scanning electron microscope at ×10,000 magnifications.

12.2.5.4 STABILITY STUDY

Stability studies were carried out by storing the ethosomal formulations at two different temperatures 4°C and 25 ± 2°C (ambient room temperature). The drug content was estimated for every 15 days for 2 months to identify any change in the EE of ethosomal formulation.

12.2.6 IN VITRO PERMEATION STUDY

A volume of 1 L PBS (pH 7.4) was prepared by dissolving 8.7 g NaCl, 1.82 g $K_2HPO_4 \cdot 2H_2O$, and 0.23 g KH_2PO_4 in 1 L of distilled water.[9,10] The pH of the buffer was adjusted to 7.4 by addition of 0.1 M NaOH and/or 0.1 M HCl. Freshly prepared PBS (pH 7.4) was used in the in vitro permeation studies. Abdominal skin from sacrificed pig is obtained from the local slaughterhouse. After removing the hair, the abdominal skin was separated from the underlying connective tissue with a surgical scalpel. The excised skin was placed on aluminum foil and any adhering fat and/or subcutaneous tissue on the dermal side of the skin was gently scrapped off using a scalpel. The skin was checked carefully to ensure that the skin samples are free from any surface irregularity such as fine holes or crevices in the portion that is used for transdermal permeation studies. The obtained skins were stored at $4.0 \pm 1.0°C$ in iso-osmotic saline solution (0.9% w/v NaCl Soln.) until use, thus avoiding any morphological modification.

12.2.6.1   DETERMINATION OF DRUG PERMEATION ACROSS THE EXCISED SKIN SAMPLE

Before the percutaneous experiments, skin samples were placed into 10 mL of PBS for 1 h at room temperature. The permeation of diclofenac sodium from an ethosomal system was measured in experiments carried out for 12 h at $32 \pm 0.5°C$ in side-by-side Franz diffusion cells. The compositions (in % w/v) of the three donor compartments, each containing 1.0% w/v diclofenac sodium were (1) 30% v/v ethanolic solution, 3.0% w/v PL, (2) ethosomal system ($F3_C$), and (3) Omnigel®. The skin was mounted between donor and receptor compartment with the stratum corneum side facing upward into the donor compartment. PBS with 20% ethanol was taken in the receptor compartment to improve the solubility of diclofenac. The formulation was applied on the skin in donor compartment which was then covered with aluminum foil to avoid any evaporation process. Samples (5 mL) were withdrawn at predetermined time intervals over 12 h and suitably diluted to analyze the drug content. The receptor medium was immediately replenished with equal volume of fresh medium to maintain the sink conditions throughout the experiment. The percentage of drug release was plotted against time to find the drug release pattern. The amount of diclofenac sodium retained in the skin was determined at the end of the 12 h in vitro permeation studies. The formulation remain in the in vitro permeation experiment was removed by washing with distilled water. The receptor content was replaced by absolute ethanol and kept for further 12 h with stirring at $32 \pm 0.5°C$, and the drug content was estimated by UV–visible spectroscopy. This receiver solution diffused through the skin, disrupting any ethosome structure and extracting deposited drug from the skin.

12.3 RESULTS AND DISCUSSION

12.3.1 COMPATIBILITY STUDY

Physical and chemical compatibility of drug and excipients are studied by peak matching in IR spectra (600–4000 cm^{-1}) of pure drug and excipients and their physical mixture. No significant shifts in the peaks of the components are observed, suggesting compatibility of the components of the formulation (Figs. 12.1–12.3).

12.3.2 HPLC CALIBRATION CURVE

The calibration curves of diclofenac sodium in HPLC are obtained by plotting the peak area for sample against the concentration of standard solution (Fig. 12.4). The R^2 of calibration curve is 0.9993 signifying the linearity of the plots. The unknown concentrations of samples from various analytical studies are obtained by replacing the values of absorbance (x) in the straight line formula for the respective calibration curves.

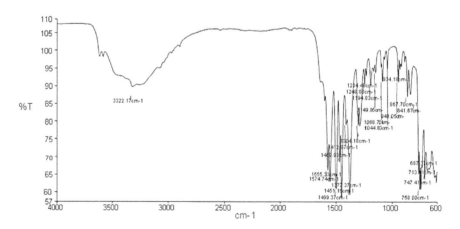

FIGURE 12.1 FTIR spectra of diclofenac sodium.

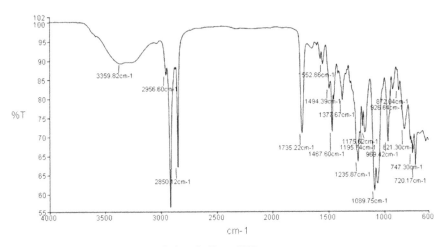

FIGURE 12.2   FTIR spectra of phospholipon 90H.

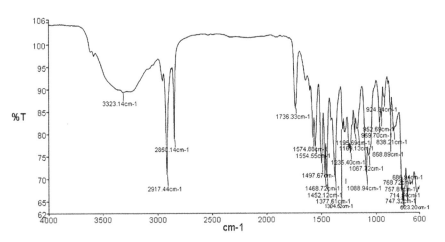

FIGURE 12.3   FTIR spectra of diclofenac sodium and phospholipon 90H.

12.3.3   DRUG ENTRAPMENT EFFICIENCY

Drug entrapment within a vesicular carrier is an important parameter to be defined to really evaluate the delivery potentiality of the system. For this reason, the EE of diclofenac sodium within the formulations $F1_A$ – $F3_C$ were evaluated in an attempt to investigate the influence of ethosome composition, that is, the quantity of ethanol and phospholipid, on the drug loading capacity (Fig. 12.5).

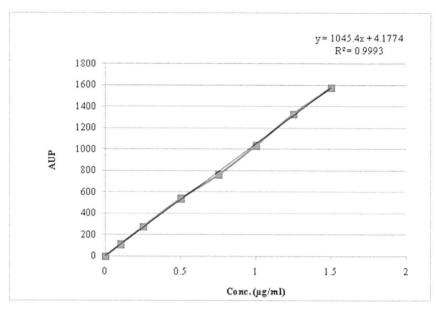

FIGURE 12.4   Calibration curves of diclofenac sodium.

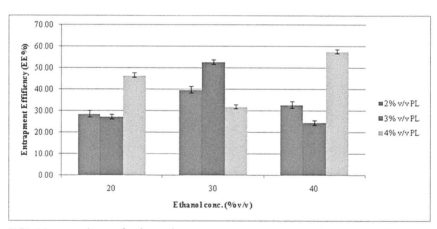

FIGURE 12.5   **(See color insert.)** Drug entrapment efficiency of ethosomes with varying concentrations of ethanol and phospholipids.

As shown in Table 12.2, both ethanol and phospholipid amounts, used for ethosome preparation, positively influenced the EE of the colloidal carrier, namely, the higher the amount of ethanol and phospholipid, the greater the diclofenac sodium entrapment within the ethosomes. This is

TABLE 12.2 Influence of the Amount of Ethanol and Phospholipon 90H (PL) Used for the Preparation of Ethosomes on Vesicle Suspension Mean Size, Polydispersity Index, and Diclofenac Sodium Entrapment Efficiency.

Formulation codea	Conc. of ethanol (% v/v)	Conc. of PL (% w/v)	Entrapment efficiency (EE%) b ± SD (n = 2)	Average vesicular size (nm)	PDI	Zeta potential (mV)
F1$_A$	20	2	28.44 ± 11.68	447.4	0.602	−3.231
F1$_B$	30	2	39.73 ± 8.02	−	−	−
F1$_C$	40	2	32.84 ± 8.15	−	−	−
F2$_A$	20	3	27.36 ± 9.64	−	−	−
F2$_B$	30	3	52.78 ± 4.01	319.2	0.473	−0.963
F2$_C$	40	3	24.64 ± 14.28	−	−	−
F3$_A$	20	4	46.42 ± 4.93	420.7	0.416	−1.462
F3$_B$	30	4	31.72 ± 7.67	−	−	−
F3$_C$	40	4	57.47 ± 3.05	463.4	0.493	−0.449

EE, entrapment efficiency; PDI, polydispersity index; PL, Phospholipon 90H.
[a]Drug concentration used in each formulation kept constant at 10 µg/mL or 1% w/v.
[b]Mean of readings from two identical batches.

in agreement to previously reported findings. The range of EE of nine ethosomal formulations was about 28.44–57.47%. Significant variations were observed in the EE of different batches of the same formulation, with F3$_C$ showing the least variation with SD of 3.05%. The maximum EE was observed in the formulation containing 4% w/v of phospholipids and 40% v/v of ethanol.

12.3.4 SEM ANALYSIS

The dependence of vesicle size on phospholipid content was determined for ethosomes containing 40% ethanol and phospholipid concentrations ranging from 2% to 4%. The average size of the ethosomes (F2$_B$) was 319.2 nm, and the polydispersity index was 0.473, which indicates that the vesicle population sizes were relatively homogeneous (Fig. 12.6). These data correlate well with the vesicular size visualized in the SEM

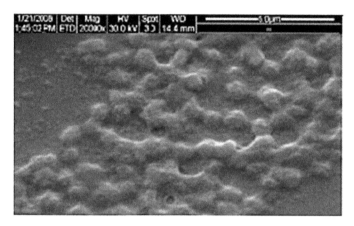

FIGURE 12.6   Scanning electron microscope (SEM) image of ethosomes (F3$_B$) (magnification ×10,000).

experiments. In the ethanol concentration range of 20–40%, the size of the vesicles increased with increasing phospholipid concentration. The largest particles in preparations containing 20% v/v ethanol and 2% w/v phospholipid were 447.4 nm and the smallest in preparations containing 30% v/v ethanol, but phospholipid 3% w/v was 319.2 nm. Ethosome size exhibits a limited dependence on ethosomes phospholipid concentration.

12.3.5   STABILITY STUDIES

The results revealed that (Table 12.3) the drug retention capacity (EE) was more with ethosomal formulation (F3$_C$) stored at 4°C than at 25 ± 2°C. The decrease in EE may be due to drug leakage from the ethosomes at higher temperature. Hence, increase in temperature decreased the drug retention capacity of ethosomes.

TABLE 12.3   Entrapment Efficiency of Ethosomal Formulation F3$_C$ When Stored at Various Temperatures for 2 Months.

Time of testing	(EE %) at 4°C	(EE %) at 25°C
Immediately after formulation (within 24 h)	57.47 ± 1.05	57.47 ± 1.05
After 15 days	53. 95 ± 0.57	51.65 ± 0.68
After 30 days	51. 27 ± 0.45	42.03 ± 0.97
After 60 days	44.97 ± 0.72	31.39 ± 0.38

EE, entrapment efficiency.

12.3.6 IN VITRO TRANSDERMAL DRUG PERMEATION AND SKIN RETENTION OF DRUG

In vitro skin permeation of diclofenac sodium (1% w/v) from ethosomes (F3$_C$), hydroethanolic solution (4% w/v PL in 40% v/v ethanol) and marketed formulation Omnigel® were studied using Franz diffusion cell with an effective permeation area of 2.54 cm^2 and a receiver compartment volume of 27.9 mL (Figs. 12.7 and 12.8). The ethosomal formulation was selected for the in vitro skin permeation on the basis of high EE and smaller vesicular size. In each study, accurately weighed/measure amount of formulation containing diclofenac sodium equivalent to 10 μg is taken in the donor compartments.

Each study was performed in triplicate ($n = 3$) under identical conditions to obtain statistically relevant data.

12.3.6.1 TRANSDERMAL PERMEATION OF DRUG

The cumulative percentage of drug release from ethosomal system (F3$_C$), hydroethanolic solution, and Omnigel® were found to be 18.76%, 17.54%,

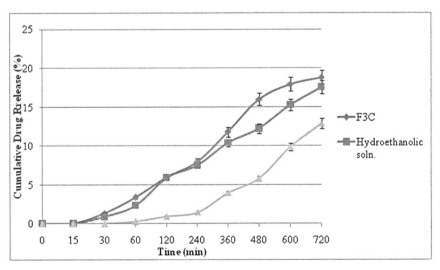

FIGURE 12.7 **(See color insert.)** Comparison of in vitro skin permeation of drug from various formulations in phosphate buffer saline pH 7.4 ($n = 3$).

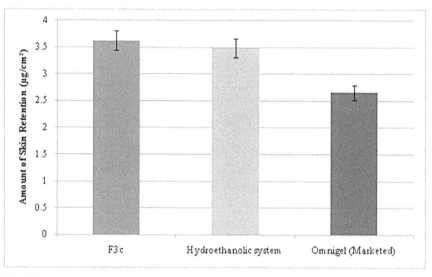

FIGURE 12.8   Comparison of in vitro drug retention in skin after 12 h ($n = 3$).

and 12.87%, respectively, at the end of 12 h study. From this study, it has been observed that the ethosomal system showed higher skin permeation of drug (more than 6%) than the marketed formulation, Omnigel®. Further, the permeation enhancement of drug from ethosomes was also greater than the hydroethanolic solution.

12.3.6.2   SKIN RETENTION OF DRUG

The ethosomal formulation (F3$_c$), hydroethanolic solution, and Omnigel® showed the skin deposition of 3.62 ± 18.3 µg/cm², 3.46 ± 7.0 µg/cm², and 2.66 ± 5.2 µg/cm², respectively, at the end of 12 h experiment. The ethosomes showed significantly higher skin deposition possibly due to combined effect of ethanol and phospholipid, thus providing a mode for dermal and transdermal delivery of diclofenac sodium.

Diclofenac sodium, a phenyl acetic acid derivative, is used for anti-inflammatory and analgesic effects in the symptomatic treatment of acute and chronic inflammatory conditions. It is completely absorbed from the GI tract but undergoes extensive first pass metabolism in the liver. Oral dose of diclofenac sodium causes an increased risk of serious gastrointestinal adverse events, including bleeding, ulceration, and perforation of the stomach or the intestines. Due to the presence of these, oral adverse effects necessitate the need for investigating other route of drug delivery of diclofenac sodium. Transdermal delivery of the drug can improve its bioactivity with reduction of the side effects and enhance the therapeutic efficacy. Therefore, the goal of the present study was to develop ethosomal delivery systems for diclofenac sodium.

The analysis of physicochemical features and technological parameters showed that ethosomes suitable for potential clinical use should be characterized by a small mean sizes and a narrow size distribution, which should be maintained also after their application on the skin.

Vesicle size is critical to topical drug delivery systems, for example, vesicles smaller than 300 nm are able to deliver their contents, to some extent, into deeper layers of the skin. In our study, the average diameters of ethosomes and liposomes were below 300 nm, indicating that these systems had potential for delivery of drug through the skin.

The surface charge distribution in colloidal vesicles generally depends on the chemical structure of phospholipids and their distributions in the vesicular bilayer. In the case of ethosomes investigated in this study, the obtained data suggested that phospholipids were organized in a vesicular bilayer with negative-charged head groups on the outer side of colloidal systems.

As concerning technological parameters, EE values depended on the composition of ethosomes; the greater the percentage of ethanol in the formulations, the higher the diclofenac sodium entrapment.

The magnitude percentages of diclofenac released from ethosomes formulation $F3_C$ during 12 h were high, and a sustained release of diclofenac, suitable for in vivo administration, was obtained. Ethosomes were previously shown to maintain a sustained release of entrapped drugs and to penetrate intact through the SC membranes when applied as topical drug delivery system. We further tested their capability to improve the percutaneous diclofenac passage. Diclofenac sodium loaded ethosomes permeated faster through the SC membranes than hydroethanolic solution of diclofenac and phospholipid and marketed topical diclofenac formulation, Omnigel®. This increase probably depended on the ethanol percentage used during the preparation procedure of the formulation $F3_C$. In fact, 40% w/v of ethanol could modify the SC lipid membrane and increase the amount of hydrophobic compounds permeated through the skin. The amount of diclofenac permeated was more than 10% lower than that observed in the case of diclofenac-loaded ethosomes at the same treatment time, thus confirming the crucial role of the ethosomal carrier for the improvement of the percutaneous passage of diclofenac. The obtained results are in agreement with other data previously reported in literature for hydrophobic drugs.

12.4 CONCLUSION

In this study, we developed and evaluated ethosomes as carriers for the topical application of an anti-inflammatory agent such as diclofenac sodium for the enhanced transdermal drug delivery. Formulation $F3_C$ was found to have the highest EE of $57.47 \pm 1.05\%$ and was selected for all further studies. The formulation ($F3_C$) was stable when stored at 4°C for 60 days with no significant reduction in drug entrapment. The cumulative percentage of drug release from ethosomal system ($F3_C$) was found to be 18.76% at the end of 12 h study. From this study, it has been observed that the ethosomal system showed higher skin permeation of drug (more than 10%) than the marketed formulation, Omnigel® which in turn it may be better alternate to oral delivery as well.

KEYWORDS

- **ethosomes**
- **diclofenac sodium**
- **transdermal drug delivery**
- **skin permeation**
- **topical delivery**

REFERENCES

1. Verma, P.; Pathak, K. Therapeutic and Cosmeceutical Potential of Ethosomes: An Overview. *J. Adv. Pharm. Technol. Res.* **2010,** *1*(Jul (3)), 274.
2. Alexander, A.; Dwivedi, S.; Ajazuddin; Gir, T. K.; Saraf, S.; Saraf, S.; Tripathi, D. K. Approaches for Breaking the Barriers of Drug Permeation Through Transdermal Drug Delivery. *J. Control. Release* **2012,** *164,* 26–40.
3. Elsayed, M. M. A.; Abdallah, O. Y.; Naggar, V. F.; Khalafallah, N. M. Lipid Vesicles for Skin Delivery of Drugs: Reviewing Three Decades of Research. *Int. J. Pharm.* **2007,** *332,* 1–16.

4. Dayan, N.; Touitou, E. Carriers for Skin Delivery of Trihexyphenidyl HCl: Ethosomes vs. Liposomes. *Biomaterials* **2000,** *21,* 1879–1885.

5. Chourasia, M. K.; Kang, L.; Chan, S. Y. Nanosized Ethosomes Bearing Ketoprofen for Improved Transdermal Delivery. *Results Pharma Sci.* **2011,** *1,* 60–67.

6. Paolino, D.; Celia, C.; Trapasso, E.; Cilurzo, F.; Fresta, M. Paclitaxel-Loaded Ethosomes: Potential Treatment of Squamous Cell Carcinoma, A Malignant Transformation of Actinic Keratoses. *Eur. J. Pharm. Biopharm.* **2012,** *81,* 102–112.

7. Maheshwari, R. G. S.; Tekade, R. K.; Sharma, P. A.; Darwhekar, G.; Tyagi, A.; Patel, R. P.; Jain, D. K. Ethosomes and Ultradeformable Liposomes for Transdermal Delivery of Clotrimazole: A Comparative Assessment. *Saudi Pharm. J.* **2012,** *20,* 161–170.

8. Touitou, E.; Dayan, N.; Bergelson, L.; Godin, B.; Eliaz, M. Ethosomes—Novel Vesicular Carriers for Enhanced Delivery: Characterization and Skin Penetration Properties. *J. Control. Release* **2000,** *65,* 403–418.

9. Karri, V. N.; Raman, S. K.; Kuppusamy, G.; Mulukutla, S.; Ramaswamy, S.; Malayandi, R. Terbinafine Hydrochloride Loaded Nanoemulsion Based Gel for Topical Application. *J. Pharm. Invest.* **2015,** *45*(1), 79–89.

10. Li, G.; Fan, Y.; Fan, C.; Li, X.; Wang, X.; Li, M.; Liu, Y.; Tacrolimus-Loaded Ethosomes: Physicochemical Characterization and In Vivo Evaluation. *Eur. J. Pharm. Biopharm.* **2012,** *82*(1), 49–57.

CHAPTER 13

TOXICITY OF NANOMATERIAL-BASED SYSTEMS IN DRUG DELIVERY

MEREENA LUKE P.[*]

International and Inter University Centre for Nanoscience and Nanotechnology, Mahatma Gandhi University, Kottayam 686560, Kerala, India

[*]*E-mail: merinaluke@gmail.com*

ABSTRACT

The uses of nanoparticles (NPs) are exclusively popular in biological applications medicine and diagnosis due to its significant properties. These NPs tend to increase an adverse toxicological effect on animals and environment, although their toxicological effects associated with human exposure are yet not comprehensive. Therefore, evaluation of NPs toxicity is necessary in biomedical applications, including drug delivery systems, gene delivery, and therapeutic applications. However, it is possible that nanomedicine in future would play a crucial role in the treatment of human diseases and also in enhancement of normal human physiology.

13.1 INTRODUCTION

The emergence and rapid growth of nanoscience and technology has led to the development of variety of nanostructured materials that are now used in pharmaceutics, biomedical products, cosmetics, and industries. Nano-technologies facilitate other technologies, and the principal outcome is the fabrication of intermediate materials because it connects different disciplines such as physics, chemistry, genetics, information and communication technologies and cognitive sciences, etc.[1] Engineered nanomaterials such as the quantum dots, carbon nanotubes (CNTs), dendrimers, fullerene and fullerene-based derivatives, single- and multiwalled CNTs, functionalized CNTs, polymer nanoparticles (NPs), carbon black, nanocoatings, etc. have diameters less than 100 nm.[2]

Recent advances in nanotechnology marked a new era in the field of nanomedicine and biomedical engineering, which includes many applications of nanomaterials and nanodevices for diagnostic and therapeutic purposes.[3] NP-based diagnostic and therapeutic techniques have been introduced as helpful "add-on" in personalized treatment of patients, and they would have huge contributions to health care.[4] They hold great promise in a range of biomedical applications, including medical imaging biosensors, prostheses, and implants and diagnostics and for delivery of therapeutic compounds.[5]

Nanostructures materials, NPs, nanospheres, nanotubes, and nanofibers produced via nanotechnologies by-products have the potential risk to raise the concern for humans, public health, consumer safety, and the health and safety of workers and the environment.[6] Nanoscale materials have novel and unique physicochemical properties than bulk materials; they also have an unpredictable impact on human health. However, despite the extensive use of NPs at present, it is essential to realize the NP-mediated toxicity. Our present knowledge of the toxicology of NPs is limited.[7] Recently, the term "nanotoxicology" has emerged in the literature, it is a new discipline of to investigate the potential adverse effects of NPs. *Nanotoxicology is defined as the study of the nature and mechanism of toxic effects of nanoscale materials/particles on living organisms and other biological systems. It also deals with the quantitative assessment of the severity and frequency of nanotoxic effects in relation to the exposure of the organisms.*[8]

Recent developments in nanotechnology have led to the progress of the new field of nanomedicine, which uses various nanomaterials for therapeutic and diagnostic applications. The field of "nanomedicine" exclusively relates to the development of all nanosized tools for the diagnosis, prevention, and treatment of disease. Nanomaterials have advantages over biomedical applications because they can be engineered to have excellent properties such as selectivity, size, shape, and biocompatibility.[9] These special properties permit the nanomaterials to impinge on the human body differently than traditional medical treatments.

Therefore, the potential risks of NPs and nanomaterial products, especially in the field of medicine, need to be evaluated properly before human adverse impacts occur. In this chapter, the toxological hazards of NPs particularly in biomedical applications are discussed. Further, present advances and limitations about the toxicity testing and risk evaluation of NPs are identified.

13.2 FACTORS OF TOXICITY

Popularity of nanotechnology in the field of medicine, their applications have been limited due to their potential toxicity and long-term secondary adverse effects. The important mediators of nanotoxicity are size and surface area, chemical composition, and shape of the NPs.

13.2.1 SIZE

Particle size and surface area are important material characteristics from a toxicological point of view. As the size of a particle decreases, its surface area increases and permits a larger proportion of its atoms or molecules to be displayed on the exterior than the interior of the material and enhances the toxicity.[10] The modifications in the physicochemical and structural properties are well defined in the case of engineered nanomaterials (size range of 1–100 nm) that could be responsible for enhanced material interactions, which may lead to toxicological effects. Additionally, particle size induces significant differences in the cellular delivery mechanism and distribution of NP. The size of the NP plays a key role in cytotoxicity. It generates cytotoxicity and inflammatory response in living organisms, due to the migration of NPs across the epithelial barrier.[11] Several toxicological studies have demonstrated that smaller NPs of dimensions <100 nm cause adverse respiratory health effects compared to larger particles of the same material. Inhaled particles of different sizes exhibit different fractional depositions within human respiratory tract. It has been observed that ultrafine particles with diameters <100 nm deposit in all regions, whereas particles <10 nm deposit in the tracheobronchial region, while particles between 10 and 20 nm deposit in the alveolar region. As a result, the translocation or distribution of NPs has been found to be size dependent, which in turn decide their toxicities issues.[12]

13.2.2   SURFACE AREA

The greater surface-area-to-volume ratio of these particles causes superior chemical reactivity and results in the formation of excess reactive oxygen species (ROS). ROS formation is one of the mechanisms of NPs toxicity which could cause oxidative stress, inflammation, and consequent damages to the proteins, cell membrane, and DNA.[13] Surface area was therefore a driver for inflammation and for these materials; the differences in severity of the response disappeared when the dose was expressed as surface area. The importance of particle size and, by implication, the amount of surface area presented to the biological system for particle toxicity. In general, the size dependent toxicity of NPs can be attributed to its ability to enter into the biological systems.[14]

13.2.3   CHEMICAL COMPOSITION

The chemical composition of the nanoparticles is another factor for the intrinsic toxic properties. For example, the effect of carbon black has been shown to be more severe toxicity than that of titanium dioxide, whereas both of them initiate the lung inflammation by epithelial damage in rat models. NPs in surrounding air can have an extremely complex composition, and constituents, such as organics and metals, can interact. Metallic iron can boost the effect of carbon black NPs, resulting in enhanced reactivity, including oxidative stress.

13.2.4   SHAPE AND SURFACE CHARGE

Shape dependent toxicity has been reported for NPs including carbon nanotubes (CNTs), silica, allotropies, nickel, gold, and titanium nanomaterials. Shape is another vital factor in the toxicity of NPs. Fibers provide a significant example of the debate about shape, especially in relation to inhalation, where the physical parameters of thinness and length appear to determine respirability and inflammatory potential. The smaller the diameter of a spherical particle, the more the surface-to-volume ratio increases. Larger surface area increases the chemical reactivity. This is particularly important for biological interactions.[15]

Surface charge of NPs definitely affects the biological systems. The changes in surface charge result in significant differences in the in vivo biodistribution of NPs. NPs, including polymeric nanocomplexes gold nanospheres and dendrimers, have the same range of size variations, different surface charges yielded noticeable difference in both distribution and uptake efficiency. Particles showed different degrees of toxicity depending on their surface charges. NPs with a positively charged surface tended to have much higher toxicity.[16]

13.3 MECHANISM OF NANOTOXICITY

The mechanism of nanotoxicity has recently been deliberate intensively. Mechanisms on the nanotoxicity with biological systems can be either chemical or physical. Chemical mechanisms include the production of ROS, dissolution and release of toxic ions, disturbance of the electron/ion cell membrane transport activity, oxidative damage through catalysis, lipid peroxidation, or surfactant properties (Fig. 13.1). Physical mechanisms at the nano–bio interface are mainly a result of particle size and surface properties. This includes disruption of membranes membrane activity, transport processes, protein conformation, and protein aggregation or fibrillation. ROS induction by NPs is considered the prime cause of nanotoxicity and has been attributed to the presence of pro-oxidant functional groups on their reactive surface or due to NP–cell interactions.[17] However, in excess, it has been found to cause severe damage to cellular macromolecules such as proteins, lipids, and DNA, resulting in detrimental effects on cells.

13.3.1 OXIDATIVE STRESS AND DNA DAMAGE

Toxicity of NPs is mainly due to oxidative stress, followed by DNA damage and apoptosis. ROS production is a normal cellular process which is involved in varied aspects of cellular signaling, as well as in the defense mechanism of the immune system. NPs can cause a wide variety of DNA damage, ranging from chromosomal fragmentation, DNA strand breakages, and the induction of gene mutations. Apoptosis is a highly complex and tightly regulated pathway involving several signaling molecules. Metal oxide NPs including TiO_2, ZnO, Fe_3O_4, Al_2O_3, and CrO_3 of particle sizes ranging from 30 to 45 nm were found to induce apoptosis.[18] In vivo studies showed that various NPs induce oxidative stress, DNA damage, and apoptosis, thereby mediating developmental and reproductive toxicity. However, the actual reasons of signaling molecules mediating apoptosis in NP-induced toxicities were sufficiently studied yet, and thus attempts to better understand this mechanism may prove useful in reducing the toxicological side effect.[19]

The toxicity of NPs can be attributed to the features as follows:

- Surface-area-to-volume ratio of the particles, it boosts their interaction with the surrounding molecules.
- Chemical composition of the nanoparticles—which is responsible for its reactivity.

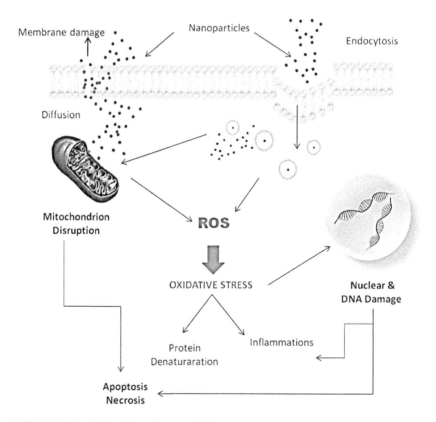

FIGURE 13.1   Mechanism of nanotoxicity.

- Surface charge of the particle is responsible for electrostatic interactions.
- Hydrophobicity and lipophilic groups may allow the particle to interact with proteins and membranes, respectively.
- Complementarity of nanostructure could cause inhibition of enzyme activity either competitive or noncompetitive.
- Oxidative stress and over production of ROS.[20]

13.4 ROUTES OF NANOPARTICLE EXPOSURE

Humans can be exposed to nanomaterials through several routes such as inhalation, injection, oral ingestion, and the dermal route. Specifically, the respiratory system, gastrointestinal tract, the circulatory systems, as well as the central nervous system (CNS) are known to be adversely affected by NPs.[18] The major sources of exposure, routes of uptake into the human body, transportation, and distribution pathways of NPs and related hazards or diseases are summarized in Figure 13.2.

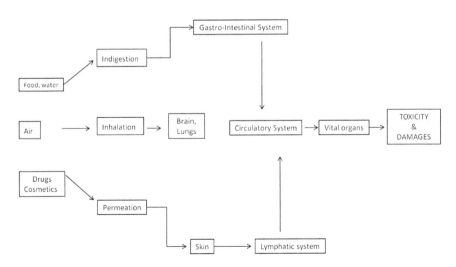

FIGURE 13.2 Routes of nanoparticle exposure.

13.4.1 EXPOSURE THROUGH RESPIRATORY SYSTEM

The lung is the organ that transports the air via the respiratory tract to the alveoli, where oxygen and carbon dioxide are exchanged with the environment. There are 300 million alveoli to facilitate this gas exchange via diffusion, encompassing a surface area of approximately 140 m^2.[21] The human respiratory tract consists of three sequential regions, assisting the filtration effect, including nasopharyngeal, tracheobronchial, and the pulmonary regions. This specific design and the defense mechanisms may protect the organism from harmful materials at the portal of entry; however, these defenses may not always be as effective for NPs. When the NPs are inhaled, their deposition, clearance, and translocation within the respiratory tract will be different from the larger particles.[22]

13.4.2 EXPOSURE THROUGH SKIN

NPs may penetrate into sweat glands and hair follicles. Skin exposure to cosmetics, sunscreens, and dusts resulted in accumulation of NPs. Baroli et al. reported that metallic NPs smaller than 10 nm could penetrate the hair follicle and stratum corneum and sometimes reach the viable epidermis. However, metallic NPs are unable to permeate the skin.[23]

13.4.3 EXPOSURE THROUGH INGESTION

Exposure of nanomaterials into gastrointestinal tract can occur after uptake of daily food, drinks, and medicines. NPs absorbed by any means can cause cytotoxicity effects. Cytotoxicity means that NPs prevent cell division, hinder cell proliferation, damage DNA, and biological system and lead to cell death by biological process called apoptosis.[24,25]

13.5 TYPES OF NANOPARTICLES USED IN CLINICAL APPLICATION

Nanomaterials can migrate through cell membranes below a critical size and are capable of crossing the blood–brain barrier (BBB). These characteristics are used to develop nanoscaled ferries, which transport high potential pharmaceutics at the site.[26] There are different kinds of NPs which are suitable to be applicable in drug and gene-delivery, probing DNA structures (Fig. 13.3). Diverse surface properties, distinctive magnetic properties, tunable absorption and emission properties, and recent advances in the synthesis and engineering of various NPs recommend their potential as probes for molecular imaging and early detection of diseases such as cancer.[27]

13.5.1 DRUG DELIVERY

The unique physicochemical properties of NPs permit the delivery of therapeutic molecules to desired sites in the body.[28] The types of NPs applied in the biomedical system include liposomes, polymer NPs (nanospheres and nanocapsules), solid lipid NPs, nanocrystals, polymer therapeutics such as dendrimers, fullerenes (most common as C60 or buckyball, similar in size of hormones and peptide a-helices), and inorganic NPs[29] (e.g., gold and magnetic NPs), etc. and are summarized in Table 13.1.

13.6 ROUTES OF ADMINISTRATION

The route of administration is the way through which the dosage form is administered into the body for treatment of various diseases and disorders.[32] Various routes of administrations play a marked role in the bioavailability of the active drug in the body. Routes can also be classified based on where the target of action is. Action may be topical, enteral, parenteral, or inhalation[33] (Fig. 13.4). The selection of routes for drug administration depends on factors like (1) characteristics of the drug, (2) utilization of

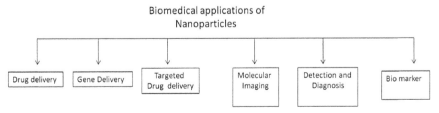

FIGURE 13.3 Biomedical applications of nanoparticles.

TABLE 13.1   Nanosystems Used in Drug Delivery.

Nanosystems used in drug delivery	Size (nm)	Characteristics	Applications
Dendrimers	<10	Highly branched, nearly mono-dispersed polymer system produced by controlled polymerization; three main parts core, branch, and surface	Controlled and targeted drug delivery
Liposomes	50–100	Phospholipid vesicles, biocompatible, versatile, good entrapment efficiency	Long circulatory, offer passive and active delivery of gene, protein, peptide, and various other drugs
Polymeric nanoparticles	10–1000	Biodegradable, biocompatible, offer complete drug protection	Excellent carrier for controlled and sustained delivery of drugs. Surface modified nanoparticles can be used for active and passive delivery of bioactive
Polymeric micelles	10–100	Block amphiphilic copolymer micelles, high drug entrapment, payload, bio stability	Drug and gene delivery
Quantum dots and nanocrystals[28]	2–9.5	Semiconducting material synthesized with II–VI and III–V column element; fluorescence bright, narrow emission, broad UV excitation, and high photo stability	Long-term multiple color imaging of liver cell DNA hybridization, immunoassay, receptor-mediated endocytosis
Fullerene[30]	>10	Carbon allotrope, 7 Å in diameter with 60 carbon atoms arranged in a shape known as truncated icosahedrons	Stimulate host immune response and production of fullerene specific antibodies and imaging
Carbon nanotubes	0.5–3 diameter and 20–1000 length	Third allotropic crystalline form of carbon sheets single layer or multiple layer (multi-walled nanotubes) strength and unique electrical properties	Functionalization enhanced solubility, penetration to cell cytoplasm and to nucleus, as carrier for gene delivery, peptide delivery
Metallic nanoparticles	<100	Gold and silver colloids High surface area accessible for functionalization, stable	Drug and gene delivery, highly sensitive diagnostic assays, thermal ablation, and radiotherapy enhancement

Nanosystems used in drug delivery	Size (nm)	Characteristics	Applications
Magnetic nanoparticles	<100	Magnetite Fe_2O_3, meghemite coated with dextran	Drug targeting diagnostics to in medicine
Ceramic nanoparticles	<100	Silica, alumina, titania	Drug and biomolecules delivery
Nanopores[3]	<100	Aerogel, produced by cell gel chemistry	Controlled release drug carriers
Silica nanoparticle[31]	10–300	Xerogels and mesoporous With biocompatibility, highly porous structure and ease of modification	Drug carriers
Ferro fluids	<100	Iron oxide magnetic nanoparticles surrounded by polymeric layer	For capturing cells and other biological targets
Nanoshells	<100	Dielectric core and metal shell	Tumor targeting

drug, (3) site of action of the drug, (4) conditions of the patient, (5) age of the patient, and (6) effect of gastric pH, digestive enzymes, and first-pass metabolism.[34]

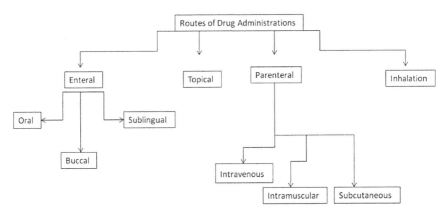

FIGURE 13.4   Major routes of drug administration.

13.6.1   TARGETED DELIVERY OF NANOMEDICINE

Nanomedicine is the application of nanotechnologies in a healthcare milieu and the majority of benefits that have already been seen involve the use of NPs to improve the behavior of drug substances. Nowadays, nanomedicines are used globally to improve the diagnosis and treatments of patients suffering from a range of diseases including cancer, kidney disease, fungal infections, cholesterol, multiple sclerosis, chronic pain, asthma, and emphysema. Irrespective of the route of delivery, once a drug-containing nanomedicine has reached the systemic circulation, it distributes into cells and tissues, with broad variability depending on the delivery system used the nanomedicine characteristics, and the variability between individual patients. A relevant factor to consider is the drug release rate and the nanomedicine stability.

The distribution of such a nanomedicine is generally very different to free drug and this has a number of important implications. Nanomedicine can effectively diffuse into tissues through well-described mechanisms, such as the enhanced permeability and retention effect. Several studies have described that high-molecular-weight drugs, prodrugs, and NPs tend to accumulate in sites of inflammation or cancer, which are tissues with an increased blood flow and a more permeable vasculature. Capillaries and blood vessels reaching certain solid tumors are characterized by large pores, ranging from 100 nm to several hundred nanometers. Therefore, nanomedicine with sizes between 100 and 500 nm can effectively enter the tumor mass.

13.7   TOXICITY TO TARGETED SYSTEMS

There are numerous applications of nanomaterials with manifold ways for humans to use and be affected by them. All NPs, on exposure to tissues and fluids of the body, will immediately adsorb onto their surface some of the macromolecules that they encounter at their portal of entry. The specific features of this adsorption process will depend on the surface characteristics of the particles, including surface chemistry and surface energy, and may be modulated by intentional modification or functionalization of the surfaces.[26]

13.7.1 NANOPARTICLE TOXICITY IN RESPIRATORY SYSTEM

The lung is the organ that transports the air via the respiratory tract to the alveoli, where oxygen and carbon dioxide are exchanged with the environment. There are 300 million alveoli to facilitate this gas exchange via diffusion, encompassing a surface area of approximately 140 m^2.[35] Foreign particles, including nanoobjects that deposit into the lung, are mostly removed by mucociliary transport as they pass the respiratory tract or bronchial tubes. Fine particles (<2.5 mm) can be transported with the air into the alveoli. When the NPs are inhaled, their deposition, clearance, and translocation within the respiratory tract will be entirely different from large particles. After deposition in the respiratory tract, translocation of NPs may potentially occur to the lung interstitium, the brain, liver, spleen, and possibly to the fetus in pregnant females.[36] Apart from exposure to the lungs, inhaled NPs can also gain access to other organs via the olfactory bulb followed by olfactory nerve in the nose. NPs with a size about 35 nm were detected in the brain olfactory bulb after inhalation exposure.[37] According to the diffusion motion, exposure to airborne NPs via the inhalation route will deposit throughout the entire respiratory tract. Many NPs are characterized by a high surface reactivity and thus exhibit a high inherent potential for induction of inflammation and generation of ROS at the spot of deposition.[38] Inhalation of NPs leads to deposition of NPs in respiratory tract and lungs. It causes lung-related disease such as asthma, bronchitis, lung cancer, pneumonia, etc. Translocation of nanomaterials therefore could lead to brain.[39] Jung-Taek Kwon acute pulmonary toxicities were caused by the inhalation of metallic silver NPs (Fig. 13.5).

13.7.1.1 TOXICITY BY PULMONARY DRUG DELIVERY

The lung is an attractive target for drug delivery due to the noninvasive nature of inhalation therapy, the lung's large surface area, localization/accumulation of drugs within the pulmonary tissue, and avoidance of first-pass metabolism, thus reducing systemic side effects.[26,40] Nanocarrier systems for pulmonary drug delivery have several advantages which can be exploited for therapeutic reasons and, thus, toxic effects of NPs by pulmonary drug delivery are intensively studied.

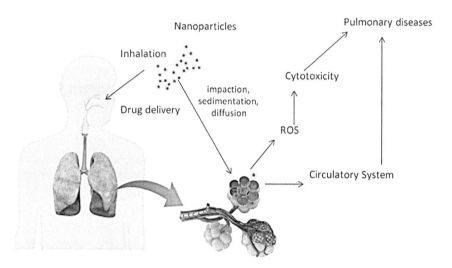

FIGURE 13.5 Toxicity of nanoparticle in respiratory system.

13.7.1.1.1 *Polymeric Nanoparticles*

Polymeric NPs are biocompatible, surface modifiable and are capable of sustained drug release. They show potential for applications in the treatment of various pulmonary conditions such as asthma, chronic obstructive pulmonary disease, tuberculosis, and lung cancer as well as extrapulmonary conditions such as diabetes. The development of synthetic polymers such as the biocompatible and biodegradable poly(lactic-*co*-glycolic acid) (PLGA) for use as drug carrier devices; drug-loaded polymeric nanosystems are effective substitutes for lung cancer treatment. PLGA NP productively enhances therapeutic effects and reduces adverse effects via sustained and targeted drug delivery. In addition, the modification of drug-loaded nanopolymeric system with biological capping materials such as chitosan or bovine serum albumin (BSA) will further reduce the toxicity and improve the biocompatibility. Romero et al. demonstrated a reduction in cytotoxicity of PLGA NPs stabilized with BSA compared to synthetic coating materials in cultured lung cancer cells.[41] Similarly albumin serum protein was found to be highly biocompatible, and it acts as a useful stabilizer for drug delivery vehicles. Similarly, chitosan stabilization resulted in near-total cellular preservation and improved pulmonary mucoadhesion in an in vivo lung cancer model.[42]

13.7.1.1.2 Carbon Nanotubes

CNTs are frequently used for in vivo inhalation models and can be subdivided into single wall carbon nanotubes (SWCNTs) and multiwall carbon nanotubes (MWCNTs) with the former being considered more cytotoxicity. This difference in toxicity has been attributed to the larger surface area of SWCNTs compared to the multilayered alternative. CNTs show biomedical potential in areas such as drug delivery, photodynamic therapy (PDT), and as tissue engineering scaffolds. Previous studies have highlighted the toxic potential of both SWCNTs and MWCNTs; after murine intratracheal instillation of CNTs, pulmonary epitheloid granulomas and interstitial inflammation with subsequent fibrotic changes were observed. Previous studies have highlighted the toxic potential of both SWCNTs and MWCNTs; after murine intratracheal instillation of CNTs, pulmonary epitheloid granulomas and interstitial inflammation with subsequent fibrotic changes were observed.[43] Cytotoxicity is thought to be mediated by the upregulation of inflammatory cytokines such as tumor necrosis factor alpha. However, a recent study by Mutlu et al. postulates that toxicity after murine intratracheal instillation of SWCNTs arises due to nanotubular aggregation rather than the large aspect ratio of the individual nanotube. This has led to the development of several methods to achieve improved nanotube dispersion. Exposure of animals to MWCNTs results in contradictory reports with some authors appending toxicities in the range of asbestos poisoning, whereas other studies found MWCNTs to be biocompatible and far from cytotoxic.[44]

13.7.1.1.3 Silica (SiO$_2$) Nanoparticles

Recently, SiO$_2$ particles have been introduced into the biomedical field as biomarkers, cancer therapeutics, and drug delivery vehicles.[45] Moderate dosage of <20 µg/mL silica NPs are considered as safe. At a dose of 25 µg/mL, silica NPs exhibited agglomerative potential. On A549 cells, both in vitro and dose-dependent cytotoxicity exhibited at a critical concentration of 50 µg/mL. Cell death was mediated by ROS induction and membrane lipid peroxidation. Silica NPs further exhibited particle size-dependent cellular toxicity with smaller diameters causing more harm than bigger ones.[46] Paradoxical results have been obtained in animal studies focusing on the exposure of lungs to silica NPs. After intratracheal exposure of silica, NPs caused an acute pulmonary inflammation and neutrophil infiltration of lung tissue with the development of chronic granulomatous changes after 14 weeks in a dose-dependent manner.[47] An enduring study with the same silica NPs results the induction of anti-inflammatory mediators and the reversibility of inflammatory as well as fibrotic changes.

13.7.1.1.4   Silver Nanoparticles

Because of their antiseptic properties, silver NPs are widely used in creams, textiles, surgical prosthesis, and cosmetics and as bacteriocides in fabrics and other consumer products.[35] Due to their widespread applications, silver NPs and their products have emerged as a massive source of ENPs to the surroundings. Both silver NPs and dissolved silver are known to have significant antibacterial properties. Kim et al. carried out a series of inhalation studies focusing on the acute, subacute, and subchronic toxicity of silver nanoparticles (AgNPs) in rats.[48] It reveals that sequential inhalation of AgNPs does not exhibit any significant changes in body weight or clinical changes more over lung function tests revealed no statistical differences between exposed and control groups. In contrast, subchronic inhalation for 13 weeks at a maximum concentration of 515 $\mu g/m^3$ (five times the limit) exhibited time and dose-dependent alveolar inflammatory and granulomatous changes as well as decreased lung function. Silver NPs can bind to different tissues (bind to proteins and enzymes in mammalian cells) and cause toxic effects, such as adhesive interactions with cellular membrane and production of highly reactive and toxic radicals like ROS, which can cause inflammation and show intensive toxic effects on mitochondrial function. The current American Conference of Governmental Industrial Hygienist's limit for silver dust exposure is 100 $\mu g/m^3$. These results suggest that while high-dose chronic exposure to AgNP has the potential to cause harm, under current guidelines and limits such excessive particle inhalation would seem unrealistic.

13.7.2   NANOPARTICLE TOXICITY IN DERMAL SYSTEM

The skin is the largest organ of the body and functions as the first-line barrier between the external environment and the internal organs of the human body. Consequently, it is exposed to an excess of nanoparticle within creams, sprays, or clothing. Topically applied NPs can potentially penetrate the skin and access the systemic circulation and exert undiserable effects on a systemic scale.[49] Permeation and distribution of nanomaterials in skin usually led to irritation of the inflammation area. Skin barrier alterations such as the wounds, scrapes, or dermatitis can act as exposures routes to NPs into the body (Fig. 13.6). Injured skin represents a fine channel for entry of nano- and microparticles (0.5–7 μm). Smaller NPs penetrated deeper into the tissue than larger ones which were mainly accumulated in the more superficial epidermis and dermis. These findings may have important implications with regards to efficient NP-based dermal drug delivery. Several studies have recently targeted the subject of NP skin interactions.[50] Possible pathways of cellular uptake of NPs are phago-cytosis, macropinocytosis, clathrin-mediated endocytosis, nonclathrin-, noncaveolae-mediated endocytosis, caveolae-mediated endocytosisor diffusion.[51] During phagocytosis, particles become engulfed via specific membrane receptors, leading to the formation of an early phagosome. Physical interactions of NPs with subcellular compartments are determinant of the type of ROS generated, it causes the DNA damage and cell death.

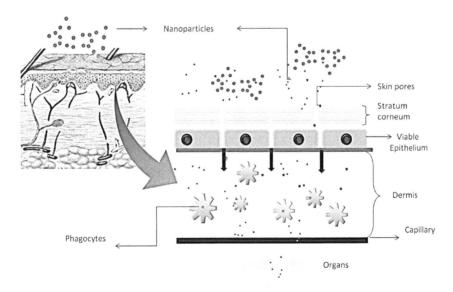

FIGURE 13.6   Permeation and distribution of nanomaterials in skin.

13.7.2.1 TOXICITY BY DERMAL DRUG DELIVERY

13.7.2.1.1 Silver Nanoparticles

AgNP possesses proven and excellent antimicrobial effects which are currently used in many products, especially in wound dressings and clothing. Thus, Ag is one of the most constantly studied NPs in terms of toxicity. Topical application of AgNPs can induce the primary state known as argyria, a gray–blue discoloration of the skin and liver caused by deposition of Ag particles in the basal laminae of such tissues.[52] Potential toxicity from dermal exposure was shown with silver NPs, which diminished human epidermal keratinocytes viability. In an in vitro study of cultured human keratinocytes, Lam et al. observed a substantial decrease in cell viability and cytotoxicity of silver nanocrystals. Furthermore, several toxicity studies based on AgNP have been carried out on cell lines including mouse fibroblast, rat liver, human hepatocellular carcinoma, and human skin carcinoma cells. All of them exhibited a rise in ROS and oxidative stress-mediated cell death and apoptosis with AgNP concentrations in the range of 2.5–200 µg/mL.[53] The level of toxicity was concentration-reliant; it may vary with surface coatings. The studies by Samberg et al. reported toxicity of uncoated AgNP on human epidermal keratinocytes in contrast to carbon-coated particles. Trop et al. observed reversible silver toxicity in a human burns patient who was treated with an AgNP-coated dressing.[54] These results suggest reversible toxicity of AgNP and a transient discoloration of exposed skin. Extensive presence of nanosilver in textiles, wound dressing, sport clothes, and other products, which come in direct contact with the skin, may cause dermal exposure and related undesirable side effects. Real mechanism of AgNP toxicity is unknown; however, the ROS generation and oxidative stress may consider as two key routes. Once excessive ROS production exceeds the antioxidative capacity of the cell, oxidative stress is induced with successive production of inflammatory mediators, followed by DNA damage and apoptosis.

13.7.2.1.2 Quantum Dots

Quantum dots may be functionalized to serve both diagnostic and therapeutic purposes. Findings by W.I. Hagens et al. have shown that quantum dot NPs skin penetration was as a result of skin abrasion thus the staff safety issues arising from working sites especially of those involved with quantum dot and other NPs manufacturing centers.[56] Another study by Zhang et al reported the penetration of quantum dot (QD621) NPs (i.e., NPs containing cadmium and selenium core with cadmium sulphite) when topically applied to weaning porcine skin. The same group also used the same QD621 and found that the quantum dot could penetrate neonatal human epidermal kerationcytes leading to inflammatory responses.[57]

A study by Tang L, et al. provides information on penetration and distribution of quantum dots in skin and its toxicity in organs. They studied the transdermal delivery capacity through mouse skin of water-soluble CdSeS quantum dots (QDs) and the deposition of these QDs in the body. QD solution was coated onto the dorsal hairless skin of male ICR mice. Experimental results indicate that QDs can penetrate into the dermal layer and are limited to the uppermost stratum corneum layers and the hair follicles. These results suggest that QDs have in vivo transdermal delivery capacity through mouse skin and are harmful to the liver and kidney.[58]

13.7.2.1.3   Titanium Dioxide Nanoparticles

Titanium dioxide nanoparticles are widely used in commercial sunscreen and cosmetics due to its UV-light blocking properties; also, it provides better transparency and esthetics to creams. TiO_2 nanoparticles cause toxicity, affecting cellular functions such as cell proliferation, differentiation, mobility, apoptosis, etc.[59] The penetrative effect of TiO_2 was studied by the repeated administration of TiO_2 containing sunscreen on the skin of volunteers. The studies revealed that TiO_2 penetrated into the open part of a hair follicle.

13.7.2.1.4   Silica Nanoparticles

Silica materials are used for implantable, oral, intravenous, and dermal drug delivery systems. Park et al. evaluated toxicity of differently sized silica NPs on cultured human keratinocytes; it confirmed that at a concentration of 50-μg/mL, silica NPs exhibited significant dose-dependent toxicity on human keratinocytes with a considerable reduction in cell viability.[60] Exposure of NPs to human skin equivalent model showed no inflammatory changes even at a maximum concentration of 500 μg/mL.[26]

13.7.2.1.5   Gold Nanoparticle

Nanogold is a widely used nanomaterial with its applications in cell imaging, cancer therapy, tissue welding, nanomedicine, etc. Sonavane et al. investigated the in vitro cutaneous penetration of different sizes (15, 102, and 198 nm) of gold nanoparticles (AuNPs) using the Franz diffusion cell method with rat skin. Gold NPs exhibited size-dependent penetration through rat model; transmission electron microscopy of the sample skin further confirmed the accumulation of smaller gold NPs (15 nm) in profound region of the skin, whereas larger particles were observed mainly in epidermis and dermis.[58,61] Pernodet et al. studied the effects of citrate/gold NPs at different concentrations and exposure times on human dermal fibroblasts found the presence of AuNPs. In intracellular NP actin stress fibers disappeared and inducing major adverse effects on cell viability. The cytological properties such as cell spreading and adhesion, cell growth, and protein synthesis to form the extracellular matrix were changed significantly, suggesting that the internal cell activities were dented.[59,62]

13.7.3   NANOPARTICLE TOXICITY IN LIVER

The liver is the largest internal human organ weighing about 1.5 kg in an adult, with many essential roles in metabolism and clearance. Bile acids are formed by the liver, which are of critical importance for the maintenance of cholesterol metabolism and intestinal lipid absorption.[63] The liver is a critical target tissue for drug delivery because many fatal conditions including chronic hepatitis, enzyme deficiency, and hepatoma occur in hepatocytes.[64] Biodegradable NPs are effective drug delivery devices. Various polymers have been used in drug delivery research as they can effectively deliver the drug to a target site and thus increase the therapeutic benefit, while minimizing side effects.[65] Nanomaterial-based drugs may overcome many of the hurdles of traditional non-nano drugs, because they bear the advantage of enabling a cell-type specific drug delivery based on binding to a specific surface structure. High toxicity and poor specificity of the therapeutic agents may lead to systemic toxicity and adverse effects that are harmful for the patient.

13.7.3.1 TOXICITY BY HEPATIC DRUG DELIVERY

Targeted drug delivery can be a highly desirable strategy to improve the therapeutic outcome, with significantly decreased toxic side-effects compared to traditional chemotherapy. For targeting of polymeric NP to liver, various ligands, such as folic acid and asialoglycoproteins, galactosyl residues, glycyrrhizin derivative, have been introduced into drug carriers.[66]

13.7.3.1.1 Gold Nanoparticles

Chen et al. studied the d size-associated toxicity of AuNPs in mice model. Animals were exposed to intraperitoneal injection with AuNP ranging from 3 to 100 nm in size at a concentration of 8 mg/kg/week for 4 weeks. Histological examination of organs indicated Kupffer cell activation in the liver, splenic white pulp diffusion, and structural deformities in lung parenchyma with gold deposition at these sites. However, particles ranging from 8 to 37 nm induced severe systemic adverse side effects, such as fatigue, loss of appetite, weight loss, and fur color changes[67] improved cytocompatibility by biological surface coatings, the primary mechanism of liver toxicity is thought to arise from acute inflammatory changes and subsequent apoptosis. AuNP toxicity depends on surface functionalization and the route of administration.

13.7.3.1.2 Silver Nanoparticles

AgNP may gain access into the circulation via various routes due to its excellent antimicrobial effects. Many studies have indicated the propensity for AgNP to accumulate within the liver and induce oxidative stress-related toxicity. The degree of toxicity depends on various parameters including particle concentration size, shape, and the ability to deplete cells of antioxidants a study by Teodoro et al. confirmed significant toxicity exerted by AgNP on BRL3A rat liver cells by evaluating the significant reduction in mitochondrial function, and simultaneous rise in lactate dehydrogenase leakage from cells, significantly lower levels of the antioxidant glutathione with an enhanced ROS concentrations.[68] Kim et al. reported the oxidative stress-dependent cytotoxicity.[69]

13.7.3.1.3   Silica Nanoparticles

Few studies have been carried out, and contradictory results regarding toxicity further necessitate systematic research. Xie et al. demonstrated the hepatotoxic potential of silica NPs causing mononuclear inflammatory cell infiltrates at the portal area with concomitant hepatocyte necrosis.[70] Such inflammatory changes are postulated to relate to frustrated phagocytosis of larger silica NPs (>100 nm) with subsequent stimulation of proinflammatory cytokine release. A size relationship has also been reported by Kumar et al. who found small silica NPs (<25 nm) to be biocompatible and nontoxic to murine liver parenchyma.[71]

13.7.3.1.4   Quantum Dots

Quantum dots may be used in a variety of biomedical applications. The general structure of QDs comprises an inorganic core–shell and an organic coating to which biomolecules can be conjugated to facilitate targeting to specific sites within the body. Cytological studies pointing on QD-induced adverse effects found toxicity by the liberation of metal ions released from the heavy metal core. A study by Yang et al. viewed that oxidative environments promote degradation and metal ion leaching. The liver is of particular importance with regards to biotoxicity because of first-pass metabolism and potential accumulation and deposition within the organ[72] QD size was also postulated to be a major parameter in organ-specific deposition with smaller particles (<20 nm) extravasating through capillary orifices that are large enough in the liver (~100 nm in size). The extensive half-life of QDs is a clear implication of hepatic toxicity. The toxicity of QDs can diminish by surface coatings in order to protect the core from degradation. Conventionally, QDs are coated with a layer of zinc sulfide (ZnS) or mercaptoacetic acid.[73]

13.7.4   NANOPARTICLE TOXICITY IN BRAIN

Most of the drugs for treatment of brain diseases are therefore limited by the well-defined protective system (BBB). As a result, various drug delivery and targeting modes are currently being developed to improve the transport and distribution of drugs into the brain.[74] Neurodegenerative diseases, such as Alzheimer's, Huntington's, and Parkinson's diseases, are also the targets of NP-based therapeutic approaches. Similarly to malignant tumors, overcoming BBB is one of the causes for the use of NPs in the treatment of neurodegenerative diseases.[75] Targeted drug delivery allows the use of smaller drug doses and hence reduces systemic adverse effects. Nevertheless, the fate of NPs after translocation into the structures of the CNS requires toxicological analyses. Recently, Oberdorster et al. demonstrated the uptake of NPs by sensory nerve endings within the pulmonary epithelium and subsequent axonal translocation into the CNS which presents a further route of exposure to potentially harmful NPs.[76]

13.7.4.1   TOXICITY BY BRAIN TARGETED DRUG DELIVERY

The brain is an important organ with many essential functions and mechanisms, and it highly protected by well-defined barrier systems. The barrier systems prevent the entry of invasive therapeutics. Hence, a number of strategies like invasive physiological disadvantages of and pharmacological approach are used for targeting drug molecules to brain.[78]

13.7.4.1.1   Silver Nanoparticles

NPs of transition metals, silver, copper, aluminum, silicon, carbon, and metal oxides can easily cross the BBB and induce damage to the barrier integrity by changing endothelial cell membrane permeability. Sharma et al. have studied that the interaction between AgNP and BBB has revealed functional disruption of the BBB and following brain edema formation in rat model.[79] Similarly, Tang et al. demonstrated AgNP-induced BBB destruction, astrocyte swelling, and neuronal degeneration in a rat model.[80] Powers et al. demonstrated the development of neurotoxin due to AgNP in zebra fish embryos that causes long-term changes in synaptic functioning. In addition, exposure to ionized silver (Ag$^+$) increases the yield of the neurotransmitters dopamine and serotonin (5-HT) that leads to Ag$^+$-induced neurotransmitter hyperactivity. The greater Ag$^+$-induced neurotransmitter hyperactivity resulted in behavioral changes consistent with a lower anxiety threshold.

13.7.4.1.2   Gold Nanoparticles

AuNP is mainly used as imaging agents, diagnosis, and therapeutic for CNS.[81] However, mild toxicity was observed with smaller AuNP (3 nm) compared to larger ones (5 nm). Previous studies in various cell lines and in vivo models have established the size-dependent toxicity. Extremely small NPs were found to be widely distributed to almost all tissues including the brain, but the accesses of larger particles to the cerebral tissues are completely restricted. This may be due to either the physical obstruction provided by the BBB or the rapid and more efficient removal of larger particles from the circulation compared with smaller particles. As a result, a lesser particle concentration would accomplish and consequently be absorbed by the BBB. However, the repeated administration of relatively large AuNP, the quantity in the brain was significantly lower than that in the liver, spleen, and kidneys.[82]

13.7.4.1.3   Quantum Dots

Functionalized QDs are widely used as diagnostic and therapeutic purposes. They as can be used as nanovehicles for brain-targeted drug delivery. Kato et al. intraperitoneally injected CdSe/ZnS core/shell-typed QDs functionalized with captopril with a diameter of 13.5 nm into mice. Six hours postinjection, organs were collected and brains were dissected into several areas including the olfactory bulb, cerebral cortex, hippocampus, thalamus, and brainstem. Even though there was a detection of unexpectedly high amounts of Cd within the brain tissues, no signs of inflammation, neuronal loss, or functional disorders were present. Small QDs may transfer through minute gaps (<20 nm) between astrocytic foot processes which constitute the BBB out of the capillary bed into the brain parenchyma.

13.8   TOXICITY BY OTHER COMPLEX NANOSYSTEMS

13.8.1   DENDRIMERS

Dendrimers are well-defined, versatile polymeric architecture with properties resembling biomolecules. Dendritic polymers emerged as outstanding carrier in modern medicine system because of its derivatizable branched architecture and flexibility and tunable structure.[83] Dendritic scaffold has been found to be appropriate carrier for a variety of drugs including anticancer, antiviral, antibacterial, antitubercular, etc., with capacity to enhance solubility and bioavailability of poorly soluble drugs.[84] The wide range of applications in pharmaceutical field, the use of dendrimers in biological system is controlled because of intrinsic toxicity associated with them.[85] This toxicity is attributed to the interaction of surface cationic charge of dendrimers with negatively charged biological membranes in vivo. Interaction of dendrimers with biological membranes results in membrane disruption via orifices formation, membrane thinning, and erosion.[86] Dendrimer toxicity in biological system is generally characterized by hemolytic toxicity, cytotoxicity, and hematological toxicity. Dendrimer biocompatibility has been evaluated in vitro and in vivo for efficient presentation of biological performance. In order to diminish this toxicity, two major strategies have been utilized; designing and synthesis of biocompatible dendrimers; and varying the peripheral charge of dendrimers by surface engineering.[87]

Cytotoxicity of dendrimers is mainly relevant for biomedical purposes. Dendrimers cytotoxicity depends on the core material and is strongly influenced by the nature of the dendrimers' surface. The presence of several surface functional groups enables a simultaneous interaction with a number of receptors; thus, it enhances biological activity.[88] The provision of surface functionalities in dendrimers offers particular ligands attachment such as folic acid, silver salts complexes, antimicrobial agents, or poly (ethylene glycol). The attached compounds can improve surface activity, the biological and physical properties of dendrimers. The size and charge of poly amido amine (PAMAM) dendrimers influence their cytotoxicity. The higher generation (G4–G8) PAMAM dendrimers exhibit toxicity due to their high cationic charge density. The cationic charge of dendrimer surface effectively altered either by neutralization of charge, for example, PEGylation, acetylating, and carbohydrate and peptide conjugation or by introducing negative charge such as half generation dendrimers. Neutral and negatively charged dendrimers do not interact with biological environment and hence are compatible for biomedical applications.[89] Chemical modification of the surface is an important strategy to overcome the toxicity problems associated with the dendrimers. Positively charged dendrimers introduced into the systemic circulation interact with blood components causing destabilization of cell membranes and cell lysis.[90] Roberts et al. observed that cationic PAMAM dendrimers caused a decrease in cell viability; however, they found no evidence of their immunogenicity in rabbits. Additionally, they studied the toxicity of cationic PAMAM Starburst® in mice, and they suggested that even high doses of low generation cationic dendrimers do not cause side effects.[91]

13.8.2 FULLERENES

Fullerenes as a class of carbon allotropes have been studied extensively for their material properties and potential technological applications, including those in biology and medicine. Since a major focus in the latter has been on drug development and formulation, in this paper we highlight some representative studies related to such a focus, including the use of fullerenes for drug-like functions and for their improving the formulation of established drugs. Also discussed are some other potential medically relevant applications of fullerenes, such as their serving as potent agents in PDT and magnetic imaging. Fullerenes as a distinctive class of carbon allotropes have been considered over extensively to their unique material properties, furthermore possibility innovative applications, including clinical applications.[92] The important focus in the clinical and biological applications has been on the drug development and formulations, including the utilization about fullerenes for drug-like system.[93]

The fullerenes are neither hydrophilic nor hydrophobic. Hence, it has strong attractive interactions with the head group layer of the membrane; moreover, it has a high solubility in the lipid core of the membrane. These unique properties give rise to an unusually high permeability of lipid membranes. This may explain how it can penetrate the BBB by passive diffusion, which is normally not the case with molecules larger than 500 Da. The molecular size of the fullerene derivative was about 3.5 nm. Exposures were for 8 and 24 h. After 8 h, the compound was mostly located in the epidermal layer in the unflexed skin, whereas the 60-min flexed skin showed a greater epidermal content and the 90-min flexed skin showed an even greater epidermal content with penetration into the dermis. After 24 h, the highest concentration of particles was in the upper epidermis with a lower concentration in the dermis. Skin flexed for 90 min showed the greatest penetration to the dermis. The compound was localized in the intracellular space between the keratinocytes.[94]

The oral administration water-soluble C60 fullerene derivative (pristine C60 suspension and particle size range 60–1650 nm) in rat models indicated radioactivity was mainly in the liver (73%), whereas blood (1.6%), lungs (1.4%), and the brain (0.2%) had a low content after 1 h. The distribution of the radioactivity 160 h after the administration showed that only 1.6% was in the liver, and no radioactivity was detected in the urine and in the brain. Only 5.4% was eliminated in the feces in addition to trace amount in the urine. Thus, excretion occurred very slowly with redistribution of the remaining activity to the muscles and the hair.

Fullerene and its derivatives showed a profound difference in cytotoxicity in cell culture C60 crystals caused cytotoxicity due to a direct ROS-generating effect that caused cell membrane damage, lipid peroxidation, and necrotic cell death. The genotoxic studies show that fullerene has no direct DNA reactive effects. However, ROS formation may promote development of cancer. At low concentrations, fullerene derivatives had antioxidant and radical-scavenging effects, preventing ROS-induced damage. However, at a high concentration, airway administration caused development of inflammation, which causes ROS formation. Fullerenes cause beneficial effects at low concentrations, but at high concentrations, they may be able to induce inflammation and if chronic, may promote development of cancer, but enduring skin painting study suggests that fullerene did not show tumor-promoting effect.

13.9 CONCLUSION

NPs have certain unique characteristics which can be and have been exploited in many biomedical applications. However, these unique features are postulated to be the grounds for NP-induced biotoxicity which arises from the complex interplay between particle characteristics (e.g., size, shape, and charge), administered dose and host immunological integrity. Recently, more emphasis has been placed onto understanding the role of the route of particle administration as a potential source for toxicity. Nanoparticle exposure especially through inhalation, and afterward translocation inside the body to far off destinations at low measurements various putative systems of toxicological harm have been distinguished, including the formation of ROS, protein misfolding, membrane perturbation, and direct physical impairment. A basic stride in nanotoxicology is to describe the nanomaterial under examination, and this is considerably more troublesome than is the situation in traditional toxicology in view of the large number of factors in the parameter space. These include particle size, roughness, shape, charge, composition, and surface coatings. Finally, it can change contingent on the matrix into which it is presented. There is a need to build up an administrative structure based on the scientific research which will restrain human exposure to undesirable engineered nanomaterials in nature to safe levels. In any case, the restorative utilization of nanomaterials in medication requires an alternate structure which adjusts the remedial benefit against the potential threat of damage. In summary, current difficulties in evaluating NP toxicity originate in the inherent discrepancies found amongst toxicity study protocols such that is has become apparent that a unifying protocol for the toxicological profiling of NPs may be required in order to achieve reliable outcomes that have realistic implications for the human usage of nanoparticles.

KEYWORDS

- nanomaterials
- toxicity
- drug delivery
- diagnosis
- health
- factors of toxicity

REFERENCES

1. Parveen, S.; Misra, R.; Sahoo, S. K. Nanoparticles: A Boon to Drug Delivery, Therapeutics, Diagnostics and Imaging. *Nanomed. Nanotechnol. Biol. Med.* **2012,** *8*, 147–166. DOI: 10.1016/j.nano.2011.05.016.
2. Whitesides, G. M. Nanoscience, Nanotechnology, and Chemistry. *Small* **2005,** *1*, 172–179. DOI: 10.1002/smll.200400130.
3. Surendiran, A.; Sandhiya, S.; Pradhan, S. C.; Adithan, C. Novel Applications of Nanotechnology in Medicine. *Indian J. Med. Res.* **2009,** *130*, 689–701.
4. Koo, O. M.; Rubinstein, I.; Onyuksel, H. Role of Nanotechnology in Targeted Drug Delivery and Imaging: A Concise Review. *Nanomed. Nanotechnol. Biol. Med.* **2005,** *1*, 193–212. DOI: 10.1016/j.nano.2005.06.004.
5. Ingle, A. P.; Shende, S.; Pandit, R.; Paralikar, P.; Tikar, S.; Kon, K.; Rai, M. Nanotechnological Applications for the Control of Pulmonary Infections. In *Volume 1 in Clinical Microbiology: Diagnosis, Treatments and Prophylaxis of Infections*; 2016, pp 223–235.

6. Bakand, S.; Hayes, A.; Dechsakulthorn, F. Nanoparticles: A Review of Particle Toxicology Following Inhalation Exposure. **2012**, *24*, 125–135. DOI: 10.3109/08958378.2010.642021.

7. Elsaesser, A.; Howard, C. V. Toxicology of Nanoparticles. *Adv. Drug Deliv. Rev.* **2012,** *64*, 129–137. DOI: 10.1016/j.addr.2011.09.001.

8. Fischer, H. C.; Chan, W. C.; Nanotoxicity: The Growing Need for In Vivo Study. *Curr. Opin. Biotechnol.* **2007,** *18*, 565–571. DOI: 10.1016/j.copbio.2007.11.008.

9. Wang, J.; Gao, W. Nano/Microscale Motors: Biomedical Opportunities and Challenges. *ACS Nano* **2012,** *6*, 5745–5751. DOI: 10.1021/nn3028997.

10. Karlsson, H. L.; Gustafsson, J.; Cronholm, P.; Möller, L. Size-Dependent Toxicity of Metal Oxide Particles—A Comparison between Nano- and Micrometer Size. *Toxicol. Lett.* **2009,** *188*, 112–118. DOI: 10.1016/j.toxlet.2009.03.014.

11. Blanco, E.; Shen, H.; Ferrari, M. Perspective Principles of Nanoparticle Design for Overcoming Biological Barriers to Drug Delivery. **2015,** *33*, 941–951. DOI: 10.1038/nbt.3330.

12. Gatoo, M. A.; Naseem, S.; Arfat, M. Y.; Dar, A. M.; Qasim, K.; Zubair, S. Physicochemical Properties of Nanomaterials: Implication in Associated Toxic Manifestations. *Biomed. Res. Int.***2014,** *2014*, 498420. https://doi.org/10.1155/2014/498420

13. Fard, J. K.; Jafari, S.; Eghbal, M. A. A Review of Molecular Mechanisms Involved in Toxicity of Nanoparticles. **2015,** *5*, 447–454. DOI: 10.15171/apb.2015.061.

14. Lovri, J.; Yan, S. B.; Fortin, G. R. A.; Winnik, F. M. Differences in Subcellular Distribution and Toxicity of Green and Red Emitting CdTe Quantum Dots. *J. Mol. Med.* **2005,** *83*, 377–385. DOI: 10.1007/s00109-004-0629-x.

15. Nel, A. E.; Mädler, L.; Velegol, D.; Xia, T.; Hoek, E. M. V.; Somasundaran, P.; Klaessig, F.; Castranova, V.; Thompson, M. Understanding Biophysicochemical Interactions at the Nano–Bio Interface. *Nat. Mater.* **2009,** *8*, 543–557. DOI: 10.1038/nmat2442.

16. Shin, S. W.; Song, I. H.; Um, S. H. Role of Physicochemical Properties in Nanoparticle Toxicity. *J. Nanomater.* **2015,** *5*(3), 1351–1365. DOI: 10.3390/nano5031351.

17. Fu, P. P.; Xia, Q.; Hwang, H.; Ray, P. C. Mechanisms of Nanotoxicity: Generation of Reactive Oxygen Species. *J. Food Drug Anal.* **2014,** *22*, 64–75. DOI: 10.1016/j.jfda.2014.01.005.

18. Khanna, P.; Ong, C.; Bay, B. H.; Baeg, G. H.; Nanotoxicity: Interplay of Oxidative Stress, Inflammation and Cell Death. *J. Nanomater.* **2015,** *5*, 1163–1180. DOI: 10.3390/nano5031163.

19. Manke, A.; Wang, L.; Rojanasakul, Y. Mechanisms of Nanoparticle-Induced Oxidative Stress and Toxicity. *Biomed. Res. Int.* **2013,** 2013. DOI: 10.1155/2013/942916.

20. Aydın, A.; Sipahi, H.; Charehsaz, M. Nanoparticles Toxicity and Their Routes of Exposures. **2012,** Recent Advances in Novel Drug Carrier Systems Ali Demir Sezer, IntechOpen, DOI: 10.5772/51230.

21. Krug, H. F.; Wick, P. Nanotoxicology: An Interdisciplinary Challenge. *Angew. Chem.* **2011,** 1260–1278. DOI: 10.1002/anie.201001037.

22. Ostiguy, C.; Cloutier, Y. Studies and Health Effects of Nanoparticles. n.d.

23. Baroli, B. Nanoparticles and Skin Penetration. Are There Any Potential Toxicological Risks? *J. Fur Verbraucherschutz Und Leb.* **2008,** *3*, 330–331. DOI: 10.1007/s00003-008-0357-1.

24. Kuhlbusch, T. A.; Asbach, C.; Fissan, H.; Göhler, D.; Stintz, M. Nanoparticle Exposure at Nanotechnology Workplaces: A Review. *Part. Fibre Toxicol.* **2011,** *8,* 22. DOI: 10.1186/1743-8977-8-22.

25. Abbott, L. C.; Maynard, A. D. Exposure Assessment Approaches for Engineered Nanomaterials. *Risk Anal.* **2010,** *30,* 1634–1644. DOI: 10.1111/j.1539-6924.2010.01446.x.

26. Yildirimer, L.; Thanh, N. T. K.; Loizidou, M.; Seifalian, A. M. Toxicological Considerations of Clinically Applicable Nanoparticles. *J. Nanotoday* **2011.** DOI: 10.1016/j. nantod.2011.10.001.

27. Nune, S. K.; Gunda, P.; Thallapally, P. K.; Lin, Y.; Laird, M.; Berkland, C. J. Nanoparticles for Biomedical Imaging. **2011,** *6,* 1175–1194. DOI: 10.1517/17425240903229031.

28. Bhatia, S. Nanoparticles Types, Classification, Characterization, Fabrication Methods. In *Natural Polymer Drug Delivery Systems*; Springer International Publishing: Cham, 2016; pp 33–93. https://doi.org/10.1007/978-3-319-41129-3_2.

29. Lehner, R.; Wang, X.; Marsch, S.; Hunziker, P. Intelligent Nanomaterials for Medicine: Carrier Platforms and Targeting Strategies in the Context of Clinical Application. *Nanomed. Nanotechnol. Biol. Med.* **2013,** *9,* 742–757. DOI: 10.1016/j. nano.2013.01.012.

30. Mudshinge, S. R.; Deore, A. B.; Patil, S.; Bhalgat, C. M. Nanoparticles: Emerging Carriers for Drug Delivery. *Saudi Pharm. J.* **2011,** *19,* 129–141. DOI: 10.1016/j. jsps.2011.04.001.

31. Kannan, R. M.; Nance, E.; Kannan, S.; Tomalia, D. A. Emerging Concepts in Dendrimer-Based Nanomedicine: From Design Principles to Clinical Applications. *J. Intern. Med.* **2014,** *276,* 579–617. DOI: 10.1111/joim.12280.

32. Verma, P.; Thakur, A. S.; Deshmukh, K.; Jha, A. K.; Verma, S. Routes of Drug Administration. *Int. J Pharm. Stud. Res.* **2010,** *1*(1), 54–59.

33. Aronson, J. K. Routes of Drug Administration: Uses and Adverse Effects: Part 1: Intramuscular and Subcutaneous Injection. *Adv. Drug React. Bull.* 2008, (253), 971–974: DOI: 10.1097/FAD.0b013e328329bb21.

34. Kwatra, S.; Taneja, G.; Nasa, N. Alternative Routes of Drug Administration—Transdermal, Pulmonary & Parenteral. *Indo Glob. J. Pharm. Sci.* **2012,** *2,* 409–426.

35. Srivastava, V.; Gusain, D.; Sharma, Y. C. Critical Review on the Toxicity of Some Widely Used Engineered Nanoparticles. **2015.** DOI: 10.1021/acs.iecr.5b01610.

36. Kunzmann, A.; Andersson, B.; Thurnherr, T.; Krug, H.; Scheynius, A.; Fadeel, B. Toxicology of Engineered Nanomaterials: Focus on Biocompatibility. *Biochim. Biophys. Acta.* **2011,** *1810,* 361–373. DOI: 10.1016/j.bbagen.2010.04.007.

37. Li, Muralikrishnan, S.; Ng, C. T.; Yung, L. Y.; Bay, B. H. Nanoparticle-Induced Pulmonary Toxicity. *Exp. Biol. Med.* **2010,** *235,* 1025–1033. DOI: 10.1258/ ebm.2010.010021.

38. Sørig, K.; Campagnolo, L.; Chavatte-palmer, P.; Tarrade, A.; Rousseau-ralliard, D.; Valentino, S.; Park, M. V. D. Z.; De Jong, W. H.; Wolterink, G.; Piersma, A. H.; Ross, B. L.; Hutchison, G. R.; Stilund, J.; Vogel, U.; Jackson, P.; Slama, R.; Pietroiusti, A.; Cassee, F. R. A Perspective on the Developmental Toxicity of Inhaled Nanoparticles. *Reprod. Toxicol.* **2015,** *56,* 118–140. DOI: 10.1016/j.reprotox.2015.05.015.

39. Thomas AJ Kuhlbusch, Christof Asbach, Heinz Fissan, Daniel Göhler, Michael Stintz Workplace Exposure to Nanoparticles: A review. *J. Part. Fibre Toxicol.* **2011**, *8*, 22, 1–18, DOI: https://doi.org/10.1186/1743-8977-8-22.

40. Singh, A. K. Mechanisms of Nanoparticle Toxicity. *Eng. Nanoparticles* **2016**, 295–341. DOI: 10.1016/B978-0-12-801406-6.00007-8.

41. Romero, G.; Estrela-lopis, I.; Zhou, J.; Rojas, E.; Franco, A.; Espinel, C. S.; Gonza, A.; Gao, C.; Donath, E.; Moya, S. E. Surface Engineered Poly(Lactide-*co*-glycolide) Nanoparticles for Intracellular Delivery: Uptake and Cytotoxicity—A Confocal Raman Microscopic Study. *Biomacromolecules* **2010**, 2993–2999, DOI: 10.1021/bm1007822.

42. Tahara, K.; Sakai, T. Improved Cellular Uptake of Chitosan-modified PLGA Nano-spheres by A549 Cells. **2009**, *382*, 198–204. DOI: 10.1016/j.ijpharm.2009.07.023.

43. Warheit, D. B.; Laurence, B. R.; Reed, K. L.; Roach, D. H.; Reynolds, G. A. M.; Webb, T. R. Comparative Pulmonary Toxicity Assessment of Single-Wall Carbon Nanotubes in Rats. *Toxicol. Sci.* **2004**, *125*, 117–125. DOI: 10.1093/toxsci/kfg228.

44. Mitchell, L. A.; Gao, J.; Wal, R. V.; Gigliotti, A.; Burchiel, S. W.; McDonald, J. D. Pulmonary and Systemic Immune Response to Inhaled Multiwalled Carbon Nano-tubes. *Toxicol. Sci.* **2007**, *100*, 203–214. DOI: 10.1093/toxsci/kfm196.

45. Hirsch, L. R.; Stafford, R. J.; Bankson, J. A.; Sershen, S. R.; Rivera, B.; Price, R. E.; Hazle, J. D.; Halas, N. J.; West, J. L. Nanoshell-Mediated Near-Infrared Thermal Therapy of Tumors Under Magnetic Resonance Guidance. *Proc. Natl. Acad. Sci. U.S.A.* **2003**, *100*, 13549–13554. DOI: 10.1073/pnas.2232479100.

46. Napierska, D.; Thomassen, L. C. J.; Rabolli, V.; Lison, D.; Gonzalez, L.; Kirsch-Volders, M.; Martens, J. A.; Hoet, P. H. Size-Dependent Cytotoxicity of Monodis-perse Silica Nanoparticles in Human Endothelial Cells. *Small* **2009**, *5*, 846–853. DOI: 10.1002/smll.200800461.

47. Cho, W. S.; Choi, M.; Han, B. S.; Cho, M.; Oh, J.; Park, K.; Kim, S. J.; Kim, S. H.; Jeong, J. Inflammatory Mediators Induced by Intratracheal Instillation of Ultrafine Amorphous Silica Particles. *Toxicol. Lett.* **2007**, *175*, 24–33. DOI: 10.1016/j.tox-let.2007.09.008.

48. Kim, J. S.; Kuk, E.; Yu, K. N.; Kim, J.-H.; Park, S. J.; Lee, H. J.; Kim, S. H.; Park, Y. K.; Park, Y. H.; Hwang, C.-Y.; Kim, Y.-K.; Lee, Y.-S.; Jeong, D. H.; Cho, M.-H.; Jones, S. A.; Bowler, P. G.; Walker, M.; Parsons, D.; Silver, S.; Phung, L. T.; Catauro, M.; Raucci, M. G.; De Gaetano, F. D.; Marotta, A.; Crabtree, J. H.; Burchette, R. J.; Siddiqi, R. A.; Huen, I. T.; Handott, L. L.; Fishman, A.; Zhao, G.; Stevens, S. E.; Ay-monier, C.; Schlotterbeck, U.; Antonietti, L.; Zacharias, P.; Thomann, R.; Tiller, J. C.; et al. Antimicrobial Effects of Silver Nanoparticles, **2007**, *3*, 95–101. DOI: 10.1016/j.nano.2006.12.001.

49. Gautam, A.; Singh, D.; Vijayaraghavan, R. Dermal Exposure of Nanoparticles: An Understanding. *J. Cell Tissue Res.* **2011**, *11*(1), 2703–2708.

50. Hirai, T.; Yoshioka, Y.; Higashisaka, K.; Tsutsumi, Y. Potential Hazards of Skin Ex-posure to Nanoparticles. **2016**. DOI: 10.1007/978-4-431-55732-6.

51. de Chastellier, C.; Forquet, F.; Gordon, A.; Thilo, L. Mycobacterium Requires an All-Around Closely Apposing Phagosome Membrane to Maintain the Maturation Block and This Apposition is Re-Established When It Rescues Itself from Phagolysosomes. *Cell. Microbiol.* **2009**, *11*, 1190–1207. DOI: 10.1111/j.1462-5822.2009.01324.x.

52. Samberg, M. E.; Oldenburg, S. J.; Monteiro-riviere, N. A. Evaluation of Silver Nanoparticle Toxicity in Skin. **2010**, *118*, 407–413. DOI: 10.1289/ehp.0901398.

53. Kawata, K.; Osawa, M.; Okabe, S. In vitro Toxicity of Silver Nanoparticles at Non-cytotoxic Doses to HepG2 Human Hepatoma Cells. *Environ. Sci. Technol.* **2009**, *43*, 6046–6051. DOI: 10.1021/es900754q.

54. Trop, M.; Novak, M.; Rodl, S.; Hellbom, B.; Kroell, W.; Goessler, W. Silver-Coated Dressing Acticoat Caused Raised Liver Enzymes and Argyria-Like Symptoms in Burn Patient. *J. Trauma Inj. Infect. Crit. Care.* **2006**, *60*, 648–652. DOI: 10.1097/01. ta.0000208126.22089.b6.

55. Yah, C. S.; Simate, G. S.; Iyuke, S. E. Nanoparticles Toxicity and Their Routes of Exposures. **2012**, *25*, 477–491.

56. R.J. Aitken, K.S. Galea, C.L. Tran, J.W. Cherrie, Exposure to Nanoparticles, in: Environ. Hum. Heal. Impacts Nanotechnol., 2009: pp. 307–356. doi:10.1002/9781444307504.ch8.

57. L.W. Zhang, W.W. Yu, V.L. Colvin, N.A. Monteiro-riviere, Biological interactions of quantum dot nanoparticles in skin and in human epidermal keratinocytes, 228 (2008) 200–211. doi:10.1016/j.taap.2007.12.022.

58. T. Lei, Z. Chunling, S. Guangming, J.I.N. Xun, X.U. Zhongwei, In vivo skin penetration and metabolic path of quantum dots, 56 (2013) 181–188. doi:10.1007/s11427-012-4404.

59. Lademann, J.; Schaefer, H.; Otberg, N.; Teichmann, A.; Blume-Peytavi, U.; Sterry, W. Penetration of microparticles into human skin, Hautarzt ,**2004**, 1117–1119. DOI: 10.1007/s00105-004-0841-1.

60. Seo, H.; Hong, J.; Park, M.; Kim, M.; Kim, Y.; Cho, K. Evaluation of Silica Nanoparticle Toxicity After Topical Exposure for 90 Days. **2014**, 127–136.

61. Sonavane, G.; Tomoda, K.; Sano, A.; Ohshima, H.; Terada, H.; Makino, K. Colloids and Surfaces B: Biointerfaces In Vitro Permeation of Gold Nanoparticles Through Rat Skin and Rat Intestine: Effect of Particle Size. *Colloids Surf. B: Biointerfaces* **2008**, *65*, 1–10. DOI: 10.1016/j.colsurfb.2008.02.013.

62. Pernodet, N.; Fang, X. H.; Sun, Y.; Bakhtina, A.; Ramakrishnan, A.; Sokolov, J.; Ulman, A.; Rafailovich, M. Adverse Effects of Citrate/Gold Nanoparticles on Human Dermal Fibroblasts. *Small* **2006**, *2*, 766–773. DOI: 10.1002/smll.200500492.

63. Plaats, A. V. D. Chapter 2 Anatomy and Physiology of the Liver. In *The Groningen Hypothermic Liver Perfusion System for Improved Preservation in Organ Transplantation*, 2005; pp 6–7.

64. Erion, M. D.; Van Poelje, P. D.; Mackenna, D. A.; Colby, T. J.; Montag, A. C.; Fujitaki, J. M.; Linemeyer, D. L.; Bullough, D. A. Liver-Targeted Drug Delivery Using HepDirect Prodrugs. **2005**, *312*, 554–560. DOI: 10.1124/jpet.104.075903.al.

65. Pandey, A.; Mishra, S.; Tiwari, A.; Misra, K. Targeted Drug Delivery (Site Specific Drug Delivery). **2004**, *63*, 230–247.

66. Li, C.; Zhang, D.; Guo, H.; Hao, L.; Zheng, D.; Liu, G.; Shen, J.; Tian, X.; Zhang, Q. Preparation and Characterization of Galactosylated Bovine Serum Albumin Nanoparticles for Liver-Targeted Delivery of Oridonin. *Int. J. Pharm.* **2013**, *448*, 79–86. DOI: 10.1016/j.ijpharm.2013.03.019.

67. Erion, M. D.; van Poelje, P. D.; MacKenna, D. A.; Colby, T. J.; Montag, A. C.; Fujitaki, J. M.; Linemeyer, D. L.; Bullough, D. A. Liver-Targeted Drug Delivery Using

HepDirect Prodrugs. *J. Pharmacol. Exp. Ther.* **2005**, *312*, 554–560. DOI: 10.1124/jpet.104.075903.

68. Teodoro, J. S.; Simoes, A. M.; Duarte, F. V.; Rolo, A. P.; Murdoch, R. C.; Hussain, S. M.; Palmeira, C. M. Assessment of the Toxicity of Silver Nanoparticles in vitro: A Mitochondrial Perspective. *Toxicol. In Vitro* **2011**, *25*, 664–670. DOI: 10.1016/j.tiv.2011.01.004.

69. Kim, S.; Choi, J. E.; Choi, J.; Chung, K. H.; Park, K.; Yi, J.; Ryu, D. Y. Oxidative Stress-Dependent Toxicity of Silver Nanoparticles in Human Hepatoma Cells. *Toxicol. in Vitro* **2009**, *23*, 1076–1084. DOI: 10.1016/j.tiv.2009.06.001.

70. Liu, T.; Li, L.; Teng, X.; Huang, X.; Liu, H.; Chen, D.; Ren, J.; He, J.; Tang, F. Single and Repeated Dose Toxicity of Mesoporous Hollow Silica Nanoparticles in Intravenously Exposed Mice. *Biomaterials* **2011**, *32*, 1657–1668. DOI: 10.1016/j.biomaterials.2010.10.035.

71. Cho, M.; Cho, W.-S.; Choi, M.; Kim, S. J.; Han, B. S.; Kim, S. H.; Kim, H. O.; Sheen, Y. Y.; Jeong, J. The Impact of Size on Tissue Distribution and Elimination by Single Intravenous Injection of Silica Nanoparticles. *Toxicol. Lett.* **2009**, *189*, 177–183. DOI: 10.1016/j.toxlet.2009.04.017.

72. Ho, C.; Chang, H.; Tsai, H.; Tsai, M.; Yang, C.; Ling, Y.; Lin, P. Quantum Dot 705—A Cadmium-Based Nanoparticle, Induces Persistent Inflammation and Granuloma Formation in the Mouse Lung. *Nanotoxicology* **2013**, *7*, 105–115. DOI: 10.3109/17435390.2011.635814.

73. Das, G. K.; Chan, P. P. Y.; Teo, A.; Loo, J. S. C.; Anderson, J. M.; Tan, T. T. Y. In vitro Cytotoxicity Evaluation of Biomedical Nanoparticles and Their Extracts. *J. Biomed. Mater. Res. Part A* **2010**, *93*, 337–346. DOI: 10.1002/jbm.a.32533.

74. de Boer, A. G.; Gaillard, P. J. Drug Targeting to the Brain. *Annu. Rev. Pharmacol. Toxicol.* **2007**, *47*, 323–355. DOI: 10.1146/annurev.pharmtox.47.120505.105237.

75. Jo, D. H.; Kim, J. H.; Lee, T. G.; Kim, J. H. Size, Surface Charge, and Shape Determine Therapeutic Effects of Nanoparticles on Brain and Retinal Diseases. *Nanomed. Nanotechnol. Biol. Med.* **2015**, *11*, 1603–1611. DOI: 10.1016/j.nano.2015.04.015.

76. Connor, E. E.; Mwamuka, J.; Gole, A.; Murphy, C. J.; Wyatt, M. D. Gold Nanoparticles Are Taken up by Human Cells but Do Not Cause Acute Cytotoxicity. *Small* **2005**, *1*, 325–327. DOI: 10.1002/smll.200400093.

77. Kumar, M.; Misra, A.; Babbar, A. K.; Mishra, A. K.; Mishra, P.; Pathak, K. Intranasal Nanoemulsion Based Brain Targeting Drug Delivery System of Risperidone. *Int. J. Pharm.* **2008**, *358*, 285–291. DOI: 10.1016/j.ijpharm.2008.03.029.

78. Sharma, H. S.; Hussain, S.; Schlager, J.; Ali, S. F.; Sharma, A. Influence of Nanoparticles on Blood–Brain Barrier Permeability and Brain Edema Formation in Rats. *Acta Neurochir. Suppl.* **2010**, *106*, 359–364. DOI: 10.1007/978-3-211-98811-4_65.

79. Tang, J.; Xiong, L.; Zhou, G.; Wang, S.; Wang, J.; Liu, L.; Li, J.; Yuan, F.; Lu, S.; Wan, Z.; Chou, L.; Xi, T. Silver Nanoparticles Crossing Through and Distribution in the Blood–Brain Barrier In Vitro. *J. Nanosci. Nanotechnol.* **2010**, *10*, 6313–6317. DOI: 10.1166/jnn.2010.2625.

80. Pathak, Y.; Thassu, D. Drug Delivery Nanoparticles Formulation and Characterization, Informa Health care, New York 2009, 191, 92–116.

81. Lasagna-Reeves, C.; Gonzalez-Romero, D.; Barria, M. A.; Olmedo, I.; Clos, A.; Sadagopa Ramanujam, V. M.; Urayama, A.; Vergara, L.; Kogan, M. J.; Soto, C. Bio-

accumulation and Toxicity of Gold Nanoparticles After Repeated Administration in Mice. *Biochem. Biophys. Res. Commun.* **2010,** *393,* 649–655. DOI: 10.1016/j. bbrc.2010.02.046.

82. Baker, J.; Majoros, I. Dendrimer-Based Nanomedicine, Pan Stanford Publishing Pte. Ltd.**2008,**25-26 DOI: 10.4032/9789814241182.

83. Kesharwani, P.; Jain, K.; Jain, N. K. Dendrimer as Nanocarrier for Drug Delivery. *Prog. Polym. Sci.* **2014,** *39,* 268–307. DOI: 10.1016/j.progpolymsci.2013.07.005.

84. Duncan, R.; Izzo, L. Dendrimer Biocompatibility and Toxicity. *Adv. Drug Deliv. Rev.* **2005,** *57,* 2215–2237. DOI: 10.1016/j.addr.2005.09.019.

85. Pan, B. F.; Gao, F.; Gu, H. C. Dendrimer Modified Magnetite Nanoparticles for Protein Immobilization. *J. Colloid Interface Sci.* **2005,** *284,* 1–6. DOI: 10.1016/j. jcis.2004.09.073.

86. Wilczewska, A. Z.; Niemirowicz, K.; Markiewicz, K. H.; Car, H. Nanoparticles as Drug Delivery Systems. **2012.**

87. Rainwater, J. C.; Anslyn, E. V. Amino-Terminated PAMAM Dendrimers Electrostatically Uptake Numerous Anionic Indicators. *Chem. Commun.* **2010,** *46,* 2904–2906. DOI: 10.1039/b925229k.

88. Hong, S.; Bielinska, A. U.; Mecke, A.; Keszler, B.; Beals, J. L.; Shi, X.; Balogh, L.; Orr, B. G.; Baker, J. R.; Banaszak Holl, M. M. Interaction of Poly(amidoamine) Dendrimers with Supported Lipid Bilayers and Cells: Hole Formation and the Relation to Transport. *Bioconjugate Chem.* **2004,** *15,* 774–782. DOI: 10.1021/bc049962b.

89. Roberts, J. C.; Bhalgat, M. K.; Zera, R. T. Preliminary Biological Evaluation of Poly-amidoamine (PAMAM) Starburst Dendrimers. *J. Biomed. Mater. Res.* **1996,** *30,* 53–65. DOI: 10.1002/(SICI)1097-4636(199601)30:1<53::AID-JBM8>3.0.CO;2-Q.

90. Bakry, R.; Najam-ul-haq, M.; Huck, C. W. Medicinal Applications of Fullerenes. **2007,** *2,* 639–649.

91. Anilkumar, P.; Lu, F.; Cao, L.; Luo, P. G.; Liu, J.-H.; Sahu, S.; Tackett, K. N.; Wang, Y.; Sun, Y.-P. Fullerenes for Applications in Biology and Medicine. *Curr. Med. Chem.* **2011,** *18,* 2045–2059.

92. Crosera, M.; Bovenzi, M.; Maina, G. Nanoparticle Dermal Absorption and Toxicity: A Review of the Literature. **2009,** 1043–1055. DOI: 10.1007/s00420-009-0458-x.

93. Nielsen, G. D.; Roursgaard, M.; Jensen, K. A.; Poulsen, S. S.; Larsen, S. T. In Vivo Biology and Toxicology of Fullerenes and Their Derivatives. **2008,** 197–208. DOI: 10.1111/j.1742-7843.2008.00266.x.

94. Orlova, M. A.; Trofimova, T. P.; Shatalov, O. A.; Orlova, M. A. Fullerene Nanoparticles Operating the Apoptosis and Cell Proliferation Processes in Normal and Malignant Cells. **2013,** *5,* 99–139.

INDEX

Milton Keynes UK
Ingram Content Group UK Ltd.
UKHW022059141024
449569UK00031B/1699

9 781774 633960